Being alive on land

Tasks for vegetation science 13

Series Editors

HELMUT LIETH

University of Osnabrück, F.R.G.

HAROLD A. MOONEY

Stanford University, Stanford, Calif., U.S.A.

Being alive on land

PROCEEDINGS OF THE INTERNATIONAL SYMPOSIUM
ON ADAPTATIONS TO THE TERRESTRIAL ENVIRONMENT
HELD IN HALKIDIKI, GREECE, 1982

edited by

N.S. MARGARIS, M. ARIANOUSTOU-FARAGGITAKI
and W.C. OECHEL

1984 **DR W. JUNK PUBLISHERS**
a member of the KLUWER ACADEMIC PUBLISHERS GROUP
THE HAGUE / BOSTON / LANCASTER

Distributors

for the United States and Canada: Kluwer Boston, Inc., 190 Old Derby Street, Hingham, MA 02043, USA
for all other countries: Kluwer Academic Publishers Group, Distribution Center, P.O.Box 322, 3300 AH Dordrecht, The Netherlands

Library of Congress Cataloging in Publication Data

(International Symposium on Adaptations to Terrestrial
    Environment (1982 : Chalkidike, Greece)
    Being alive on land.

    (Tasks for vegetation science ; 13)
    1. Plants--Adaptation--Congresses. 2. Botany--
Ecology--Congresses. I. Margaris, N. S. II. Arianoustou-
Faraggitaki, M. III. Oechel, W. C. IV. Title. V. Series.
QK901.2.I58 1982   581.5'264   84-7191
ISBN-13: 978-94-009-6580-5

ISBN-13: 978-94-009-6580-5   e-ISBN-13: 978-94-009-6578-2
DOI: 10.1007/978-94-009-6578-2

Cover design: Max Velthuijs

Series editor's forword

The 'Kassandra symposia' for biological topics concerning
the problems in the Mediterranean region have become an
established activity through the efforts of Dr. Margaris.
Not all of these symposia fit into the T:VS series. This
volume, edited by Drs. Margaris, Arianoutsou-Farragitaki,
and Oeschel, is to a certain extend the continuation of the
T:VS volume 4 , edited by Drs. Mooney and Margaris.

The somewhat unusual title for a scientific book must be
understood in that respect. It is difficult for the orga-
nizers of symposia to receive enough contributions for a
particular topic and at the same time find the best contri-
butors from a wide geographical distribution.

The table of contents shows that adaptive mechanisms of
plants were the main concern during the symposium during
which this volume was generated. The majority of the papers
deal with Mediterranean type and Tundra ecosystems.

The topics dealt with during the meeting were so extensive that the papers could not be expected to cover everything. There are, however, substantial contributions to the
individual problems presented by several authors.

The ecosystem concept applied in this symposium makes it
possible that a variety of plant groups are discussed. While
this may appear as a collage of papers at first sight, it becomes clear when all the Kassandra symposia are considered
that Dr. Margaris intends to compile as much ·ecosystem information as possible for the Phryganic ecosystem type.
Much basic information needs to be compiled before an ecosystems model can be designed.

As the editor of the T:VS series I see it as one objective that much of the material needed for the modelling of
the Mediterranean type ecosystems will be accumulated in
this series.

One of the major problems for the editors was the editing
of the manuscripts from authors outside of the English speaking countries. Since the editors themselves are not from the
English tongue, the team of the series editor tried to copy-
edit as much as possible. We hope that we have caught most
of the errors. It was not possible, however, to change complete passages or pages in a manuscript which was delivered in
camera ready form.

In spite of the linguistic shortcoming I believe that the
volume presents a substantial contribution to the analysis
of adaptation of plants to stress habitats..

Osnabrück, February 1984   H. Lieth

CONTENTS

Part Seven: Pollution

INTRODUCTION

The present volume includes papers presented in the International Symposium on Adaptations to Terrestrial Environment, held in Halkidiki, Greece from September 26th to October 2nd, 1982, as well as some invited ones from well known scientists working in the same field.

It seemed rather optimistic to deal just in the same volume with such a variety of organisms (micro-organisms to higher plants) on the basis of their adaptive strategies for survival on land. It would appear as the entire ecology ought to be included. It was a challenge for us. We undertook this challenge hoping that the output would not be unsuccessful.

The Editors allowed the authors of the accepted papers great leeway in terms of thoroughness of their contributions. The quality of the papers included is high while some of them had to be rewritten in order to include valuable comments developed during the Symposium discussions. We have tried to include many papers from Eastern Europe since generally, because of the language problem, they do not get widely known. The Editors wish to express their thankfulness to UNESCO for sponsoring the Symposium in the frame of Man and Biosphere Program; to all scientists who have contributed papers in this volume; and to Mrs. A. Karamanli-Vlahopoulou for her patient and skillful typing of part of the manuscript.

N. S. Margaris
M. Arianoutsou-Faraggitaki

Division of Ecology
Department of Biology
Faculty of Sciences
University of Thessaloniki
GREECE

W. C. Oechel

Systems Ecology Research Group
College of Sciences
San Diego State University
California 92182
U.S.A.

PART I

SYSTEMS' ADAPTATIONS

CONVERGENCE AND NON-CONVERGENCE OF MEDITERRANEAN TYPE COMMUNITIES IN THE OLD AND THE NEW WORLD

A. Shmida and R. H. Whittaker[*] (Department of Botany, Hebrew University, Jerusalem, Israel and Section of Ecology and Systematics, Cornell University, Ithaca, N.Y., USA)

The fathers of phytogeography observed that, wherever on a continent one has a given type of climate, one could expect to find in response to that climate the same structural type, the same plant formation, as expressing in vegetation the impact of the environment on the evolution of plants (Schimper, 1903; Grisebach, 1884). It is a natural idea, particularly it is a natural idea of people for western Europe and the eastern United States (most of whose ancestors were European), where we have the deciduous forest, to perceive the deciduous forest as a conveyor of convergent communities. Also, there has been a development quite recently in the U.S., founded by the International Biological Program (IBP), for a more systematic, second generation approach to this question. This has involved people like Mooney, Solbrig, Orian, Pianka and others in two international biome projects: one comparing Mediterranean communities and the other comparing desert communities of the world (Orian, Solbrig, 1977; Cody, Mooney, 1978).

The Mediterranean communities are a nice case to study ecological convergence because there are five or six of these (depending on whether or not you count the Mallee). They have certain things in common. They are in a climate which we call the Mediterranean, by which it is meant a maritime climate with dry summers, the rain concentrated in the winter months, and not very much rain at that, and warm temperatures. So all these share those characteristics and the expectation would then be that one would find convergence in the plant life and perhaps also in the animal ways of life in these communities. One can indeed find such convergence and a great deal has been written about this in the IBP project (Mooney, 1977). One can ask, specifically for vegetation, what

kind of convergence occurs, to what degree should convergence be expected, in what way does convergence not occur and why, when this does not appear. The approach taken, first of all, is to give a series of short travelogues, or descriptions of particular areas of the Mediterranean vegetation in these areas, commencing with California. These specific vegetation types of all these Mediterranean communities is what is called Maquis (-Chaparral-Matorral-espinal-Mallee, see Shmida, 1981a for details). This is a type dominated by shrubs whith sclerophyll leaves. By this of course is meant evergreen, usually rather small, often spiny leaves, of which the *Quercus calliprinos* in the Mediterranean Basin and *Q. dumosa* in California are perfect representatives. In California one calls such a community the chaparral; this again is dominated by such sclerophyll shrubs. The vegetation there is part of a pattern, along climatic gradients: one can go from a mixed evergreen forest, through the sclerophyll forests, to woodlands, to harsh chaparral, to soft charappal (coastal sage definitive comparable with the Phrygana (Batha) in the Mediterranean Basin) (Shmida and Whittaker, 1981c). All these areas have some such spectrum or gradiation of vegetation as this. This kind of gradient is useful for studies of birds (Cody, 1975; Terbor Terborgh, 1973). It is therefore important to think what one is comparing with what when talking about different Mediterranean areas (Shmida, 1981b). For example, there is little point in comparing hard chaparral in one case with coastal sage in another (Shmida, 1981a).

The community we are primarily concerned with is the sclerophyll-dominated shrubland, represented in the Mediterranean Basin by Maquis and in California by Chaparral. The major leaf types in the chaparral are: first these broad sclerophyll leaves and secondly, in California, there is a needle leaf plant which is represented by the

---

*Deceased October 19, 1982

Chamise (*Adenostoma* sp.) which has tiny leaves (leptophylls). In California, by the large, the more mesophytic, the more humid phase of the hard chaparral gives way to Chamise as an intermediate stage to the coastal sage. This is unduplicated in any of the other Mediterranean areas (Shmida, Whittaker, 1981b).

The next thing one hears about California is the fire cycle. The chaparral is combustible: it grows up and it burns in 10 or 20, sometimes 30 or 50 years, but there is always a characteristic alternation of plants. After a fire there is a flush of annuals and other herbaceous plants which are gradually overtopped or replaced by a canopy of shrubs. These shrubs in time begin to decline and open somewhat. This cycle again is unlike the patterns in other communities but is intrinsic to the Californian chaparral (Shmida, 1981). If one wants to think of a climax in this community, the climax perhaps _is_ the cycle; it is the full flora throughout this pattern of alternation of types after fires. The notable thing about species diversity in the Californian chaparral is that it is low (Shmida, Whittaker, 1981; Shmida, 1981a). We have counted 25 different species in one dunams of a typical chaparral (average of 59 "Whittaker diversity sample" in Southern California).

Another point about the Californian chaparral was observed by Edgar Anderson, who saw the interesting genetic mixture of the chaparral. Many of the species he observed show strong evidence of introgressive hybridization with other species. They are part of hybrid swarms and are connected with other species that are not themselves part of the chaparral and it is difficult to draw a boundary between them. This genetic aspect of the Californian chaparral suggests something about its evolution. This is an assemblage of shrubs that are primarily of Pliocene and Pleistocene ages which have spread extensively in Pleistocene times with the expansion of aridity. They are not sharply differentiated by the longer term processes of evolution, as is the case in some other shrublands in the Mediterranean ecosystems of the Southern Hemisphere. The glacial fluctuations of the Pleistocene period are then another important factor in the pattern of species evolution of the Californian chaparral and the Mediterranean Basin Maquis.

In addition, much recent extension of the Mediterranean shrubland vegetation has been caused by fire. Because the chaparral burns easily and the fire has a tendency to push the chaparral into the woodland and the sclerophyll forest, a drier type of vegetation is pushed into an environment that would otherwise be occupied by a moister type of vegetation. This indeed is true in the overall Mediterranean also. Now from this type of vegetation, which to an American and a European is a typical Mediterranean community, let us turn to something which is atypical to it, at the other end of the world, and talk about South Africa. The community in question is a Fynbos. This is Afrikaans for fine bush. It describes a community of densely mixed small shrubs with narrow branches which are intertwined, one individual species with another, and bearing very large numbers of tiny leaves, or leptophyll leaves. The fynbos is dominated by these leptophylls. The Sonebouth graduates from woodland with an overlayer of protaceous plants (Proteaceae), sometimes three to four meters tall, to dwarf fynbos that are only about a quarter of a meter high. These are affected by the environment, particularly moisture conditions and fire. The long green slopes of the fynbos communities on mountains are reminiscent, in a way, of Scotland. One can argue that in terms of structure and ericoid composition, they can be compared to European heath (Specht, 1979). In many fynbos communities one can observe an overtopping layer of *Protea* sp. and then the mixed layer of small densely interlocking shrubs of the Rosaceae and the Restionaceae. The Restoid is a Southern Hemisphere family which is loosely related to grasses and gives some "grassy" appearance to the fynbos. The striking feature here is not the structure of the fynbos, but the richness of perennial species and their flowers.

The fire succession in South Africa does not look at all like the California vegetation. Restoids tend to be dominant early in the succession and persist right to the end. The leptophyll shrubs tend to appear quite early and rise up to a dominance which they then maintain. They belong to the Ericoid shrubs, because some of them are in the Heath family and they look much like it. The third group, the Proteioids, come up more slowly and become part of the final structure of the community in many cases. This, then, is quite a different fire pattern. The plants by and large are persistent throughout the pattern rather than involving a flush of annuals, particularly herbs as is the case in California. (The annuals are, in fact, practically absent from the fynbos).

Then there is the matter of floral display. The Californian chaparral is beautiful in flower as are phases of the Maquis in Israel (Mediterranean Basin), but the fynbos is so much more so. It compels the attention of the northern biologist. Not only do the flowers seem larger, more colorful, more striking, buth there are flowers which are more peculiar or outlandish to a northern hemisphere biologist.

Another characteristic is the matter of bird and flower relationships. Both areas, California and South Africa, have coevolution of birds with flowers, but the patterns of these are quite different. In North America, this increasing interaction of flowers and birds led to the familiar Hummingbird-red flower to which they are primarily attracted (Grant, Grant, 1968). In South Africa, the evolution led in quite a different direction: it led towards sunbirds and huncreepers in the protea and other flowers and to quite different manners of pollination of the plant.

There can be an evolutionary diversion in some of these cases, given a different fauna to begin with, and the effect of these interactions through evolutionary time. The South African situation has also produced an extraordinary evolutionary explosion in some genera. The genus *Erica* (heath), for example, has about 600 species in South Africa and there are many other genera that are with some hundreds of species. What is the meaning of this evolution of richness? It involved a long evolution on land surfaces that have been in a relatively continuous climate. California was buffetted in the Pleistocene period by glacial displacements, by desertification and by the elevation of mountain ranges along the coast (Wright, 1967). South Africa, while it has by no means been free of climatic change, has been in relatively more consistent climates throughout (Van Zinderen-Bakker, 1978). The effect in South Africa of the alternation of climates, seen as Pleistocene glaciations in the North, was not glaciation but alternate periods of greater drought and humidity. These climatic changes brought on migration of floras into and out of the mountain landscape. Often species migrated out of the center and differentiated and migrated back sometimes into other ranges. Given this complex mountain landscape with alternating climates, the result is, apparently, the extraordinary elaboration of species in some genera such as *Erica*. There is also a remarkable degree of geographic diversity differentiation (Shmida, 1981b).

Another factor which contributes to the species richness is the parent material. If one looks at the western U.S. and thinks of parent material as limestone or nonlimestone, then a large share of the species can be shared between these two, not all, but quite a few. This is true in the glaciated part of the Northwest, i.e. the Wasatch Mountains where the flora is clearly postglacial. In the oldest parts of California, the floras of the different parent materials may overlap a bit, but not strikingly; the bulk of species are specialized for limestone and nonlimestone soils. In California there might be a relationship of 40% sharing of species in a given area between the soils. In South Africa there is very little sharing between the floras on the two soil types (about 5% is a figure quoted). An implication of this would seem to be that during evolutionary time, species that had occurred on one parent material or the other had first become ecotypes and, second, had diverged into species. During the progressive enrichment of flora in South Africa, the species specialized to one soil type or another

and their distribution became very narrow.
The third area of a Mediterranean climate to be
mentioned is that of the Australian heath. The
Australians call these heaths, though many of
them are not heaths in the true sense. By heathlands we mean a shrubland that is dominated by
plants of Ericaceae*, or more broadly speaking
the Heath tribe-Ericaceae. The Australian sand
heaths are quite different. They are a coastal
fringe in Australia centered in the southwest
part of the continent, occurring here and there
along the southern coast and around to the south-
east coast of the continent. Physiogenically, the
Australian sand heaths, for which there is a
rather elegant name proposed -Kwongan- is rather
similar to the fynbos in South Africa. They have
the same dominance by leptophyllous leaves belonging to the Restioidiae and Proteaceae. They have
the same basic structure and the same high species
diversity. They also have the same gradient of
vegetation types from woodland dominated by overtopping Proteaceae, to a tall shrubland, to a
sparse shrubland, to a semi-desert dwarf shrub.
They have many families in common and quite a
few genera in common. There is then a long-term
evolutionary relationship between these two. The
relationship is not a recent one, it goes back
to the break of Gondwanaland and it is expressed
primarily on higher taxonomic levels (that is, a
subfamily represented in one area and an equivalent
subfamily in the other area, and so on) (Raven,
Axelrod, 1974). Much of the Australian heath is
drier than the fynbos and is characterized by a
kind of triple floristic assemblage of quite different sets of species. There are very few species
shared between the three major kinds of soils in the
area (sandstone, limestone and fossil latterite).
The fossil latterite is presumably a relic of a
woodland or forest that existed in the past more
humid climate and through the dessication of the
Plio-Pleistocene period it became the environment

---

*Some others identified heath as leptophyll shrubby
formation dominated on acid soils (Specht, 1979)

of the Australian heath now.
Australia includes another community that some of
the plant geographers have put in this cluster of
Mediterranean formation, and that is the Malee
(Schimper, 1903; Raven, 1971). The Malee is again a
shrubland dominated by species of *Eucalyptus*. The
shrubs are giant shrubs of which individual shoots
are several inches in diameter, at chest height, and
several meters high. The shoots arise from enormous
lignotubers, a vast underground woody potato, which
sends up these shoots. When they burn off in a fire,
a new set of shoots is sent up again. The Malee type
covers extensive tracts of southern Australia, particularly with the Malee scattered in a rather open
growth where the underlying vegetation is sometimes
shrubland or grassland. One can think of this as
shrubland on a differentsize scale of plants, but
different from any other shrublands in the world. It
is essentially an indigenous Australian formation and,
regardless of the principle of convergence of vegetation in response to climate, cannot really be mapped
on any other continent. The Malee appears monotonous,
but it is actually rather rich in species. On a tenth
of a hectare   50-60 species have been counted.
It is not rich on a scale with the fynbos or with
the sandheath.
The last Mediterranean are to be discussed is Israel
as an example of Old World Mediterranean landscape.
The Maquis of Israel with its sclerophyll types,
is very mucy like the chaparral of California in
its overall structure. Disturbances push Maquis
into Garigue and Garigue into Batha (~Phrygana ~
Coastal sage, Shmida, 1981a), etc. If one considers
the disturbance in the history of Israel, one notes
that the vegetation in its original aspects is very
much like that of California. We have counted in a
typical open Maquis 143 species (Mean of 23 "Whit-
taker-diversity-sample", each of them in 1000 m^2
area). (The maximum of 189 different species -the
highest number of plant species per hectare recorded
in the world!!).
Israel has, however, been occupied by man for a very
long time. Pastoral societies go back some millenia.
The area has been longer and more intensively occu-

pied and pressed upon by man than any other part of the world (Shmida, Whittaker, 1981a; Naveh, Dan, 1973). An inference from this might be that we would expect Israel to be something like California, but with a very degraded vegetation and poor species diversity, because in California our experience is that disturbance by grazing reduces species diversity (Shmida, Whittaker, 1981c). What is found, though, is in a rather stunning contrast to what was expected: The Israeli vegetation is fantastically rich in species compared with temperate zone standards.

There is some basis for interpreting this extraordinary richness in Israel. The soil has mostly gone and what is left are eroded rocky limestone surfaces and in many cases these surfaces offer a complex pattern of different sorts of microsites, different sizes of soil pockets, different relationship of rain draining into pockets and different degrees of shelter from different kinds of grazing and browsing animals. The complexity of microsites in which different kinds of plant species can grow is undoubtedly one of the reasons for the high diversity. Along with this is the fact that the plants are small. They are small because they are grazed and it "doesn't pay" to maintain a high profile that is easily grazed. The grazing then, by keeping much of the flora low and lowering the competitive exclusion between species (Paine, 1966; Shmida, Ellner, 1981), contributes to the species diversity. The weeds, mostly annuals are affected by a range of grazers that includes goats, sheep, cattle, donkeys and camels. It appears that diversity may be highest in some of the pastures that are subjected to these various types of "plant predators". Their grazing prevents strong dominance by any species and where grazing pressure is not at its most extreme, plants escape the stress in different ways (Shmida, Whittaker, 1981a).

Israel is situated in the Fertile Crescent where the earliest civilizations appeared and has been affected by human beings since the middle Pleistocene and has been affected severely by man for 10,000-30,000 years. We postulate that, at that time, this extraordinarily rich vegetation, rich in species in a given stand, has appeared (Shmida, Whittaker, 1979). This is in direct contrast with American observations where, when stock is introduced (mostly sheep and cattle) into native prairies and mountain meadows, the usual experience is an abrupt and rapid decrease in species diversity in the natural vegetation because of grazing pressure.

The question then is why are the effects different in the two cases? Why does grazing in the United States drive species diversity down and in Israel grazing has produced the extraordinary richness of species? The answer might work as follows:

In the United States, one starts with communities that are adapted to a modest amount of grazing such as bison and staghorned antelopes. When the grazers are removed entirely, the turf closes in somewhat and there is some drop in species diversity. By and large, more grazing means fewer species right on down until one ends up in a weed field in which there is practically nothing left. With evolutionary time, species that are adapted to grazing will come in from other areas. Some species are fire adapted and are therefore partly preadapted to grazing, some species will be introduced from other continents and are perhaps brought in by caravans. Species accumulate in this range of relatively intense grazing pressure. As a consequence, while there may be no increase in species in the ungrazed part of this spectrum of communities, there is a progressive enrichment of species and ecotypes of existing species which are adapted to grazing and which accumulate through evolutionary time. Carry on this process through a few millenia, as is the case in Israel, and one has a community in which the greatest richness of species is under what would be considered very heavy grazing. The evolutionary origin of these species is a question of interest that one cannot speculate on too far. Some of the plants have probably been recruited from the Arab lands that surround the Levant; some of them probably come from interior Asia. In any case, there has been a great accumulation of species that are adapted to extremely severe grazing pressure, although the more severe pressure, again, can drive species diversity down. The circumstances indicate that in America, under light grazing, there is a high species diversity and in Israel there is extremely high species diversity under heavy grazing.

Some conclusions for this: much evidence has been brought that is nonsupportive of the idea of convergence and one can believe, in fact that the idea of convergence of the disjunct Mediterranean ecosystems has been overstressed in some of the literature. The treatment of convergence has been too simple and one should consider what converges and why and what has not converged and the reasons for that (Shmida, 1981a,b). Convergence research is not completely disputed (e.g., the work of Mooney, Pianka and others). For example, it has to be true that sclerophyll leaves will have similar functions in Australia, California and in Chile, but on the broader scale of communities (which is where the main concern lies) the idea of convergence should be limited. There is some convergence in some respects but then not in others.

There are other factors which contribute to nonconvergence, one of which is the significance of heritage and the historical factors. One knows about the Founder Effect as exerting a long-term genetic influence on a population. When whole communities are considered, one thinks of heritage factors that involve taxa that were available in a given geographical area when the course of evolution began. Different biotas have started with different biological assemblages and throughout the course of evolution different canalizations (in the sense given by Waddington, 1957) occurred so, although there are similar environmental conditions, they diverged in many ecological features. It is known, for example, that Australia has some very bizarre mammals and plants. The Australian fauna, in fact, has a very different spectrum of mammals than the spectrum of mammals that has evolved in the Northern Hemisphere. There are types of mammals in Australia that are unre-

presented in the Northern Hemisphere. These, in particular, are the series of animals that live in trees but browse on leaves. The converse is also true that there are types of animals in the Northern Hemisphere that are not represented in Australian mammalian fauna (Keast, 1972).

This heritage idea could then imply not only that there is a lasting long-term evolutionary implication of the broad taxa with which the evolution on a given continent started, but that the continent diverged through evolutionary time because of the manner in which these taxa interacted with one another.

In summary, the features of nonconvergence between the different Mediterranean ecosystems have been elaborated here in order to compensate for the convergence features which have been so emphasized in the last years. On the whole there is a big difference between the Mediterranean ecosystems of the Northern Hemisphere (including California, the Mediterranean Basin and Chile) and those of the Southern Hemisphere (Sout Africa and Australia). Whereas the southern ones are enormously rich in woody species wich are the product of a long, relatively moderate evolution, the northern ones are richer in annuals derived mainly through the coevolution of human culture and disturbance.

REFERENCES

Cody ML (1975) Towards a theory of continental species diversities, bird distributions over Mediterranean habitat gradients. In Cody ML, Diamond JM, eds. Ecology and Evolution of Communities, pp. 214-257, Harvard University, Cambridge.

Cody ML and Mooney HA (1978) Convergence and dissimilarities of Mediterranean-climate ecosystems, Ann. Rev. Ecol. Sys.

Grant KA and Grant V (1968) Hummingbirds and their flowers, Columbia University Press.

Grisebach AHR (1884) Die Vegetation der Erde, Auf. 2, Leipzig.

Mooney HA (ed.) (1977) Convergent evolution in Chile and California Mediterranean climate ecosystems. US/IBP Synthesis Series 5, Academic Press.

Naveh Z and Dan J (1973) The human degradation of Mediterranean landscapes in Israel. In Di Castri F and Mooney HA, eds. Mediterranean Ecosystems, pp. 373-390. Springer-Verlag.

Orians GH and Solbrig OT (1977) An evolutionary approach to ecosystem. In Orians GH and Solbrig OT, eds. Convergent Evolution in Desert Ecosystems.

US/IBP Synthesis Series 7, Academic Press.

Orians GH and Solbrig OT (eds.) (1977) Convergent evolution in warm deserts. US/IBP Synthesis Series 8, Pennsylvania, Dowden, Hutchinson and Ross Inc.

Paine RT (1966) Food web complexity and species diversity, Am. Nat. 100, 65-76.

Raven P and Axelrod DI (1974) Angiosperm biogeography and past continental movements, Ann. Miss. Bot. Garden 61, 539-673.

Schimper AFW (Trans. Fisher WR) (1903-1904) Plant geography on a physiological basis, Oxford.

Shmida A (1981a) Comparison between the Mediterranean vegetation of Israel and California. Submitted to Ann. of the Missouri Bot. Garden.

Shmida A (1981b) Biogeography of the desert floras of the world. To appear in Even-Ari M and Noy-Meir I, eds. Hot Deserts: Ecosystems of the World, Vol. 12, Elsevier.

Shmida A and Ellner S (1981) On the existence of ecological equivalents. Manuscript.

Shmida A and Whittaker RH (1979) Convergent evolution of arid regions in the New and Old Worlds. In Vegetation and History, ed. Tuxen. Ber. Sym. Int. Vereing. Vegetationskunde, Rinteln, pp. 437-450.

Shmida A and Whittaker RH (1981a) Spiny plants: An adaptation to arid climate? Oecologia (in press).

Shmida A and Whittaker RH (1981b) Pattern and biological microsite effects in two shrub communities, southern California, Ecology 62, 234-251.

Shmida A and Whittaker RH (1981c) Growth-form strategies and species diversity trends along a rainfall gradient -a comparison between Israel and California. Manuscript.

Specht RL (1979) Heathlands and related shrublands of the world -a descriptive study. Amsterdam, Elsevier Scientific Company.

Terborgh J (1973) On the notion of favorableness in plant ecology, Am. Nat. 107, 481-501.

Van-Zinderen-Bakker Jr. EM (1978) Quaternary vegetation changes in southern Africa. In Werger JA ed. Biogeography of South Africa, pp. 131-148.

Waddington CH (1957) The strategy of the genes. London, George Allen and Unwin.

Whittaker RH (1977a) Evolution of species diversity in land communities, Evol. Biol. 10, 1-65.

Wright HE (1967) Late quaternary vegetational history of North America. In Turekian KK, ed. The Late Cenozoic Glacial Ages. Yale University Press.

Wright Jr. HE (1976) The environmental setting for plant domestication in the Near East, Science 194, 385-388.

FUNCTIONING OF TUNDRA VEGETATION IN RELATION TO LIFE FORMS
F.E. WIELGOLASKI (Bot. Inst., Univ. of Oslo, Blindern, Oslo 3, Norway)

1. INTRODUCTION

Vegetation is often classified in
groups according to morphological adaptations to some specific critical
factors. These morphologically similar
groups are called life forms. Probably
the most widely known life form classification is that originally proposed by
Raunkiær (1934). This is based on the
position of the renewal buds or organs,
and the corresponding protection provided during unfavourable cold or dry
periods. Terms such as trees, shrubs,
forbs (dicotyledonous herbs), grasses
and other monocotyledons, bryophytes
(mosses) and lichens are also widely
used for a broad classification of vegetation. Here this terminology is used,
but shrubs are subdivided into deciduous
and evergreen; forbs are classified in
the nearly mat-forming cushion plants,
forbs with appressed rosettes and erect
forbs; and monocotyledons are subdivided
into single shooted and tufted and also
according to their root forms.

The vegetation found in regions with
similar climate will generally be of
the same life forms all over the world
because of the adaptation to the climate. In tundra the growing seasons are
short and decrease both with latitude
and altitude (Fig. 1). Here all the vegetation is very low and no trees are
found. Both in alpine, arctic and
antarctic tundra dwarf shrubs and mat-
forming forbs are common, although the
species need not be related at all. Because of such ecological equivalence,
the use of life forms is rational in

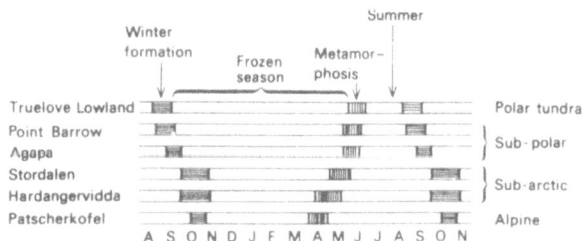

FIGURE 1. Hydrological seasons of tundra
sites. The growing season in summer is
shortest at Truelove Lowland in high
arctic Canada, somewhat longer at the low
arctic sites at Point Barrow in Alaska and
Agapa in Siberia. Stordalen in North
Sweden and Hardangervidda in alpine
Southern Norway may be looked upon either
as subarctic or low alpine sites, while
Patscherkofel in Austria with the longest
growing season clearly is an alpine site.
(Rydén, 1981).

studies on functioning of vegetation in
various parts of the world.

2. CLASSIFICATION

International classification and ordination of tundra plants (Wielgolaski, 1980)
show that monocotyledons survive very well
in extremely low summer temperatures, but
have a low cover percentage under dry
tundra conditions (Fig. 2). Under extreme
conditions, e.g. when it is extremely wet,
most monocotyledons are single shooted
with rhizomes. When the soil is somewhat
drier, the dominating monocotyledonous
species are tufted and have fibrous roots.
Forbs in tundra are found mostly in moderately moist soil (Fig. 2). It is,
however, necessary to distinguish between
different types of forbs to determine

their temperature relationships. Such
subdivisions have shown that most
cushion plants and species with appressed leaf rosettes tend to occupy the
lowest temperature zones, for instance
in the high arctic, normally in relatively nutrient rich soil. Erect forbs,
on the other hand, are found under less
extreme temperatures and normally
in somewhat moister soils. Shrubs are
also generally absent from the regions
with the lowest summer temperatures
(Fig. 2), and they are normally not
found in high arctic or high alpine
regions. Evergreen shrubs tend to dominate in the most nutrient poor communities of low arctic and low alpine
regions, and dwarf shrubs occur on
extremely dry and windswept ridges. Of
the cryptogams bryophytes are common in
the wetter tundra (Fig. 2), particularly pleurocarpous species, while
lichens are found primarily in dry
tundra. Bryophytes are abundant at
both high and low summer temperatures.
Insufficient data, however, are available to draw clear conclusions regarding temperature requirements for
lichens (Wielgolaski, 1980), because
really dry and nutrient poor soil,
where lichens normally predominate, are
seldom found at the coldest sites in
this tundra study.

3. FUNCTIONING

The ability to maintain positive net
assimilation under low temperatures is
one of the most important adaptations
to polar and alpine environments. Many
cryptogams and phanerogams in tundra vegetation have positive net assimilation
far below the freezing point. A temperature compensation point of -5°C

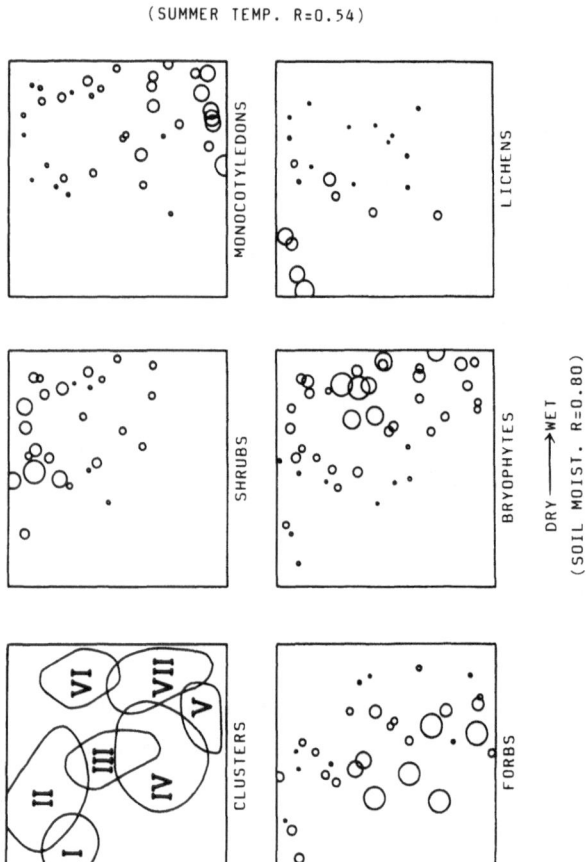

FIGURE 2. Ordination of 52 tundra stands
for 5 main life forms in relative cover
analysis (cover indicated by the size of
the circles). I = Lichen heath, II =
Mesic dry heath, III = Shrub meadow heath,
IV = Forb meadow, V = Monocot. meadow,
VI = Dwarf shrub bog, VII = Wet meadow.
(Wielgolaski, 1980).

has been found in the Alps for Oxyria
digyna (Pisek et al., 1967), Ranunculus
glacialis (Moser, 1969) and Loiseleuria
procumbens (Larcher et al., 1975).
Tieszen et al. (1980) found positive net
photosynthesis down to -4°C in the grass
Dupontia fisheri in arctic North America.
Ranunculus glacialis is also one of the
phanerogams found at the highest eleva-

tions. In Norway Resvoll (1917) found this species to accumulate photosynthates and nutirents over several years in the vegetative parts until the surplus was sufficient for flower production. This may be seen as an adaptation to the extremely short growing season in the high alpine zone snow beds, where they normally grow.

The temperature minima for net photosynthesis in mosses are also low in tundra. Kallio and Kärenlampi (1975) observed net photosynthesis down to nearly -10°C in their experiments in Finland for Rhacomitrium lanuginosum, Dicranum elongatum and Pleurozium schreberi. They further report that lichens as a group may continue net photosynthesis at lower temperatures than any other plants. A compensation point of -20°C was found for Cetraria nivalis in sub-arctic Finland. Several other lichen species showed minima for CO_2 uptake below -10°C in laboratory experiments (Lange, 1965; Atanasiu, 1971; Kallio, Kärenlampi, 1975). These experiments suggest a more clear relationship of lichens to low temperature than indicated in the ordination analysis (Fig. 2).

In addition the optimum temperature for photosynthesis is often relatively low in tundra plants. For vascular plants this may be particularly true in polar regions as an adaptation to the generally low summer temperatures in the areas. Tieszen et al. (1980) observed temperature optima for leaf photosynthesis between 10 and 15°C for several higher plant species in arctic North America. The maximum temperature for

net assimilation, however, seems to be surprisingly high for tundra plants, often above 30°C, particularly for alpine plants. In alpine tundra there are greater diurnal fluctuations in temperature than in polar regions, and leaf surface temperatures above 30°C are not uncommon on sunny days at high altitudes in temperate zones.

Mosses also have relatively low optimum temperatures for photosynthesis. Tieszen et al. (1980) report optima from 10 to 19°C for several tundra species, but stress that photosynthesis is only slightly reduced at 5°C. Pogonatum photosynthesizes at 55% of the maximum rate at 0°C. Kallio and Kärenlampi (1975) report optimum temperatures of 10°C for the mosses Dicranum elongatum and Polytrichum juniperinum, and 5°C for the moss Rhacomitrium lanuginosum sampled from various parts of the world. Lowest optimum net photosynthesis temperatures are observed in lichens. Kallio and Kärenlampi (1975) report optimum values close to 0°C in Stereocaulon paschale as well as for some other lichen species. Lange (1965) also found an optimum temperature close to 0°C in a Spitsbergen strain of Stereocaulon alpinum. Alpine strains of this species and other alpine lichen species show somewhat higher optimum temperatures, up to about 10°C for some Cladonia and Cetraria species.

Leaf structure can influence light interception, heat interception from the ground, and heat loss from the plant, and is an important link between life form and ecological function. This may be especially true in tundra vegetation, which is sensitive to small differences in temperature.

Greatest solar interception per unit leaf area and highest leaf temperatures are generally found when radiation reaches the leaves at right angles. In many parts of the world leaf temperatures are then too high for optimal growth. At high latitudes, however, the solar angle is low and temperatures normally are below optimum, at least in moist situations. This means that presence of vertical leaves in polar tundra plants may increase their production.

FIGURE 3. Steeply inclined leaves increases the solar interception at low solar angles (high latitudes) (left), while more horizontal leaves increases the interception at higher solar angles (right).

Many tundra species do have nearly vertical leaves, especially monocotyledons (Fig. 3, left). It is not surprising that high net photosynthesis is found in single shooted monocotyledons at high latitudes (Tieszen et al., 1981), where the shading effect is moderate even for the inner, nearly vertical leaves. As this life form in tundra normally grows in very wet conditions, the leaf temperature seldom is overoptimal, even on the nearly vertical unshaded leaves, because of cooling by transpiration. This may explain why this life form is favoured by the low solar angles at high latitudes when the soil moisture is high enough.

Some tundra plants have as mentioned nearly horizontal leaves. This means a decreased solar interception at the low sun angles at high latitudes compared to lower latitudes (Fig. 3, right). Leaves near the soil surface, however, always receive a heat surplus from the ground during the day, particularly if the soil is dry. Most tundra plants in such environments are protected against overheating by accumulation of several dead leaves between the living ones. This is the case in Dryas spp. and in other cushion plants, in most plants with appressed rosettes, and in tufted monocotyledons. However, if the plants grow in moist soil and/or the most important growth period takes place in early spring just after snow melt (i.e. before the soil has dried), then an increased temperature may be favourable to the tundra plants. Species such as Oxyria digyna and Ranunculus glacialis are examples of tundra plants with leaves which are nearly horizontal, particularly in spring.

Many other tundra plants are also adapted to the short growing season by having very rapid growth in spring as soon as the snow melts off. In some cases growth starts even before snow has melted because of penetration of sunlight through up to about 50 cm snow cover, dependent on the consistency of the snow. Plant growth under the snow may take place only if the upper soil layer is not frozen, but this is often the case in late winter after a good snow cover. The growth explosion in spring is mostly an elongation of small leaves initiated the previous fall and protected by bud scales and sheaths through the winter. This elongation demands a rapid and abundant supply of both

energy and nutrients from winter
storage organs.

TABLE 1. Average ratios of tundra phytomass in some vegetation types. (Wielgolaski et al., 1981.).

Region	Shoot to root (live)	Green to non-green live vascular plants	Live to dead above-ground vascular plants
Desert and semi-desert	1:0.9	1:2.3	1:1.9
Wet sedge meadows	1:21	1:23	1:1.6
Mesic-dry meadows	1:5.0	1:7.7	1:0.8
Dwarf shrub tundra	1:3.1	1:12	1:0.6
Low shrub tundra	1:2.0	1:19	1:0.2
Forest tundra	1:0.8	1:15	1:0.1

The adaptation in tundra plants to a
short growth period, often less than 8
weeks, and a low average temperature
even in this period, often less than
5°C, may, therefore, explain the large,
living non green storage biomass in relation to green often found in tundra
regions (Table 1). In monocotyledons
and forbs in low arctic and low alpine
tundra, energy and nutrients are stored
in roots and other below-ground organs.
This means a great root mass compared
to the tops. In woody plants as much is
stored in above-ground parts as below-
ground. Very strong translocation takes
place in the fall and early spring, to
and from the storage organs. This transport and transformation also demand a
lot of energy.

In extreme conditions sufficient energy
may not be available to support both
the storage and translocation systems.
Plants, for instance in high arctic
polar desert, therefore, often are adapted to reduce the translocation by
keeping the green leaves all the year.
This is the case for many of the polar
desert forbs and monocotyledons which do
not have the high root mass compared to
tops as observed in the less extreme
tundra areas (Tab. 1). Instead the energy
and nutrients are stored in the leaves
(always near the ground) which are nearly
evergreen, even if most of the leaf tips
are dead. Svoboda (1977) observed that in
the Canadian high arctic tundra the lower
parts of leaves in the middle of densely
tufted Carex nardina live 4 to 6 years.
The leaves of Saxifraga oppositifolia are
also mostly evergreen in these polar deserts, and the leaves of Dryas live
more than one year. In alpine areas, for
instance in Norway, similar adaptations
are found in Juncus trifidus and Kobresia
myosuroides growing on exposed ridges.

Evergreen dwarf shrubs are common in dry,
poor and wind swept ridges both in low
arctic and alpine regions. It is observed
that leaves remain active for 2-3 years
on these plants (Flower-Ellis, 1975;
Callaghan, Collins, 1981). Shrubs are
close to their border for survival in
tundra. They have to be low and creeping
not to be too strongly exposed to wind
and to be protected by some snow cover
against the lowest temperatures. In the
high arctic the conditions are normally
too extreme to the above-ground reserves
in woody parts and buds of even the evergreen dwarf shrubs, and all shrubs are
normally missing in that region.

Evergreen tundra plants generally have a
slow metabolic rate. As they are green in
spring, they typically do not have as
rapid growth in the spring as deciduous
species. Relatively little energy is used
for translocations, although nutrients may
be moved from old leaves to the new growth
places when this is needed. This internal

nutrient circulation may be seen as an adaptation in evergreen plants to grow on nutrient poor soil.

TABLE 2. Species of vascular plants in alpine southern Norway stratified according to altitudinal limits.(Dahl, 1975).

Altitudinal limit	Number of species	Life form						
		Phanero-phyte	Chamae-phyte	Hemi-crypto-phyte	Geo-phyte	Helo-phyte	Hydro-phyte	Thero-phyte
> 2000 m	29		44.8%	55.2%				
1800–1999 m	39		25.6%	66.7%	2.6%	2.6%		2.6%
1600–1799 m	75	8.0%	32.0%	48.0%	4.0%	5.3%		2.7%
1400–1599 m	63	4.8%	9.5%	65.1%	11.1%	1.6%	3.2%	4.8%
1200–1399 m	138	6.5%	10.1%	51.4%	10.1%	10.1%	2.9%	8.7%
1000–1199 m	109	2.8%	5.5%	55.0%	10.1%	11.9%	5.5%	9.2%

The environmental factors within tundra thus select for various life forms through adaptation and specialization to the conditions found. Previous studies (e.g. Dahl, 1975) have shown that life forms with overwintering buds just above the soil surface, chamae-phytes according to Raunkiær (1934), are favoured at increasing climatic seve-rety (Table 2), particularly tempera-ture, with altitude. In addition, life forms seem to vary with latitude and with nutrients and soil moisture.

REFERENCES

Atanasiu L (1971) Photosynthesis and respiration in some lichens in relation to winter low temperatures, Revue Rou-maine de Biologie, Ser. Bot. 16, 105-10.

Callaghan TV and Collins NJ (1981) Life cycles, population dynamics and the growth of tundra plants. In Bliss LC, Heal OW, Moore JJ, eds. Tundra Eco-systems: A Comparative Analysis, pp. 257-84. Cambridge, University Press. xxxvii + 813 pp., illustr.

Dahl E (1975) Flora and plant socio-logy in Fennoscandian tundra areas. In Wielgolaski FE, ed. Fennoscandian Tundra Ecosystems. Part 1. Plants and Microorganisms, pp. 62-7, Ecological Studies, 16. Berlin, Heidelberg, New York, Springer. xv + 366 pp., illustr.

Flower-Ellis JGK (1975) Growth in populations of Andromeda polifolia on a subarctic mire. In Wielgolaski FE, ed. Fennoscandian Tundra Ecosystems. Part 1. Plants and Microorganisms, pp. 129-34, Ecological Studies 16. Berlin, Heidelberg, New York, Springer. xv + 366 pp., illustr.

Kallio P and Kärenlampi L (1975) Photo-synthesis in mosses and lichens. In Cooper JP, ed. Photosynthesis and produc-tivity in different environments, pp. 393-423. Cambridge, University Press. xxiv + 715 pp., illustr.

Lange OL (1965) Der CO_2-Gaswechsel von Flechten bei tiefen Temperaturen, Planta 64, 1-19.

Larcher W, Cernusca A, Schmidt L, Grabherr G, Nötzel E and Smeets N (1975) Mt. Patscherkofel, Austria. In Rosswall T, Heal OW, eds. Structure and Function of Tundra Ecosystems, pp. 125-39, Ecol. Bull., 20. Stockholm, Swedish Natural Science Research Council. 450 pp., illustr.

Moser W (1969) Die Photosyntheseleistung von Nivalpflanzen, Berichte der Deutschen botanischen Gesellschaft 82, 63-4.

Pisek A von, Larcher W and Unterholzner R (1967) Kardinale Temperaturbereiche der Photosynthese und Grenztemperaturen des Lebens der Blätter verschiedener Spermato-phyten. I. Temperaturminimum der Netto-assimilation, Gefrier- und Frostschadens-bereiche der Blätter, Flora (Jena) Abt. B 157, 239-64.

Raunkiær TR (1934) The life forms of plants and statistical plant geography. Oxford, Clarendon Press.

Resvoll TR (1917) Om planter som passer til kort og kold sommer, Archiv for Mathematik og Naturvidenskab 35, 1-224.

Rydén BE (1981) Hydrology of northern tundra. In Bliss LC, Heal OW, Moore JJ, eds. Tundra Ecosystems: A Comparative Analysis, pp. 115-37. Cambridge, Universi-ty Press. xxxvii + 813 pp., illustr.

Svoboda J (1977) Ecology and primary production of raised beach communities, Truelove Lowland. In Bliss LC, ed. Truelove Lowland, Devon Island, Canada: A High Arctic Ecosystem, pp. 185-216. Edmonton, Alberta, University of Alberta Press. xxi + 714 pp., illustr.

Tieszen LL, Miller PC and Oechel WC (1980) Photosynthesis. In Brown J, Miller PC, Tieszen LL, Bunnell FL, eds. An Arctic Ecosystem, the Coastal Tundra at Barrow, Alaska, pp. 102-39. Stroudsberg, Pennsyl-vania, Dowden, Hutchinson & Ross, Inc. xxv + 571 pp., illustr.

Tieszen LL, Lewis MC, Miller PC, Mayo J, Chapin FS III and Oechel W (1981) An ana-lysis of processes of primary production in tundra growth forms. In Bliss LC, Heal OW, Moore JJ, eds. Tundra Ecosystems: A Comparative Analysis, pp. 285-356. Cambridge, University Press. xxxvii + 813 pp., illustr.

Wielgolaski FE (1980) Klassifikasjon
og ordinasjon av tundraplanter inter-
nasjonalt. In Baadsvik K, Klokk T, Rønn-
ning OI, eds. Fagmøte i Vegetasjonsøko-
logi på Kongsvoll, 16.-18.3.1980, pp.
232-43. Trondheim, K. norske Vidensk.
Selsk. Rapp. Bot. Ser. 1980-5, 279 pp.

Wielgolaski FE, Bliss LC, Svoboda J
and Doyle G (1981) Primary production
of tundra. In Bliss LC, Heal OW, Moore
JJ, eds. Tundra ecosystems: A Comparative Analysis, pp. 187-225. Cambridge,
University Press. xxxvii + 813 pp.,
illustr.

ADAPTIVE PLANT EVOLUTION IN THE ALPINE ENVIRONMENT OF THE GREEK MOUNTAINS

K. Papanicolaou and S. Kokkini (Laboratory of Systematic Botany, Department of Biology, School of Sciences, University of Thessaloniki, Greece).

1. INTRODUCTION

The Greek flora is estimated to comprise nearly 6.000 species of flowering plants, of which 500-600 occur in the alpine region. As less information is available on the Greek flora and vegetation than for the countries in central and northern Europe, much emphasis is now put on the floristic exploration as well as the study of variation and differentiation, and many areas in Greece - including islands and mountain tops - have proved eminently suited for the study of isolation and evolutionary phenomena. Whereas the flora of northern Europe was wiped out by the Pleistocene glaciations, the Greek flora remained *in situ* and is thus sufficiently old and rich to be of great interest for students of evolution. The extraordinary topographic and geological diversity of the country has speeded up evolution by producing geographically and ecologically isolated small populations. We are thus presented with a great natural laboratory for the study of many fascinating aspects of evolution and the origin of species.

Greece is a mountainous country. A great backbone of mountains stretches from north to south and further via Crete and the south Aegean island arc to Asia Minor. The highest mountain, Olympus, which is situated on the borders between Thessaly and Macedonia reaches an altitude of 2917 m, and there are numerous summits above 2.000 m (Fig. 1). The complicated geological history and the multitude of substrates -limestone, schist, granite, serpentine- are of great importance for the diversity of the flora and vegetation.

In very general terms three altitudinal zones can be distinguished in the Greek mountains: 1) the zone of the macchies from sea level up to 600-800 m, 2) the forest zone, dominated either by coniferous or deciduous trees, from 600-800 m up to 1800-2000 m, and 3) the alpine zone above the tree line. Whereas the flora of the macchie is mainly composed of species with a wide mediterranean distribution, and the forest zone contains a large central European element, the alpine flora has a high proportion of endemics, i.e. species with a restricted distribution ranging from a single mountain to maybe most of the Balkan Peninsula. The high percentage of endemics indicates a strong and long-continued isolation of the flora of the mountain tops. From an evolutionary standpoint they are islands, at least to those species that are unable to cross interjacent lowland areas.

Although the flora in the alpine belt of the Greek mountains is small when compared with that of lower altitudes (Fig. 2), the data presented do show that vascular plants are adapted to survive in very extreme environments at high altitudes.

2. THE CLIMATE IN THE GREEK MOUNTAINS

The climate of the lowest foothills in Mediterranean, i.e. with hot summers and cool wet winters, although the temperatures decrease from northern to southern Greece. The maximum precipitation occurs in November, with usually a fair amount of rain falling until May.

In Greece the annual rainfall decreases in a general northwest-southeasterly direction. Olympus is drier than Pindhus, and because of their proximity to the sea northern and eastern slopes receive more moisture than those in the south and west and thus have a more luxuriant vegetation. With increasing altitude precipitation is more evenly distributed throughout the year. Moisture from fog or low clouds is an important factor especially in the valleys at montane levels. In the alpine zone the summer climate is dry and sunny, and precipitation varies within wide limits from year to year (Table 1). In Mt Olympus the maximum precipitation recorded during the period 1966-73 for the months July to September is 542 mm and the minimum 86 mm (Cf. Strid, 1980). The variation is explained by the fact that most of

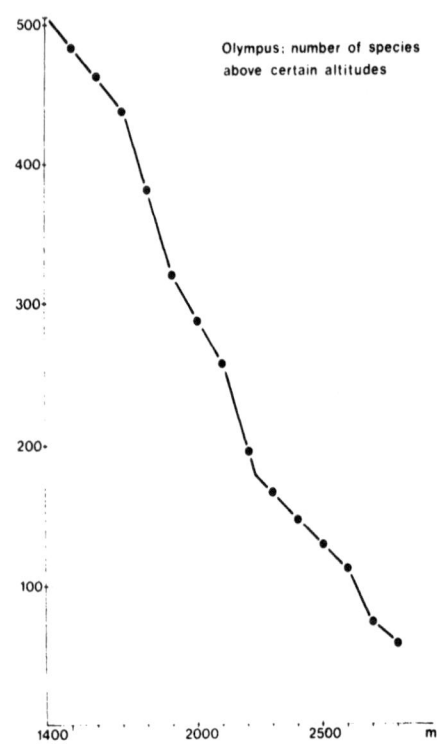

FIGURE 2. Mt Olympus. Number of species above certain altitudes (from A. Strid, pers. commun.).

TABLE 1. Main air temperature (°C) and precipitation (mm) at the summit of Mt Olympus (2815 m) (from A. Strid, pers. commun.).

	Air temperature			Precipitation (mm)		
	JUL	AUG	SEP	JUL	AUG	SEP
1966:	6.3	7.1.	3.5	16.1	44.4	177.4
1967:	5.3	7.1	3.8	235.2	66.6	242.1
1968:	6.4	5.1	3.8	7.7	36.1	88.0
1969:	3.8	6.2	4.5	24.9	13.6	48.8
1970:	6.1	6.4	3.4	148.8	25.4	112.7
1971:	4.5	7.1	2.2	78.6	82.6	93.7
1972:	5.5	4.9	2.5	192.1	95.2	41.7
1973:	6.6	4.7	4.5	111.3	56.2	110.1

Note the extreme year-to-year differences in summer rainfall (543.9 mm in 1967, 131.8 mm in 1968).August is generally drier than July and September.

FIGURE 1. View of three Greek mountains; A:Mt Taigetos (S Greece), B:Mt Athos (NE Greece), C:Mt Xerovouni (Euboea).

the rain falls erratically in thunderstorms. Melting of snow accounts for some moisture throughout the summer, but the limestone is well jointed and most of the melt-water dissappears quickly into cracks and fissures, so that the ground is usually dry only a short distance below a patch of melting snow. In schistose mountains, i.e. Belles and Kajmakcalan there are numerous seepage areas and wet meadows. Phytogeographically the alpine flora of these mountains is remarkable for its high proportion of central and northern Balkan endemics, some of which extend to the Carpathians.

3. ALPINE HABITATS

Above the tree-line the type of vegetation is determined by the harshness of the climate and the prevailing drought conditions. Summer precipitation is erratic, and most of the rain in the limestone mountains falls in violent thunderstorms. Showers of snow or hail may fall throughout the summer, but water disappears quickly into deep cracks and fissures, and the mountain-sides are dry and desolate, covered with frost-shattered rock debris. The amount of moisture obtained from fog or low clouds is difficult to estimate, but is probably of less importance in alpine habitats than in the valleys at lower altitudes.

There is no clear differentiation into a low-alpine and a high-alpine zone. Species of screes and rock crevices occurring in the summit area can, with few exceptions, be found lower down where conditions are suitable. In Mt Olympus typical species of snow-bed meadows are usually restricted to altitudes of between 2300 and 2700 m. Roughly speaking four types of habitats can be distinguished (cf. Strid loc. cit. 1980): 1. Snow-bed Meadows. In shallow depressions, as well as on flat or slightly sloping ground where the snow does not melt until midsummer and a layer of fine soil has accumulated a characteristic plant community has developed, generally dominated by the grass *Alopecurus gerardi* which forms a dense verdant turf (Fig. 3). 2. Grassy Moors with Solifluction. Vast areas of moderately sloping ground

are covered with a vegetation dominated by species of *Sesleria*, *Festuca* and *Carex*. The substrate is a mixture of fine soil, gravel and coarser material. In the early spring when the uppermost layer is saturated and the underlying layers are still frozen there is a slow downward creep of the surface soil material. This has resulted in often conspicuous terracing with tussocks of wiry grasses growing along the edges of the terraces the horizontal "steps" being almost devoid of vegetation (Fig. 4). On flat ground the action of frost may result in the formation of polygones, the coarser material moving to the periphery (Fig. 5). 3. Screes. Slopes that are too steep for any fine soil to be retained are covered with stones and screes. There is very little vegetation and few species are able to grow in the more extreme habitats afforded by mobile screes with particles of uniform size (Fig. 6). Several of them are, however, endemics or are otherwise of interest. 4. Rock Crevices. Several species are typical chasmophytes, i.e. adapted to grow in cracks in the rock. This category, too, includes some of the most remarkable endemics of the Greek mountains.

FIGURE 3. Snow-bed meadow in Mt Olympus (in the saddle between Ag. Antonios and Bara, c. 2350 m) (from A. Strid, 1980, pl. 88).

FIGURE 4. Grassy Moor with Solifluction dominated by *Festuca cyllenica* in Mt Kyllini, c. 2100 m.

FIGURE 5. Polygones caused by action of frost on flat ground in Mt Olympus (from A. Strid, 1980, pl. 93).

FIGURE 6. Screes. A slope in Mt Giona which is very steep for any fine soil to be retained is covered with stones and screes.

3.1. Snow-bed Meadows

One of the most characteristic and well-defined plant communities on most of Greek mountains is found on flat or slightly sloping ground where patches of snow remain until the end of June. Phytosociologists use the term "chionophilous" plant communities. These snow-bed meadows are usually small and sharply delimited from the surrounding vegetation, standing out as smooth verdant patches. Good examples are found on Mt Olympus. The dominant grass is practically always *Alopecurus gerardi*. Another common grass, *Poa pumila*, can be recognized by the small tufts of short broad leaves and the culms several times the length of the leaves. Less common are *Anthoxanthum alpinum*, *Phleum alpinum* and *Nardus stricta*, three grasses with an alpine or boreal-alpine distribution. The first species to flower at the edge of melting snow patches are usually *Crocus veluchensis*, *C. sieberi* (Fig. 7) and the blue *Scilla nivalis*. *Corydalis densiflora* and *C. parnassica* may be found in similar places. Among the most characteristic constituents of snow-bed meadows are the Greek endemics *Trifolium parnassi* and *Herniaria parnassica*. Two more interesting little herbs with a similar life-form, *Beta nana* and *Trinia guicciardi*, are usually found at the

FIGURE 7. *Crocus sieberi* Gay flowering at the edge of melting snow in Mt Dhirfys.

edge of snow-bed meadows, i.e. beyond the very dense *Alopecurus* turf. They both have a stout tap root which may be regarded as an adaptation to the prevailing drought of late summer. Phytogeographically the flora of snow-bed meadows is far from uniform. It tends, however, to comprise a larger number of species with a wide distribution than does the flora of screes or rock crevices. Widespread arctic-alpine or boreal species are *Anthoxanthum alpinum*, *Nardus stricta*, *Sagina saginoides* and *Gnaphalium supinum*. Some occur in the mountains of central and southern Europe as, for instance, *Alopecurus gerardi*, *Phleum alpinum*, *Trifolium pallescens*, *Arenaria biflora* and *Gnaphalium hoppeanum*; others are Balkan or Greek endemics (e.g., *Luzula pindica*, *Corydalis parnassica*, *Trifolium parnassi*, *Herniaria parnassica*, *Erigeron epiroticus* and *Beta nana*) and a few have connections with southern Italy (*Ranunculus sartorianus*).

3.2. Grassy Moors with Solifluction

On moderate sloping ground where solifluction has resulted in some degree of terracing and the plant cover is between 20 and 50 per cent the vegetation can usually be characterized as a dry grassy moor.

This type of vegetation covers extensive areas and is not as well defined as that of snow-bed meadows. The dominant grasses are usually *Sesleria*, *Festuca* and to a lesser degree *Poa*. The graminids grow in tussocks, often along the edges of shallow terraces formed by solifluction, whereas the interjacent more or less flat ground is often too mobile to support any plants at all (Fig. 4). On horizontal ground frost-heaving occasionally results in the formation of polygones, i.e. polygonal or roughly circular areas 1/2 - 2 m in diameter composed of finely comminuted soil, separated by narrow intervening areas comprised of larger stones (Fig. 5). The grassy moors are generally poor in species, and the plants that do grow there can often also be found in other habitats. The following species may be mentioned: *Minuartia verna*, *Cerastium banaticum*, *Paronychia rechingeri*, *Silene ciliata*, *Iberis sempervirens*, *Acinos alpinus*, *Thymus boissieri*, *Carduus armatus*, *Centaurea pindicola* etc.

3.3. Alpine Screes

Slopes that are too steep to retain significant quantities of fine soil are covered with frost-shattered rubble supporting a sparse vegetation (Fig. 6). From a distance they may look completely barren but in fact this type of habitat contains a number of handsome and interesting species, including several endemics. Most species found on the screes have slender straggling stems or rhizomes that creep between the mobile stones to reach what little soil and moisture there may be available underneath. Grasses are generally insignificant, and the vegetation covers less than 30 per cent of the ground. Among the most characteristic are *Cardamine carnosa*, *Sedum magellense*, *Saxifraga glabella*, *Carduus armatus* and *Doronicum columnae*. Others prefer more open, but somewhat damp gravel or scree, often near snow patches. A plant of such habitats is *Ranunculus brevifolius*, a tiny buttercup with broad, toothed, glaucous leaves. Drier fell-fields provide an even more extreme habitat to which few species have become adapted. In Mt Olympus among these are at least five of the local endemics, however, viz.

Cerastium theophrasti, *Alyssum handelii*, *Ryncho-sinapis nivalis*, *Viola striis-notata* and *Achillea ambrosiaca*. In addition to the local endemics there are some other interesting species such as *Euphorbia capitulata*, *Linaria alpina* and *Galium degenii*.

3.4. Rock Crevices in the Summit Area

A few species are adapted to grow in cracks of rocks at alpine levels. The most conspicuous chasmophyte of the highest peaks is *Helichrysum sibthorpii* (Fig. 8), *Campanula oreadum*, *Silene orphanides*, *Aethionema orbiculatum* and *Viola delphinantha*.

In Mt Olympus the most widespread species of *Saxifraga* above 2700 m is *S. spruneri* which forms dense smooth mats or cushions in rock crevices or occasionally on stabilized screes. It can be distinguished from *S. scardica* (which is less common in the summit area) by the obtuse, softly hairy leaves and snow-white flowers on short stems.

S. sempervivum is easily recognized by the spike of purple flowers; it is fairly common at alpine levels. The similar *S. grisebachii*, which is usually found at lower altitudes, has much broader, spathulate leaves.

FIGURE 8. *Helichrysum sibthorpii* Rouy growing in a rock crevice in Mt Athos.

4. ENVIRONMENTAL FACTORS IN THE ALPINE BELT IMPORTANT TO PLANTS

Flowering plants are especially notable for their great phenotypic plasticity, and this characteristic is strongly developed in many arctic and alpine species. Although most flowering plants are strongly mobile as species, the individuals are stationary. Because they cannot retreat from environmental vicissitudes, plasticity in both structure and physiology are of obvious value to plants in a severe and variable climate. Unfortunately, our awareness of such pronounced phenotypic variation may cause us to overlook very similar genetic variation. The latter may not be recognized unless the plants are grown in a controlled environment. The exact causes of most phenotypic changes are obscure and may prove to be complex. Indeed the complexity may prevent us from recognizing them as being nongenetic, and thus lead us into taxonomic difficulties. The most important characteristics of the environment in the alpine belt are: 1. Soils, 2. Light intensity, radiation and quality, 3. Temperature, 4. Wind, and 5. Snow cover.

5. MORPHOLOGICAL ADAPTATIONS IN PLANTS OF THE ALPINE BELT

The life-form spectrum of a flora can be used as a rough measure of the prevailing climate and the selective action of the climate on those plant forms best adapted for survival.

Looking at the alpine flora of the Greek mountains in the field one soon discovers a number of morphological features that are unknown or rare at lower levels on the mountains and in their surroundings. The most conspicuous of these are dense grass tussocks, acaulescent growth, dense cushion growth, and sclerophylly (Fig. 9), and we shall offer a few examples and observations for each of these.

5.1. Dense Grass Tussocks

A very characteristic feature of the vegetation in the alpine belt is the appearance of the grasses, most of which form large and dense tussocks of a dull brownish-green colour (Fig. 4). A closer inspection reveals that most leaves and culms of such a tussock are dead and decaying. The living parts tend to be largely concentrated to the centre of the tussocks, where they are protected by a felt-like mass of decaying leaf- and culm-bases which provide excellent temperature insulation for the innovation shoots in the centre.

5.2. Acaulescent and Prostrate Growth

A second peculiarity of the alpine flora is that in a number of species of diverse groups the elongation of the flowering stems is impeded so that they become acaulescent. A good example is afforded by *Beta nana* Boiss. & Heldr. (Fig. 10), the leaves and flowering stems of which are spreading on the ground and curving downwards when the plant is uprooted. The same habit occurs in occasional species of other genera, such as *Carlina acaulis* (Boiss. & Heldr.) Drude (Asteraceae), the capitulum of which is sessile, borne in the centre of a large rosette of spiny spreading thistle-like leaves, and *Trinia guicciardi* (Boiss. & Heldr.) Drude (Apiaceae). These plants evidently exploit the congenial day temperature at the soil surface at the same time as the conduction of water to their inflorescences is facilitated by the brevity of the stem and by its protected situation in the

centre of the lead-rosette. The prostrate stems occuring in species of amongst others *Trifolim parnassi* Boiss.& spruner (Fig.11), *Campanula orphanidea* Boiss., *Herniaria parnassica* Heldr.& Sart. and *Asperula suberosa* Sibth.& Sm. are apparently analogous adaptations to exploit the favourable day temperatures at the soil-surface.

5.3. Dense Cushions

Certainly the cushion growth-form of many chamae-phytes is a form well adapted to take advantages of the higher surface temperatures, reduced wind speeds as well as being adapted to greater survival in areas of little snow cover in winter. This grow-form has evolved in unrelated plant families in the alpine regions and is a good example of parallel evolution of a form well adapted for survival in the severe tundra environments (Bliss, 1956). This ecological parallel to the acaulescent and prostrate growth-form mentioned above is provided by species like *Saxifraga sancta* Griseb., *Minuartia stellata* (E.D. Clarke) Maire & Petitmengin (Fig. 12), *Astragalus taygeteus* Strid & Pers. and *Potentilla deorum* Boiss. & Heldr. These form dense cushions consisting of richly branched interwoven stems and roots with the surface covered by dense aggregated leaves (Fig. 11).

5.4. Sclerophyllous Growth

Most of the woody plants in the alpine belt are thin-stemmed, sclerophyllous shrubs. Species with this growth-form are fairly numerous and occur in a number of genera, such as *Vaccinium*, *Satureja*, *Genista*, *Prunus* and *Daphne*. The thin stems of these plants have no efficient protection against frost, but since their leaves are small and xeromorphic they may evidently stand nightly freezing with complete cessation of water conduction without suffering any damage.

6. ALPINE LIFE-FORMS OF PLANTS IN THE GREEK MOUN-TAINS

Raunkiaer (1934) reported a chamaephyte climate for the arctic and alpine tundras (treeless expanses beyond climatic timberline) in which chamaephytes

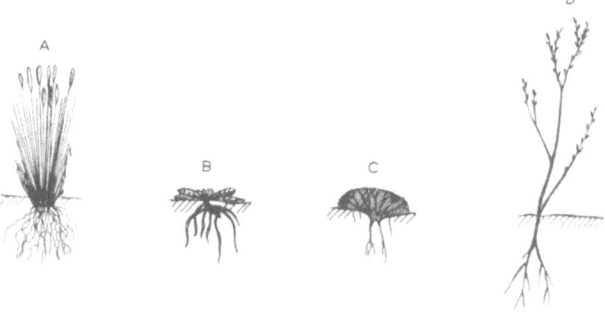

FIGURE 9. Diagramatic drawing of the four most important phanerogamic life-forms in the alpine belt of the Greek mountains. A. Tussock grass, B. Acaulescent rosette plant, C. Cushion plant, D. Sclerophyllous shrub. (from Hedberg, 1964b).

FIGURE 10. *Beta nana* Boiss. & Heldr. A characteristic acaulescent rosette plant. Note the stout tap root (to the left) and the rosette leaves (to the right) (from A. Strid, 1980, pl. 92).

FIGURE 11. *Trifolium parnassi* Boiss. & Spruner. Specimen grown in the Copenhagen botanical garden, maintaining under these conditions its prostrate habit. Material from Mt Grammos.

FIGURE 12. Dense cushion formed by *Minuartia stellata* (E.D. Clarke) Maire & Petitmengin in Mt Killini.

constitute at least 20 per cent of the life form
spectrum. Phanerophytes, plants with perennating
buds more than 30 cm above the ground, and Thero-
phytes or annuals are either lacking or are at
least quite minor in the tundra. The most important groups are the chamaephytes and hemicryptophytes. As a generalization, the percentage of
chamaephytes gradually increases as the climate
becomes more severe with increase in latitude-
arctic, and an increase in altitude-alpine (Raunkiaer loc. cit. 1934). Students of climatic adaptations in plants have often attempted to classify
the species studied into physiognomic life-forms,
each of which is believed to result from parallel
climatic adaptations in different taxonomic groups
of plants. The above mentioned Raunkiaer's life-
form system was based upon "the adaptation of the
plant to survive the unfavourable season, having
special regard to the protection of the surviving
buds or shoot-apices" (Raunkiaer loc. cit. 1934,
p. 112). As this Raunkiaer's life-form system is
not so easily applied to the Greek alpine climate
where the unfavourable season is neither winter
nor dry season but obviously the night, we classified the species in morphological types described
by Hedberg (1964a,b) which represent four distinct
life-forms well adapted to alpine conditions in
the Greek mountains, namely tussock grasses, acaulescent rosette plants, cushion plants, and sclerophyllous shrubs. From a physiological point of
view these adaptations apparently represent different ways of maintaining the water balance in
a climate with recurrent night frosts all the year
round.

A few of the relevant species may have been preadapted to this type of climate before they became
members of the Greek alpine flora as some of the
above mentioned life-forms may be equally matched
also in other types of climate, as tussock grasses.
But several of the species concerned belong to
groups endemic to the Greek mountains or at least
have their closest relatives at lower levels on
the same mountains, and may hence be expected to
have developed their alpine adaptations in this

part of the world. A few examples will be given below for each of the life-forms concerned.

7. EXAMPLES OF PRESUMED AUTOCHTHONOUS ORIGIN OF LIFE-FORMS IN THE GREEK ALPINE BELT

7.1. Tussock Grass

Many good examples of progressive adaptation to the
Greek-alpine condition are found among the tussock
grasses, the most impressive being *Festuca olympica*
Vetter characterized by culms 10-25 cm long; leaves
strongly glaucous; panicle 2.5-4.5 cm long; spikelets
6.5-8 mm long, and lemma 4.5-5.5 mm long whereas
its closest relative *F. graeca* (Hackel) Markgr.-
Dannenb. from lower levels is less densely tufted;
with longer culms, spikelets and lemma; leaves green,
and less dense panicle.

7.2. Acaulescent Rosette Plants

Good examples of acaulescent rosette plants likely
to have arisen in this environment are *Beta nana*
Boiss. & Heldr. (Chenopodiaceae-Fig. 8), the only
species of the genus entirely restricted to the
Greek alpine belt, and *Trinia guicciardi* (Boiss. &
Heldr.) Drude (Apiaceae), which is known only from
a few high mountains in Greece. There are also good
examples of progressive adaptation involving prostrate growth. Thus *Aubrieta erubescens* Griseb. and
Corydalis parnassica Orph. & Heldr. are entirely
prostrate whereas their closest relatives, *A. deltoidea* (L.) DC. and *C. bulbosa* (L.) DC. are suberect
to erect. *Trifolium parnassi* Boiss. & Spruner, a
rare Greek-alpine endemic, is of particular interest
as combining prostrate growth with sessile inflorescences (Fig. 11).

7.3. Cushion Plants

The *Trifolium* species mentioned above form a transition to the cushion plants, which are characterized
by profusely branched stems with densely compressed
branches, each of which has very brief internodes
and is terminated by a leaf rosette. A prerequisite
for good cushion growth is a particular kind of
profuse branching with branches of comparable dimensions (Rauh, 1939, p. 483). A very good example of

progressive adaptation involving cushion growth is *Saxifraga sempervivum* C. Kock, endemic to the Balkan Peninsula. Its close relative, *S. grisebachii* Degen & Dörfler occurs in lower altitudes, with loosely caespitose shoots, leaves spathulate, obtuse, and longer flowering stems. Also in *S. scardica* Griseb., endemic to S part of the Balkan Peninsula, there is a considerable ecotypic variation, plants at low altitudes forming loosely caespitose mats whereas plants of exposed alpine habitats form dense hard cushions of much smaller leaves (cf. Strid loc. cit. 1980).

7.4. Sclerophyllous Shrubs

The least exclusive of the life-forms in the Greek alpine belt is the sclerophyllous shrub, since morphologically almost identical types occur in several other types of climate. For most Greek alpine plants belonging here it is therefore difficult to establish whether this habit has developed on the Greek mountains or elsewhere.

In the Cretan mountain flora there are many examples of progressive evolution involving the above four life-forms. Greuter (1972, Table 2) has pointed out that "the most obvious cases of evolution within the Cretan flora are those where two ecologically vicarious taxa coexist on the island. Most of these cases arise from the differentiation of mountain ecotypes from lowland species... . It is interesting to note that, in the examples of Table 1, differentiation is always directed from the lowland to the mountain taxa. The former are either found outside Crete, and are often rather widespread, or, if they are endemic (as in the cases of *Teucrium*, *Scutellaria*, *Phlomis*, *Asperula* and maybe *Taraxacum* and *Leopoldia*), they are represented elsewhere by vicarious lowland taxa. The mountain ecotypes are all endemic to Crete, with two exceptions (*Satureja biroi* and *Crepis mungieri*) where similar forms have evolved, independently as it seems, on the nearby island of Karpathos. Parallel evolution of apparently identical mountain ecotypes on the different, isolated

TABLE 2. Vicarious lowland and mountain taxa in the Cretan flora.
(Binomials are used wherever they exist, irrespective of the appropriate status of the taxa)(from Greuter, 1972).

Lowland taxa	Mountain taxa	References
Paronychia macrosepala Boiss.	*P. macrosepala* var.	Greuter (1965)
Hypericum empetrifolium Willd.	*H. empetrifolium* var. *tortuosum* Rech. f.	Rechinger (1943b)
Linum arboreum L.	*L. caespitosum* Sm.	
Sedum creticum Boiss. & Heldr.	*S. hierapetrae* Rech. f.	
Trifolium uniflorum L.	*T. uniflorum* var. *breviflorum* Boiss.	
Erica monipuliflora Salisb.	*E. manipuliflora* var.	
Teucrium gossypium Rech. f.	*T. alpestre* Sm.	Strid (1965)
Scutellaria sieberi Bentham	*S. hirta* Sm.	
Phlomis lanata Willd.	*Ph. lanata* var. *biflora* Hal.	
Satureja thymbra L.	*S. biroi* Jav.	Rechinger (1943b)
Coridothymus capitatus (L.) Reich. f.	*C. capitatus* var. *albospinosus* (Bald.) Rech. f.	
Asperula incana Sm.	*A. incana* var.	
Galium heldreichii Hal.	*G. samothracicum* Rech. f.	
Galium amorginum Hal.	*G. incurvum* Sm,	
Bellis sylvestris Cyr.	*B. longifolia* Boiss. & Heldr.	
Tragopogon sinuatus Ave-Lall.	*T. lassithicus* Rech. f.	Rechinger (1943b)
Taraxacum sect. *Scariosa* Hand.-Mazz. (2 taxa)	*T.* sect. *Scariosa* (2 taxa)	Richards (pers. com.)
Lactuca alpestris var.	*L. alpestris* (Gand.) Rech. f.	
Crepis fraasii Schulte Bip.	*C. mungieri* Boiss. & Heldr.	Rechinger(1943ab)
Colchicum pusilum Sieber	*C. cretense* Greuter	Greuter (1967)
"*Muscari cousturieri*" Gand., "*M. creticum*" Vierh.	*Leopoldia spreitzenhoferi* Heldr. (=*M. amoenocomum* Rech. f.)	Bentzer (ined.)
Crocus cartwrightianus Herb.	*C. oreocreticus* Burtt	Rechinger (1949)
Dactylis hispanica Roth	*D.rigida* Boiss. & Heldr.	Borrill (1961)

mountain groups of Crete seems, by the way, to
have happened in several instances. This supports
the view that evolution from lowland to mountain
ecotypes is strongly adaptive, following almost
identical lines under given climatic and ecological conditions".

REFERENCES

Bliss LC (1956) A comparison of plant development in microenvironments of arctic and alpine
tundras, Ecol. Monogr. 26, 303-337.

Hedberg O (1964a) Features of Afroalpine plant
ecology, Acta Phytogeogr. suec. 49, 1-144 (Uppsala).

Hedberg O (1964b) Études écologiques de la Flore
afroalpine, Bull. Soc. Bot. Belg. 97, 5-18.

Greuter W (1972) The relict element of the flora
of Crete and its evolutionary significance. In
Valentine DH (ed.), Taxonomy, Phytogeography and
Evolution, London and New York

Rauh W (1939) Über polsterformigen Wuchs. In
"Nova Acta Leopoldina" N.F. 7(49), 267-508, Taf.
23-44, Halle (Saale).

Strid A (1980) Wild Flowers of Mount Olympus,
Kifissia, Greece.

HISTORY AND GENETICS OF TOLERANCE TO THE NATURAL PHYTO-TOXIC MEDIA OF WESTERN BALKANS

A-Z. LOVRIĆ (CIM-IRB, YU-41000 Zagreb, Buliceva 8-V, Yugoslavia)

1. INTRODUCTION

The so-called "serpentines" have often been
studied by botanists (cf. surveys in Whittaker
et al., 1954, and Gams, 1975) but their actual
knowledge is far from being satisfactory, especially concerning Balkans, where for example those in
Croatia have not been studied at all. On the other
hand, studies on salt sprays are rather scarce
and almost preliminary with few ecologixal details,
and the most informative data on this topic are
given by Boyce (1954); Clayton (1972); Doutt
(1941); Edlin (1943); Malloch (1972); Moss (1940);
Parsons, Gill (1968); and Wells (1939); concerning
the Adriatic: Buljan, Marinković (1949); Gračanin
(1934, 1935); Anderlind (1912); and recently
Lovrić (1971a, 1971b, 1975a, 1975b, 1976, 1977a,
1977b, 1979, 1980, 1981a, 1981b, 1982a, 1982b,
1982c)., have given some data.

FIGURE 1. Salt crust (white) at 120m alt. after a
heavy Bora storm with xerohaline fumarea in the
Senj Archipelago.

2. AREA, MATERIAL, AND METHODS

Three areas in Adriatic, known for their strong
coastal winds and the air-borne salinity these
winds cause, have been studied: Dubrovnik coast
in SE Adriatic, Vis Archipelago in central pelagic
area, and the Senj Archipelago in NE Adriatic
being the most exposed area to the heavy Bora
storms, and the one presenting the most interesting phenomena has been analysed in detail (Figs.
1,2). The ophiolites have been studied ecologically in detail on their westernmost Balkanic
locality in middle Croatia, Banija county
(especially in mount Zrinska gora), but comparatively also in minor and less rich areas of
N Croatia and W Bosnia. The ecology of gypsum
has been studied in numerous related localities
of central Dalmatia and Lika highland, and the
specific limonite dunes in Krk island. The
methods of field studies have been summarized in
Table 1. The related taxa in edaphic floras have
been studied by standard methods, but in some

FIGURE 2. Bare rockfields of the Senj Archipelago,
eroded by iterative Bora hurricanes and heavy xerohaline fumarea.

TABLE 1. A simple determination key for the field separation of phyto-toxic substrata.
Elementary equipment: HCl + portable flamelet (indications in parentheses are only auxiliary ones but not decisive alone).

1. OPHIOLITES s. lat.	2. EVAPORITES
1.1. Peridotite: stable in flame (surface opaque, blackish, rugged, covered by a patina of reddish soil).	2.1. Halite (NaCl): easily melting in flame giving them an intense yellow-gold or orange colour of Na, dissolvable in H_2O (taste salty).
1.2. Serpentinite (sensu stricto); in flame darking and crepitating of H_2O (grey-green to black-green, surface rounded-scaly, covered by a patina of grey clayish soil)	2.2. Epsomite ($MgSO_4$): not melting in flame but in hot state giving a splendent silvery light of Mg, dissolvable in H_2O (taste bitter).
1.3. Gabbroids: melting in flame, surface often splendent-crystalline (rugged-granulate, covered by a grey clayish soil as in serpentinite).	2.3. Gypsum ($CaSO_4.nH_2O$): burned in flame melting and whitening, giving to flame an orange-reddish colour of Ca, insolvable in H_2O (fragile rock, cutting possible).

interesting cases, especially in *Centaurea*, also the caryotypes are included (Lovrić, 1982c).

3. OLIGOTOXIC HABITATS

This group includes the less specific substrata where the mineral composition of soil seems to be secondary, and the dominating effects are caused by physical factors as soil aridity, pH, etc. Such habitat types are the standard salt marshes, the usual salt spray of humid maritime winds, ferrugineous soils, gypsum, and the true serpentinite. Among them, the less known habitats in our area were gypsum and ferrugineous soils, which now have been studied in detail. Their characteristic flora is presented in Table 2.
The gypsum areas and related vegetation of Dalmatia are among the most typical in Europe. The characteristic related soil of Dalmatian mainland gypsum is the xerorendzina, but in insular areas on gypsum occurs also the semidesertic yerma soil with a carbonate crust (caliche), there called "gnjilac" (fragile soil). The presence of relict thermophilic genera in gypsum as *Lycium, Capparis,* and *Ficus* indicates the southern, Gondwanian origin of the related flora, but the exclusive taxa and endemics there are not numerous.
The ferrugineous soils and related flora in the Adriatic area are the most typical in the coastal

sand dunes with limonite nodules, developed in SE Krk island, NE Adriatic. This is also the unique European locality with specialized siderophytes, and the second one studied in world, besides the Isla de Pinos in W Cuba (Alain, 1946; Lötschert, 1957, 1969). Here occurs a specific soil locally called "šupelnica", and the related flora is very rich, more than 76 taxa, but with few specific elements, except several herbaceous endemics of infraspecific or hybridogenic value (Table 2).

4. METALOPHYTES IN PERIDOTITE

Due to certain confusion in distinguishing the true petrological serpentinite from gabbro and peridotite the so-called "serpentines" of botanists provoked a series of problems and contradictions because they have been often different from the same scientific name in geology, and therefore cf. determination key in Table 1. This confusion has been especially evident in related Balkanic areas where it has been stressed by the justified criticism of Lämmermayer (1926, 1928, 1929) on the problematic existence of so-called Balkanic "serpentinophytes" and "serpentinomorphoses". Despite the opposition of a series of authors to this opinion, the most recent studies in Yugoslavia just confirmed he was almost right, while many others were wrong, except in the case of the selecting mechanism, in which none of both was right cf. Ritter-Studnička(1962), Lovrić (1980, 1981a,

TABLE 2. Characteristic flora in the oligotoxic substrata of E Adriatic coast.

GYPSUM (Middle Dalmatia)	LIMONITE dunes (Krk island, NE Adriatic)
Brassica frutescens (Vis.) Schl.& Vuk.	*Leucanthemum platylepis* Borb. ssp.
Anthyllis aegaea Turrill	*platylepis* s.str.
Colutea oriantalis Boiss.	*Edraianthus tenuifolius* ssp. *pichleri* M.G.
Capparis sicula Duh.	*Alyssum montanum* ssp. *pagense* (Baumg.)Hay.
Juniperus lycia L.	*Allium montanum* ssp. *visianii* M.G.
Lycium europaeum L.	*Artemisia biasolettiana* Vis.
Lycium intricatum Boiss.	*Iris X marchesettii* Pamp.
Artemisia arborescens L.	(=*I. illyrica X pallida*)
Suaeda vera var.*deserti* Zoh.	*Asperula X borbasiana* Kor.
Cornus australis C.A.Mey	(=*A. dalmatica X canescens*
Quercus virgiliana Ten.	
Acer campestre ssp. *marsicum* (Guss)Hay.	
Ulmus canescens Melv.	

1982a), Tatić et al. (1961).
The actual state of the ecology of Balkanic ophiolites may be summarized as following: the true serpentinites in W Balkans have a few specific floras also as other similar areas in Europe. Opposite to this, the Balkanic "serpentinophytes" are just restricted to the eruptive peridotite and not to metamorphic serpentinite nor gabbro, i.e. this specialized flora really exists but it is not serpentinic, and therefore a series of taxonomical and phytosociological names produced in Balkans by the attribute "serpentinicum" are almost unconveinable, presenting a permanent document of this imprecision. On the other hand, the ancient hypothesis on the decisive role of disproportion of abundant Mg and lacking Ca is little justified: in some Yugoslav periodoties with very diversified flora, Ca is well presented (to 4%), and a series of related studies recently confirmed that the main selecting factor on flora in ophiolites may be the toxic heavy metals: Cr, Co, Ni, Mn etc. (Antonovics et al., 1971; Brooks, 1972, 1977; Cole, 1971; Dierschke, 1975; Ernst, 1974, 1974a, 1975 ; Jaffré, 1976; Lötschert, 1969; Lovrić, 1982a,b; Malyuga,1964; Ritter-Studnička, 1962). Therefore the ophiolites are one of the most toxic natural habitats, and their specialized flora is an excellent indicator of the presence of heavy metals in soil. Among different ophiolites, the serpentinite is dominating in European local-ities and the peridotite is there very rare and restricted, but in Balkans it is more frequent and this is the main reason for floristic differences in European and Balkanic ophiolites. The Balkanic ophiolites present the western end of a great ophiolitic series (cf. Fig. 3, and Belostotskiy, 1969; Heissleitner, 1951-1952; Maxwell,1970; Steinmann, 1926; Temple and Zimmerman, 1969), beginning in Northwestern by Mt. Zrinska gora in central Croatia, and prolonged across Bosnia, Serbia, Albania, Greece, to Asia Minor and Persia. After the recent theory of "New global Tectonics" the ophiolites and especially peridotite originate from the deepest layers of magma and this provokes also their ecological speciality against other siliceous substrata (ultrabasic character, heavy metals, etc.). After the studies in Croatia and Bosnia, the true serpentinite is ecologically less basic, producing a gentle relief and grey, acid clayish soils more similar to other siliceous areas. On the contrary, in rugged relief with abundant skeleton on peridotite occurs a mountain brown soil, and in lower belts a reddish-orange soil resembling laterite, and being very peculiar in continental Croatia. The related flora of specialized metalophytes is rather specific, presenting in Banija the westernmost localities of many Balkanic "serpentinophytes" (cf. Table 3), some of them being local endemics or disjunctive relics (cf. Kišpatić, 1889).

36

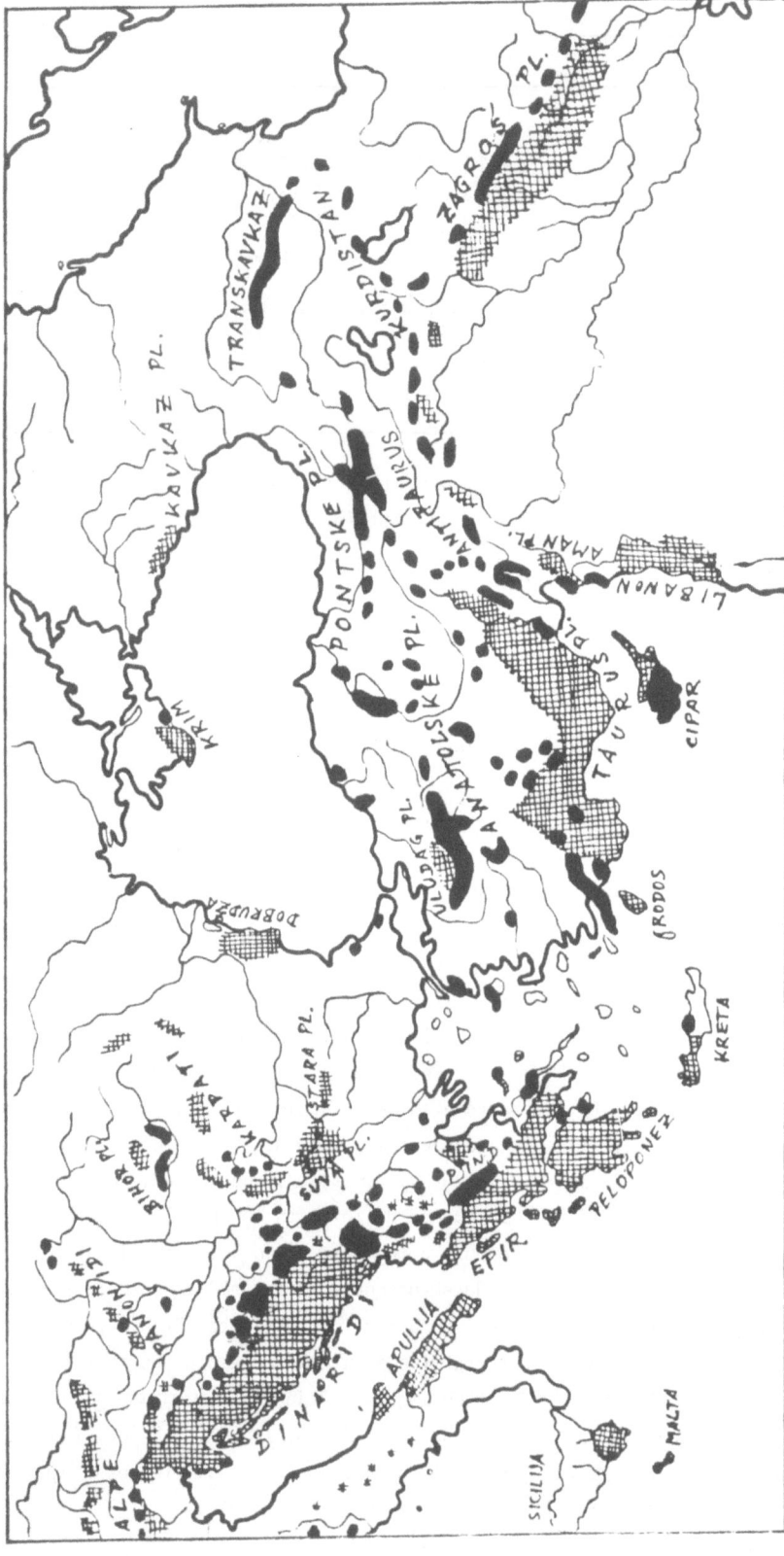

FIGURE 3. Important areas of ultrabasic ophiolites (■), and carbonate karst substrata (xxxxx), with the related specific floras in Balkan peninsula and SW. Asia.

TABLE 3. Characteristic flora in hypertoxic habitats of S.R. Croatia.

SPECIALIZED TOXICOPHYTES (endemics and disjunctive relics)	FACULTATIVE TOXICOPHYTES (with larger areas and edaphic amplitudes)

1. Peridotite metalophytes in Mt. Zrinska gora (Banija county, central Croatia)

1.1. Lignescent metalophytes:	*Carpinus caucasica* Grossh.
	Quercus dalechampii Ten.
Quercus cerris ssp. *pseudocerris* Boiss.	*Quercus virgiliana* Ten.
Rubus zvornikensis Fritsch	*Pinus nigra* ssp. *illyrica* Vidak.
	Cytisus heuffelii Wierz.
1.2. Herbaceous metalophytes:	
	Notholaena marantae (L.)Desv.
Euphorbia gregersenii Maly	*Asplenium adulterinum* Milde
Silene staticifolia Sb. Sm.	*Asplenium cuneifolium* Viv.
Cerastium moesiacum Friv.	*Centaurea micranthos* Gmel.
Polygonum moesiacum M.G.	*Plantago serpentina* Vill.
Verbascum bosnense Maly	*Ceterach officinarum* ssp. *officinarum*
Stachys recta ssp. *baldaccii* Maly	*Dianthus croaticus* Borb.
Centaurea nigra ssp. *aterrima* (Hay)Hay.	*Sedum orientale* Pold.
Centaurea nigrescens ssp. *smolinensis* (Hay.)Dost.	*Asperula montana* Borb.
Sesleria rigida Heuff.	*Thymus jankae* Cel.
Sedum glaucum W.K.	*Satureia variegata* D.C.
	Medicago rigidula L.
	Poa pannonica Kern.
	Viola beckiana Fiala

2. Borraphytes in xero-sulphatic fumarea areas of Senj Archipelago (Kvarner, NE Adriatic)

2.1. Lignescent borraphytes:	
Carpinus grandis (Ung.) Dinić	*Pinus nigra* ssp. *dalmatica* (Vis.)Franco
Pinus urani (Ung.) Ett.	*Quercus virgiliana* Ten.
Astragalus dalmaticus Bunge	*Ulmus canescens* Melv.
Astragalus vegliensis Sadl.(s.s.non auct.)	*Acer campestre* ssp. *marsicum* (Guss.)Hay.
Aurinia media (Host)Schur.	*Cornus australis* C.A.Mey.
Centaurea procellaria Lov.	*Rhamnus alaternus* ssp. *myrtifolia* (Willk)Maire
Peucedanum crassifolium Hal.& Zahl.	*Euphorbia wulfenii* Hoppe (non *E. veneta* auct.).
Plantago wulfenii Rchb.	*Helichrysum litoreum* Guss.
Prunus spinosa ssp. *istriaca* M.G.	*Frangula rupestris* (Scop.) Schur.
Fraxinus ornus ssp. *argentea* (Lam.)M.G.	*Onosma dalmaticum* Scheele s.str.
Camphorisma nestensis Turrill	*Cephalaria mediterranea* ssp. *scopolii* (Vis.) Lov.
Halimione graeca (Moq.)Lov.	*Limonium anfractum* (Salm.) Salm.
Lonicera etrusca ssp. *stabiana* (Guss.)M.G.	*Cotinus coggygria* Scop.
Ficus carica L. ssp. *caprificus*	
Rumex suffruticosa Gay.	
2.2. Herbaceous borraphytes:	
Asperula dalmatica M.G.(= *A. fragilis* Wol.)	*Phyllitis hybrida* (Milde) Christ.
Centaurea rossiana Wagn.	*Drypis jacquiniana* Murb.& Wettst.
Allium horvatii Lov. ssp. *horvatii*	*Iris rotschildi* Deg.
Senecio fluminensis Simk.	*Centaurea aliena* Wagn.
Hyoscyamus muticus L.	*Centaurea tommasinii* ssp. *petteri* (Rchb.)Lov.
Centaurea approximata Rouy	*Centaurea substituta* Tzvel.
Peltaria alliacea ssp. *crassifolia* (Mort.) Lov.	*Allium commutatum* Guss.
Chaenorrhinum litorale ssp. *aschersonii* (Simk.)Lov.	*Carlina fiumensis* Simk.
Anthyllis vulneraria ssp. *tournefortii* (Schult.)M.G.	*Scorzonera crispa* M. Bieb.
Salsola kali ssp. *pontica* (Pall.) Iljin	*Juncus litoralis* C.A.Mey.
	Colchicum kochii Parl,
	Festuca illyrica Mgf.-Dfg.
	Minuartia capillacea (All.) Graeb.
	Suaeda maritima ssp. *salsa* (L.) Soo
	Vincetoxicum croaticum Jord.& Four.

5. BORRAPHYTES AND FUMAREA SALT STORMS

Contrary to many other aerosaline coasts in world, in Adriatic two very different types of air-borne salinity are distinguishable. The wet maritime Sirocco and similar winds there provoke a classical, humide salt spray, by its chemical structure being similar to seawater with dominating chlorides, and the related flora of sprayed areas is similar to salt marshes with usual halophytes but without specific indicators of air-borne salinity, because there dominates the edaphic effect of salinized soils.

A quite different type of air-borne salinity occurs at arid continental storsma of Bora and foehn types. Due to air dryness, the quantities of dispersed seawater in air are not limited by saturation as in humid Sirocco, but also the salts are not transported in air as a wet fog but they are quickly transformed by dessication in a splendent silvery smoke presenting the crystallized salt powder that provokes in atmosphere the curious optic phenomena ressembíling a polar light. Moreover, this one is even chemically different from the usual salt spray and just therefore it becomes more toxic: due to complex phenomena of superficial tension of seawater, the crystallizing effect of different salts in the air is selective resulting in an evident stratification. In the excessively salinized area at Senj Archipelago, only the lowest 60-90 m of air above sea have a wet spray with dominating chlorides; above this they are decreasing and at ca. 150-250 m in the islands sulphates predominate in storms. During the intense Levantara hurricanes in the Senj area, the upper limit of a visible air-borne salinity is at 300-450 m. This curious and complex phenomenon of fumarea (local name "slana" i.e. salinizing) begins at a Bora of ca. 7⁰ Beaufort. Its ecological effects are quite different from usual salt spray and more similar to artificial air pollution by industrial sulphates and halogenoids. Its immediate and direct air effect on plants is the most important, while that on the soil is a secondary action. The salt crystals there interfere with the motion of stomatal pores, and by penetration in stomatal cells provoke an intoxication in non-resisting species (borraphobes), resulting in plasmolysis, permanent opening of stomatal systems and at the end a lethal desiccation by arid wind. This mechanism provokes a severe elimination of borraphobes from the Senj area, resulting in their disjunctions and local complete absence as of *Carpinus orientalis, Acer monspessulanum, Laurus nobilis, Viburnum tinus, Ligustrum, Evonymus, Corylus, Castanea* etc. being there also quite unsuitable for artificial cultivation, and therefore the postulated climax of these taxa here are very absurd and contradicting to their ecology. Among exotics the most sensitive are *Pinus halepensis, Atriplex halimus* (!), *Aesculus, Cedrus, Paulownia* etc., and very resistent ones are *Pittosporum, Robinia, Sophora, Yucca, Rosmarinus, Morus, Carpobrotus, Nerium, Broussonnetia, Agave, Poinciana*. This specific area is filled with palaeoendemics (endemism 6%, highest in Yugoslavia), including also "living fossils", as *Pinus urani* and *Carpinus grandis* from the Tertiary.

6. DISCUSSION

Two adaptation patterns, both in evaporites and in ophiolites are registered. The oligotoxic media present almost the ubiquitarians with large amplitudes responding by phenotypic modifications and compensatory post-adaptation, being often hybrid polyploids related to diploids of non-toxic areas, as confirmed in *Centaurea* (Lovrić, 1982d). Therefore they are also suitable for atoxic cultivation, as the facultative or oligotoxicophytes of recent and Quaternary origin. The hypertoxic peridotite and fumarea have true specialized toxicophytes as primitive diploids or isolated paleopolyploids, having the evolved polyploide relatives in atoxic areas. Some species have similar tolerance or genetic and family levels indicating the paleoadaptations of early edaphic specializing: the Balkanic metalophytes in peridotite may be the epibiotic from volcanic islands of Tethys Ocean, and the borraphytes from fumarea area as the relict palaeo-halophytes of Tethyan coasts. Sometimes, therefore, in atoxic cultivation, they show a vegetative luxuriance, while the sterility they appear is aliminated by pollutant addition. Therefore they are suitable for polluted areas, especially those connecting with diverse toxic habitats: *Quercus virgiliana, Cornus australis, Ulmus canescens, Cotinus, Vitex* etc.

REFERENCES

Alain H (1946) Notas taxonomicas y ecologicas sobre la flora de Isla de Pinos. Mus. Hist. Nat. Colegio de la Salle, Contrib. Ocas. 7, 1-115, La Habana.

Anderlind N (1912) Über des Verhalten mehrerer Holzarten gegen den Salzgehalt der Luft an den Klippen des Quarneros, Allgem. Forst. Jagdzeitung Wien 88, 236-239.

Antonovics J et al. (1971) Heavy metal tolerance in plants, Adv. Ecol. Res. 7, 1-85.

Belostotskiy II and Kolbantsev AV (1969) The problem of ophiolites of the Dinarids, Internat. Geology Rev. 12, 358-369.

Boyce SG (1954) The salt spray community, Ecol. Monog. 24, 29-67.

Brooks RR (1972) Geobotany and Biogeochemistry in mineral exploration. Harper and Row, New York.

Brooks RR et al. (1977) Detection of nickel-iferous rocks by analysis of herbarium specimen of indicator plants, J. Geochem. Expl. 7, 49-57.

Buljan M and Marinkovic M (1952) Results of an investigation into cycling salts on the east coast of the Adriatic (Split), Acta Adriat. 4(9), 319-352.

Clayton JL (1972) Salt spray and mineral cycling in two California coastal ecosystems, Ecology 53 (1), 74-81.

Cole MM (1971) The importance of environment in biogeographical/geobotanical and biogeochemical investigations, Can. Inst. Min. Metal. Spec. vol. 11, 414-425.

Dierschke H, ed. (1975) Vegetation und substrat. Berichte Int. Verein. Veget. Symp., J. Cramer, Vaduz.

Doutt JK (1941) Wind pruning and salt spray as factors in ecology, Ecology 22(2), 195-196.

Edlin HL (1943) A salt storm on the south coast, Quart. Jour. Forest. 37, 24-26.

Ernst W (1974) Schwermetallvegetation der Erde. Stuttgart, Fisher-Verlag.

Ernst W (1974a) Mechanismen der Schwermetall-resistenz, Verhandl. Ges. Ökol. Erlangen 1974, 189-197.

Ernst W (1975) Physiology of heavy metal resistance in plants. Proc. Int. Conf. Heavy Metals in Environment, Toronto, pp. 121-136.

Gams H (1975) Vergleichende Betrachtung euro-päischer Ophiolith-Floren, Veröff. Geobot. Inst. ETH 55, 117-140.

Gračanin M (1934) Die Salzböden des nordöstlichen Adriagebietes als klimatogene Bodentypen, Soil Res. 4(1), 20-40.

Gračanin M (1935) A pedological study of Pag island (in Croat), Ann. Exp. Forest. Zagreb 4,107-187.

Hiessleitner G (1951-1952) Serpentin- und Chrom-erzgeologie der Balkanhalbinsel und eines Teils von Kleinasien, Jahrb. Geol. Bundesamt. Wien suppl. 1(1-2), after Krause W et al. (1963), Bot. Jb. 82 (4), p. 400.

Jaffré T (1976) Composition chimique et conditions de l'alimentation minérale des plantes sur roches ultrabasiques. Cah. ORSTOM,ser. Biol. 9(1),53-63.

Kišpatic M (1889) Prolongation of the Bosnian serpentine zone into Croatia (in Croat), Rad Yug. Acad. Sci. Zagreb 139, 44-73.

Lämmermayer L (1926-1927) Materialien zur Systematik und Okologie der Serpentin-Flora, I-II, Sitzber. Akad. Wiss. Wien, Abt. I, 135, 369-407 and 136, 25-69.

Lämmermayer L (1928) Weitere Beiträge zur Flora der Magnesit- und Serpentinböden, Sitzber. Akad. Wiss. Wien, Abt. I, 55-99.

Lämmermayer L(1929) Ubereinstimmung und Unterschiede in der Pflanzendecke über Serpentin und Magnesit, Mitt. Naturw. Ver. Steiermark 71,41-62.

Lötschert W (1957) Die Savanne der Kiefern-Insel Vegetationsbilder aus Westcuba, 6. Natur und Volk 87, 194-201.

Lötschert W (1969) Pflanzen an Grenzstandorten. Stuttgart, Fischer-Verlag.

Lovrić AZ (1971a) Nouveautés de la flore halophile du Littoral croate, Osterr. Bot. Zeit. 119, 567-571.

Lovrić AZ (1971b) Coenodynamique et pédodynamique du mode battu par rapport à la Bora et l'abrasion, Thalassia Jugosl. 7(1), 195-200.

Lovrić AZ (1975a) Bora salt storms and biocenoses of the Senj archipelago (in Croat). Thesis reg. no. YUAA-9975, Inst. Bot. Univ. Zagreb.

Lovrić AZ (1975b) Végétation des milieux aéro-salins du Karst littoral croate, Rapp. Comm. Int. Expl. Sci. Medit. 23(3), 45-46.

Lovrić AZ (1976) Modèles écozonaux de la dégradation et réconstitution des paysages maritimes. 3es Journées Etud. Pollut. pp. 133-136, Monaco.

Lovrić AZ (1977a) Importance of halophytic flora in biogeography of Croatia (in Croat, Summary). Proc. 4th Symp. Biosyst. Yug. pp. 93-94, Univ. Novi Sad.

Lovrić AZ (1977b) Sempervirent plantations resisting to Bora salt storms (in Croat), Bilten Baška, Rijeka 2(3), 15-17.

Lovrić AZ (1979) Expansion océanique évolution des halophytes de Yougoslavie, Rapp. Comm. Int. Expl. Sci. Medit. 25-26(3), 171-172.

Lovrić AZ (1980) Phytogeography of dolomite, ophiolite, limonite, and gypsum substrata in W. Balkans (summary). 3rd OPTIMA Meeting Proc., II, 2p., Univ. Complutense, Madrid.

Lovrić AZ (1981a) Biocenotical ecozonation in

Mediterranean Karst of W. Dinaric Alps, Adriatic
islands, and sea bottoms, Documents Cartogr. Ecol.
24, 69-78.

Lovrić AZ (1981b) Aerosaline borraphytes in the
Senj Archipelago. Adriatic endemics 5. Rapp. Comm.
Int. Expl. Sci. Medit. 27(9), 73-76.

Lovrić AZ (1982a) Wind effects in paleoendemic
halocenoses, and ecosystems of coastal Karst (in
Croat). Diss. reg. no. YUAA-1177, Zagreb-Sarajevo
(in press).

Lovrić AZ (1982b) Asteraceae, Centaurea. In Löve
A, ed. Chromosome number reports LXXVII, Taxon
(in press).

Lovrić AZ (1982c) Petrographic substrata and
vegetation (in Croat). Forest Encyclopedia II, 4p.,
Yug. Lexic. Inst. 2nd ed., Zagreb (in press).

Lovrić AZ (1982d) Ecology of Sirocco and other
maritime winds in Adriatic (in Croat). Forest
Encyclopedia, II, 5p. Yug. Lexic. Inst. 2nd ed.,
Zagreb (in press).

Lovrić LM and Lovrić AZ (1982) Morphoanatomical
syndroms in phyto-indicators of extreme stormy
habitats at NE. Adriatic, in the same volume.

Malloch AJC (1972) Salt-spray deposition on the
maritime cliffs of the Lizard Peninsula, J. Ecol.
60, 103-112.

Malyuga DP (1964) Biogeochemical methods of pro-
specting. New York, Consultants Bureau N.Y.

Maxwell JC (1970) The Mediterranean ophiolites,
and continental drift. In Johnson H, ed. The
megatectonics of continents and oceans. New Bruns-
wick, N.J., pp. 167-193, Rutgers Univ.

Moss AE (1940) Effect of wind-driven salt water,
Jour. Forest. 38, 421-425.

Parsons RF and Gill AM (1968) The effects of
salt spray on coastal vegetation at Wilson's
Promontory, Victoria, Australia. Proc. Roy. Soc.
Victoria 81(1), 1-10.

Ritter-Studnička H (1962) Die Pflanzendecke auf
Serpentin in Bosnien (in Serbocroat + Zusammen-
fass.), Godišnjak Biol. Inst. Sarajevo 16, 91-204.

Steinmann A (1926) Die ophiolithischen Zonen
in den mediterranischen Kettengebirgen. Proc. 14th
Int. Geol. Conqr. sect. 2, pp. 638-667.

Tatić B et al. (1981) Contribution to the study
of serpentine flora of Yugoslavia, Acta Biol. Yug.
ser. G, Biosistematika 7(2), 123-135.

Temple PG and Zimmerman J (1969) Tectonic significance of Alpine ophiolites in Greece and Turkey, Geol.
Soc. Am. Abs., Ann. Programs 1979(7), pp. 221-222.

Wells BW (1939) A new forest climax: the salt
spray climax of Smith Island, North Carolina, Bull.
Torrey Bot. Club 66, 629-634.

Whittaker RH et al. (1954) The ecology of serpentine soils, Ecology 35(2), 258-288.

MORPHO-ANATOMICAL SYNDROMES IN PHYTO-INDICATORS OF EXTREME STORMY HABITATS IN THE NORTHEASTERN ADRIATIC

L.M. LOVRIĆ and A.Z. LOVRIĆ (University of Zagreb, Buliceva 8-V, YU-41000)

1. INTRODUCTION

The northeastern Adriatic coast and adjacent isles are among the most stormy areas in world coasts, especially the habitats exposed to the Bora storms. The climatological, meteorological, and synoptical characteristics of Bora have been studied in detail during more than a century (cf. exhaustive bibliography in Yoshino, 1972-1973), but its ecological effects are little known; this problem has more often been described by analogy with other winds than by original field studies. The most useful ecological data on this topic are given by Ercegović (1932), Gračanin (1962), Rogić (1957), and recently by the present authors, especially on effects in cryptogams (L.M. Lovrić, 1979, 1981; L.M. Lovrić, A.Z. Lovrić, 1981, 1982) and also in phanerogams (A.Z. Lovrić, 1971, 1974, 1975, 1977, 1980, 1982a, 1982b, 1982c). This paper gives a synthesis of the topic, including new recent results.

2. AREA, MATERIAL AND METHODS

The general studies cover the eastern Adriatic coasts from Istra peninsula to Boka Kotorska Gulf, nearly all the major islands of the Adriatic archipelago, and some maritime mountains (Mts. Bitoraj, Senjsko Bilo, Resnik, Poštak, Zavelin, Rilić, and Sniježnica Konavoska). More intensive studies were made of the lithophytic microecology of cyanobacteria and lichens, and also of spermatophyta, except fungi and mosses that are rare and few of which are specific to these extreme xeric habitats. Besides the standard morphoanatomical and phytoecological procedures, multivariate analyses were applied to obtain a better knowledge of evolutionary syndromes in related characteristic ecoforms (for technical details cf. A.Z. Lovrić, 1975, 1982a). The most detailed analyses have been focused on the Senj Archipelago,

Kvarner gulf in the northeastern Adriatic, the most stormy area of the Yugoslav coast.

3. DOMINATING WINDS AND STORMS

The warm-humid maritime winds from the Mediterranean predominate in the outer and southern islands of the archipelago, and on the southernmost coasts. Among them the most important is the Sirocco (SE to S), in Yugoslavia called "jugo" (South wind), reaching maximum force of only $10-12^{o}$ Beaufort. The major part of the continental coast, and also the northern and interior islands are exposed chiefly to continental winds (Bora sensu lato). Included is the periodical, daily, fresh and dry wind, variable in local direction, but blowing generally from North-Northeastern to Northeastern, called "burin" (little bora), and not exceeding $7-8^{o}$ Bf. The true Bora is often a cold and moderately humid cyclonic storm from Northeastern to East-Northeastern, reaching $10-11^{o}$ Bf, and in the most stormy areas blowing nearly all year.

In the Northeastern coast of Kvarner Gulf occurs also another stronger type, the Levantara hurricane ("East storm") the strongest Adriatic storm, often exceeding 7° Bf and periodically reaching the most severe known hurricane force of $16-17^{o}$ Bf (50-60 m.sec^{-1}) on the coast, and at altitudes above seven kilometers even up to 110 m.sec^{-1}. Its main direction on the coast and in the islands is Eastern-Northeastern to Eastern, but with altitude it changes to Northern to Northwestern. It dominates in winter, with coastal snowstorms, temperatures to $-18^{o}C$, and sea icing. It is very turbulent with iterative violent gusts, in severe winters lasting 1-5 weeks. (Fig. 6).

The last type of continental wind is a true foehn from Eastern-Southeastern to Southeastern called "oštrun" (strong wind). It is a continuation of the Mediterranean Sirocco, blowing from Dalmatia above

Lika highlands and then descending on the Northeastern coast of Kvarner Gulf as a dry-hot
foehn (Lukšić, 1980). The frequency of
these continental winds is greatest in the Senj
coast and adjacent archipelago, covering two thirds
to three quarters of the year. Their main effects
in related stormy ecosystems are air dryness, soil
erosion, and damage to vegetation. On the coast
and islands there is also air-borne salinity (cf.
the related paper in the same volume by A.Z. L
Lovrić, 1982c). The ecological effect of these
storms is more intense than in tropical hurricanes
because similar force but multiple frequency gives
them a greater dynamic effect. (For comparison
with other hurricanes, cf. Blüthgen, 1964; Miller,
1958; Nordhoff and Hall, 1936; Schubert, 1942;
Sloane, 1956; Tannehill, 1959; for those of the
Adriatic especially, Žgur, 1954, 1956; concerning
ice, Goldberg, 1940; and Stipaničić, 1955, 1956;
Lukšić, 1969, 1975; Degan, 1936; Yoshino, 1972-
1973, 1976). Studies on the Sirocco in the Adriatic are more scarce, but cf. Eredia (1932) and
Lukšić (1980).

4. EFFECTS ON CRYPTOGAMS
In lithophytic cryptogams,both the storms and the
carbonate karst substrata are decisive. The cryptogams are most abundant and diversified in the
purest crystallized limestones, which are easily
dissolved by boring species. With increasing
exposure, the active sheltering of cyanobacteria
in the substratum becomes evident, presenting
three degrees:
(1) The well-known epilithic genera such as *Microcoleus, Phormidium, Rivularia, Brachytrichia,* etc.,
are restricted to the very surface of rocks and to
sheltered habitats (cf. also Danin et al., 1982).
(2) Microchasmophytes (pseudoendoliths) use a
passive shelter, being submerged in the preexist-
ing fissures, especially *Dalmatella.*
(3) The endolithic genera characterize the extreme
stormy habitats, being active borers. This phenomenon is confirmed to be independent of light
and is present both in sunny cliffs and in dark

caves if exposed to wind (L.M. Lovrić, 1979, 1981).
The characteristic indicators of such extreme Bora-
exposed coastal rocks are *Hyella caespitosa* Bor. &
Flah., *Mastigocoleus testarum* Lagerh., *Entophysalis
granulosa* Kütz, *Kyrtuthrix dalmatica* Erc., *Solentia
foveolarum* Erc., *Dalmatella polyformis* Erc., etc.
Several of these, such as *Mastigocoleus, Solentia,*
and *Kyrtuthrix* are also rarely present in shelters,
and then in epilithic forms (Fig. 5).
Therefore the microfloristic spectrum and ecoforms
of cyanobacteria are very useful indicators of the
local exposure in stormy coasts. At higher and less
aerosaline coastal levels, the insolation becomes
decisive, presenting a few specific, xeric micro-
floras (*Gleocapsa sanguinea, G. kuetzingiana, G.
compacta,Tolypothrix byssoidea,* etc.).
Among coastal lichens, in stormy areas the fruticu-
lose forms are nearly absent and the crustose
species dominate. In Bora-exposed coasts the
specific indicators are *Solenopora marina* Zahl.,
Lecania quarnerica Zahl., and *Verrucaria quarnerica*
Zahl. In those more exposed to Sirocco, they are
Dirina repanda Fr., *Caloplaca calcicola* Zahl.,
Lecanora latzelii Zahl., *L. adriatica* Zahl.,
Verrucaria baumgartneri Zahl., etc. The most extreme
hurricane stricken habitats of Bora are indicated by
peculiar ecoforms of endolithic "Fensterflechten".

5. EFFECTS ON PHANEROGAMS
In lignescent spermatophyta there occur as well
some reputed phenetic modifications and reversible
anemomorphoses, generally absent in sheltered and
cultivated specimen. These include nanism, con-
densed semiglobose cushions, and espalier-shrubs as
heath. The crown deformations commonly reported,
especially in geographic studies, are really a more
complicated phenomenon than usually presented (cf.
simplification in Weischet Bartsch, 1963; Yoshino,
1973). The reputed flag-trees there are really restricted to mountains and to cold-humid inland
areas where the mechanical damage is decisive. In
dry-warm Mediterranean coast and island areas with
dominating effects of wind desiccation, the sempervirent maquis, coastal conifers, and xeric decidu-

ous trees in wind produce chiefly the peltate-
umbellate crowns, and in extreme cases the horizontal fusiform ones, but no flags if they are
inland types. Such deformations in the Adriatic
are typical for wind-resisting trees, such as
Pinus nigra ssp. *dalmatica*, *P. pinea*, *Quercus
virgiliana*, *Fraxinus ornus* ssp. *argentea*, *Ulmus
canescens*, *Acer campestre* ssp.*marsicum* etc. The
relative reduction of leaf limb (microphylly) is
also a characteristic anemomorphose (Figs. 1, 2).
In contrast, other hereditary adaptations and true
ecoforms of anemophytes are conserved when
cultivated in shelter. The most picturesque anemophytes occur in S Adriatic coasts exposed to
Sirocco: the endemics *Brassica cazzae* Ginz. and
Seseli palmoides Lov. being columnar palmoid treelets with majestic monocarpic inflorescences
before their death (Fig. 7). Another peculiar
ecoform of anemophytes are the shrubby hyperpachycauls with bottle-like succulent and lignescent
short trunk-like minute baobabs: these indicate
the most extreme strom stricken habitats, exposed to
Bora and Levantara, as are *Astragalus dalmaticus*
Bunge,*Peucedanum crassifolium* Hal.& Zahl., and
Centaurea lungensis Ginz. (Fig. 8). The usual
candelabriform pachycaul shrubs occur as different
taxa both in Bora- and Sirocco-exposed habitats,
and especially in cliffs, and therefore the stormy
E Adriatic coasts on the European continent are
the richest in such pachycaul forms; besides a
dozen widespread Mediterranean pachycauls, there
occur also thirty endemic ones (cf. Figs.7, 8, and
Table 1). These taxa are almost more primitive in
other characters than their other relatives in
the Adriatic and are therefore relics of great
evolutionary importance, especially *Aurinia*
(Alyssinae), *Campanula*, *Centaurea* (sect. *Veltis
et Pterolophus*), *Brassica* (sect. *Cramboxylon*) and
Seselinae (subg. *Seselignum*).

Other interesting but rare forms from stormy
areas are the robust columnar megaphorbias with
a tall herbaceous stem 1-2 m high. The local
endemics are *Campanula pyramidalis*, *Lilium*

FIGURE 1. *Philyrea latifolia*; fusiform crown deform-
ed by strong Bora and salt.

FIGURE 2. *Senecio fluminensis*; succulent annual of
extreme stormy rockfields.

cattaniae, *Muscari speciosum*, *Iris dalmatica*, *Centaurea adriatica* etc. For the taxonomical and evolutionary significance of all such peculiar ecoforms
cf. Arber (1928) and Mabberley (1974, 1980). Such
ecoforms are pre-adaptations, by their structure
very suitable for stormy habitats, actually representing the last refugia of these epibiotics
from Palaeo-Mediterranean floras.

The stormy mountain peaks and coastal exposed slopes
there include also the well-known xeric forms of
tragacanthic echinate cushions that are also well
adapted to storms (cf. Pignatti et al., in Dierschke,

FIGURE 3. *Drypis jacquiniana* Murb.& Wettst.:
pungent echinate cushion of tragacanthic type.

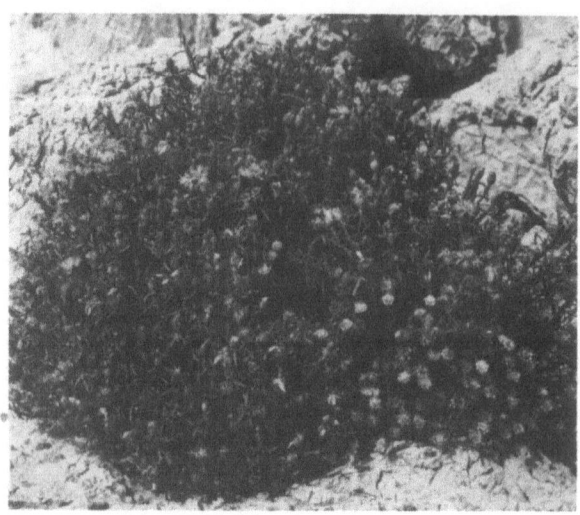

FIGURE 4. *Centaurea rossiana* Wagn.: a pungent tragacanthic cushion of tragacanthic type.

FIGURE 5. Very exposed cliffs (ca. 70 m) eroded
by strongest Bora: almost endolithic microvegeta-
tion (cyanophyta, lichens).

FIGURE 6. Winter ice (1-2 m thick) in the Senj coast,
NE Adriatic, after a heavy Bora storm.

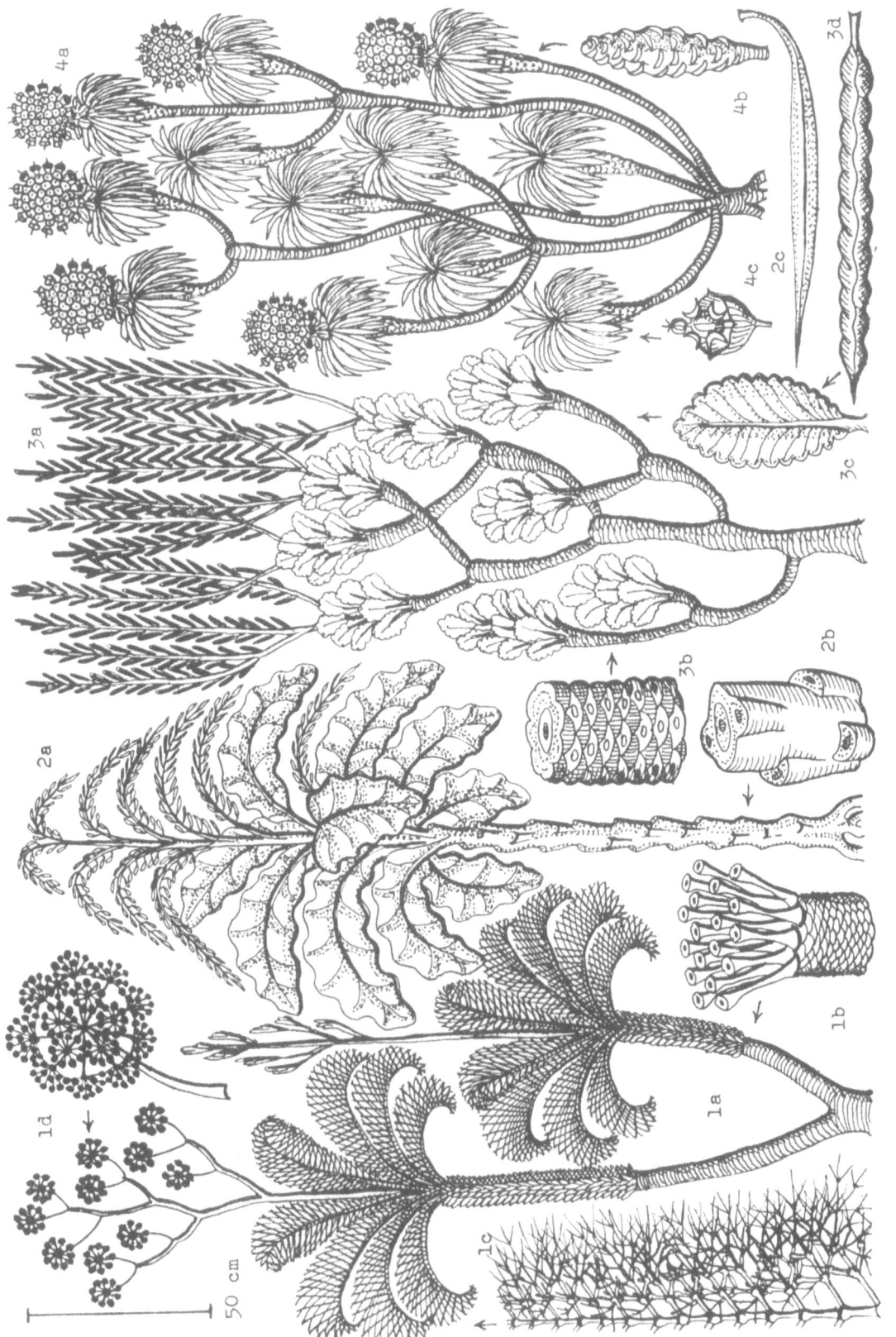

FIGURE 7. Endemic lignescent pachycauls of stormy coasts in E Adriatic: 1. *Seseli palmoides* Lov., 2. *Brassica cazzae* Ginz.,
3. *Brassica frutescens* (Vis.) Schl.& Vuk., 4. *Euphorbia wulfenii* Hoppe.

46

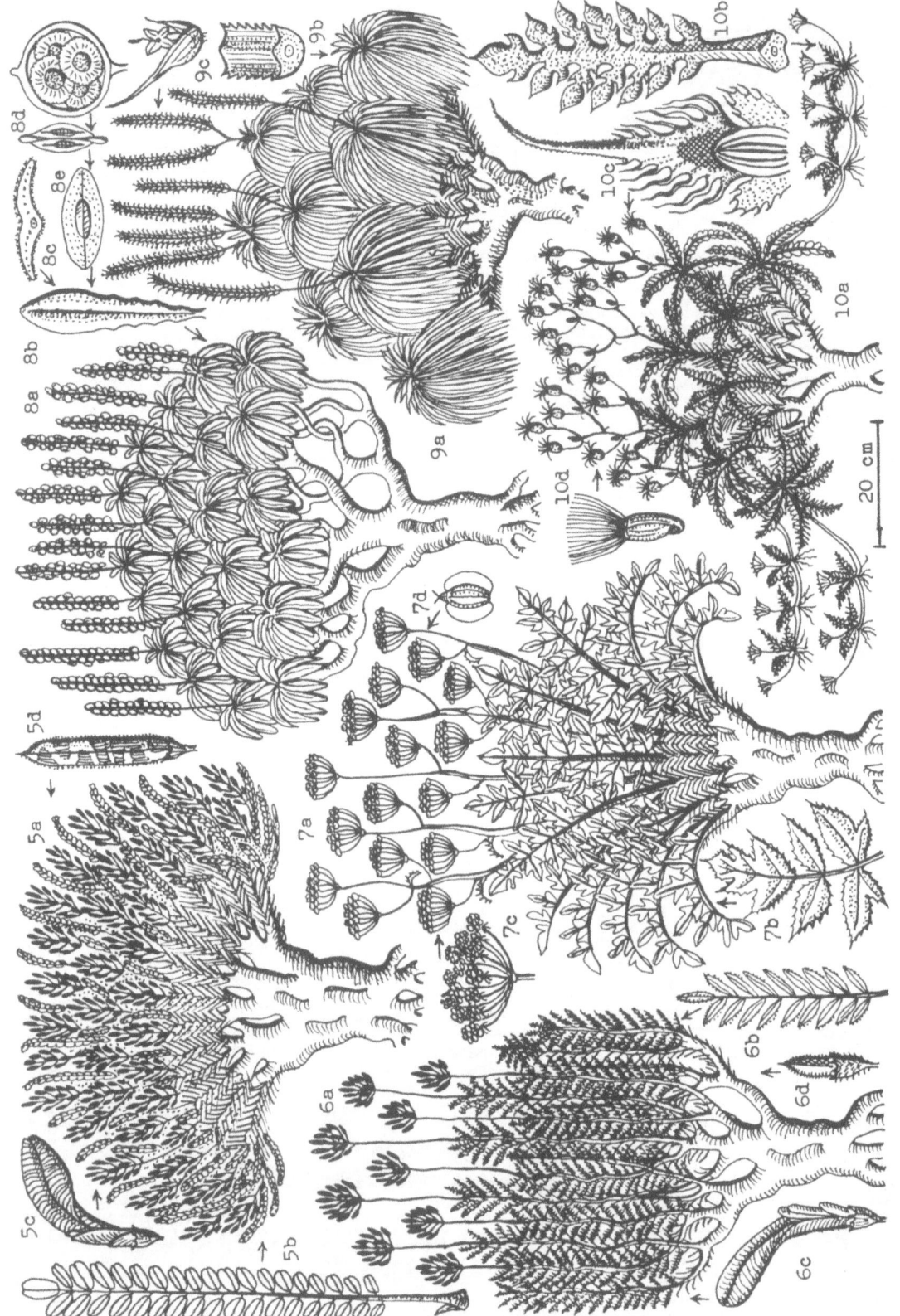

FIGURE 8. Endemic hyperpachycauls of extreme hurricanic habitats in Kvarner Archipelago,NE Adriatic: 5. *Astragalus dalmaticus* Bunge, 6. *Astragalus vegliensis* Sadl.s.str., 7. *Peucedanum crassifolium* Hal.& Zahl., 8. *Aurinia media* (Host)Schur., 9. *Plantago wulfenii* Rchb., 10. *Centaurea procellaria* Lov.

TABLE 1. Characteristic life forms in spermatophytes of extreme stormy habitats in the NE Adriatic coast and islands.

---

1. FRUTESCENT PACHYCAULS

1.1. Columnar palmoid treelents
(specific to Sirocco-exposed coasts);

Brassica cazzae Ginz.
Seseli palmoides Lov.

1.2. Trunk-succulent hyperpachycauls
(specific to Bora-exposed coasts):

Astragalus dalmaticus Bunge
Peucedanum crassifolium Hal.& Zahl.
Centaurea lungensis Ginz.

1.3. Candelabriform shrubs and bushes
(in areas of both Bora and Sirocco winds)

Aurinia media (Host.) Schur.
Astragalus vegliensis Sadl.(s.s. non auct.)
Brassica botterii Vis.
B. frutescens (Vis.) Schl.& Vuk.
Centaurea ragusina L. s. str.
C. fridericii Vis.
C. procellaria Lov.
Cephalaria mediterranea ssp *scopolii* (Vis.) Lov.
Convolvulus cneorum L.
Anthyllis aegaea Turrill
Camphorosma nestensis Turrill
Achillea abrotanoides Vis.
Euphorbia dendroides DC.
E. wulfenii Hoppe (non *E. veneta*)
Helichrysum litoreum Guss.
Halimione graeca (Moq.) Lov.
Inula methanaea Hausskn.
Limonium anfractum (Salm.) Salm.
Moltkia petraea (Tratt.) Gris.
Onosma dalmaticum Scheele s.str.
Rumex suffruticosa Gay.
Plantago wulfenii Rchb.
Campanula staubii Uechtr.

2. ECHINATE TRAGACANTHIDS

2.1. Mountain tragacanthids

Drypis spinosa L.
Astragalus angustifolius ssp. *biokovensis* Kuš.
Centaurea ceratophylla Ten.
Chamaecytisus spinescens ssp. *alaventi* Radic
Prunus prostrata Lab.
Genista michelii Guss.
Berberis croatica ssp. *dinarica* Kuš.
      (= *B. illyrica* Fuk.)

2.2. Coastal tragacanthids

Drypis jacquiniana Murb.& Wettst.
Centaurea rossiana Wagn. (= *C. cristata* ssp.
                              curictana Lov.)
Asperula dalmatica M.G. (= *A. fragilis* Wol.)
Carlina fiumensis Simk.
Salsola kali ssp. *pontica* (Pall.) Iljin
Prunus spinosa ssp. *istriaca* M.G.
Crataegus brevispina Ktze (*C. inzengae* Bert).
Calycotome infesta (Presl.) Guss.

3. COLUMNAR MEGAPHORBIAS
(robust herbs 1-2 m high)

Allium commutatum Guss. (*A. bimetrale* M.G.)
Campanula puramidalis L.
Muscari speciosum March.
Centaurea adriatica Lov.
Reichardia macrophylla Vis.& Panč.
Lilium cattaniae Vis.
Iris dalmatica (Pamp.) Bald.

1977) (Figs. 3, 4). Some of them are the western-most outposts of Oriental tragacanthic such as *Astragalus angustifolius*, or their vicarists as in *Genista* and *Prunus*, but in other cases, such as *Drypis*, *Centaurea*, *Asperula* (Table 1) they re-present Adriatic endemics of quite original aspects. These plants show other astonishing convergence in sclerotical and lignified structures, especially the so-called dracomorphic bracts (spinose-ciliate-alate in parallel),as in *Drypis*, *Centaurea rossi-ana*, *Salsola pontica*, *Carlina fiumensis* etc. With regard to types of foliage, in the stormy woods the usual sempervirent and winter-deciduous taxa are secondary and restricted to more quiet

localities.

The Bora-exposed forests are especially characterized by the semisempervirent spring-deciduous trees, such as *Quercus virgiliana*, *Fraxinus argentea* and *Petteria ramentacea*. The areas with strong Sirocco, on the other hand, are peculiar in presenting shrublands of summer-deciduous (winter-green) taxa, such as some species of *Lycium*, *Brassica*, *Capparis*, *Hali-mione* etc.

Another specific phenomenon of stormy areas is that there are the persistent leaf petioles forming protecting collars against desiccation or freezing, and these are especially frequent and well developed in the pachycaul endemic shrubs and treelets. Among

diverse types of such collars, the most primitive
and ancient may be the lignocollar type; seconda-
rily lignified petioles are best developed in
Seseli palmoides. The more evolved collar types
include spinocollars with pungent petioles, as in
some tragacanthics, then the rare filamentose
fibrocollars, herbaceous siccicollars frequent
in half-shrubs, and the most frequent, reduced
discollars, a relic of just-disappearing palaeo-
adaptations.

5. DISCUSSION

The preceding analyses have diverse theoretic and
practical implications: these extreme stormy areas
represent the refugia of primitive taxa with a very
peculiar and specific flora. On the other hand,
endolithic lichens and cyanobacteria, like hyperpachycauls and tragacanthids, are useful indicators
of exposure and of local wind dynamics, for the
distinction between sylvatic and asylvatic areas.

REFERENCES

Arber A (1928) The tree habit in Angiosperms,
its origin and meaning, New Phytol. 27, 69-84.
Blüthgen J (1964) Allgemeine Klimageographie.
In Obst E, ed. Lehrbuch der allgemeinen Geographie,
II. Berlin, De Gruyter.
Danin A et al (1982) Patterns of limestone and
dolomite weathering by lichens and blue-green
algae and their palaeoclimatic significance,Palaeo-
geogr. Palaeoclimat. Palaeoecol. 37, 221-233.
Degen A (1936) Flora velebitica, vol. I. Die
klimatischen Verhältnisse des Velebitgebirges, pp.
144-178. Budapest, Acad. Sci.
Dierschke H, ed. (1977) Vegetation und Klima.
Berichte Int. Verein. Veget. Symp. Rinteln. Vaduz,
Cramer.
Eredia F (1932) Lo scirocco in Italia, Ann. Uff.
Presagi, 5, 155-172.
Ercegovic A (1932) Etudes écologiques et socio-
logiques des cyanophycées lithophytes de la côte
yougoslave de l'Adriatique, Bull int. Yug. Acad.
26, 33-56.
Goldberg J (1940) The Arctic aspects of the
Port Senj in Adriatic (in Croat), Priroda (Zagreb)
30, 150-152.
Gračanin Z (1962)Verbreitung und Wirkung der Boden
erosion in Kroatien, Osteuropastudien, ser. I, 21,
1-333.
Lovrić AZ (1971) Coenodynamique et pédodynamique
du mode battu par rapport à la Bora et l'abrasion,
Thalassia Jug. 7(1), 195-200.
Lovrić AZ (1974) Ecological effects of the Bora
storms at Senj, and related flora and vegetation

(in Croat, Summary). Proc. 1st Congr. Ecol. Yug.
pp. 54.
Lovrić AZ (1975) Bora salt storms and biocenoses
of the Senh Archipelago (in Croat). Thesis reg. no
YUAA-9975, Inst. Bot. Univ. Zagreb.
Lovrić AZ (1977) Sempervirent plantations resist-
ing to Bora salt storms (in Croat), Bilten Baška,
Rijeka 2(3), 15-17.
Lovrić AZ (1980) Ecology of Bora storms and similar
continental winds in Adriatic (in Croat). Forest
Encyclopedia v.I, pp. 225-228, Yug. Lexic. Inst.
2nd edn. Zagreb.
Lovrić AZ (1982a) The Bora storms at Port of Senj
(in Croat), Priroda (Zagreb) 71, in press.
Lovrić AZ (1982b) Wind effect in paleoendemic
halocenoses, and ecosystems of coastal Karst (in
Croat). Diss. reg. no YUAA-1177, Sarajevo, in press.
Lovrić AZ (1982c) History and genetics of tolerance
to the natural phyto-toxic media of W. Balkans. In
the same volume.
Lovrić LM (1979) Zonation des microlithophytes et
hydrodynamisme orageux dans l'Adriatique supérieure,
Rapp. Comm. Int. Expl. Sci. Medit. 27(9), 67-68.
Lovrić LM (1981) Azonal microhabitats of litho-
phytic Cyanobacteria in stormy coastal caves of NE
Adriatic, Rapp. Comm. Int. Expl. Sci. Medit. 27(3),
37-38.
Lovrić LM and Lovrić AZ (1981) Cyanobacteria and
lichens in Karst shores. Adriatic endemics, 2, Rapp.
Comm. Int. Expl. Sci. Medit. 27(9), 67-68.
Lovrić LM and Lovrić AZ (1982) Petrophilic micro-
and macrovegetation in brackish estavelles and canyon
estuaries (in Croat, Summary). Proc. 6th Congr. Biol.,
Yug. I p, Univ. Novi Sad.
Lukšić I (1969) An interesting example of the Bora
at Port Senj (in Croat), Vijesti, Dept. Hydromet.
SR. Croat. Zagreb 19(3-4), 19-24.
Lukšić I (1980) Special effects of the cyclonic
Sirocco at Port Senj (in Croat). Proc. Conf. Weather
Agroclim. Mount Yug. v. I, 159-164, Dept. Hydromet.
Beograd.
Lukšić I (1975) Bora storms at Port Senj (in Croat).
Zbornik Mus. Senj 6, 467-494.
Mabberley DJ (1974) Pachycauly, vesselements,
islands and the evolution of arborescence in
"herbaceous" families, New Phytol. 73, 977-984.
Mabberley DJ (1980) Pachycaul plants and islands.
In Bramwell D, ed. Plants and islands, pp. 259-277,
London.
Miller BI (1958) On the maximum intensity of
hurricanes, J. Meteor. 15, 184-195.
Nordhoff CB and Hall JN (1936) The hurricanes,
257 p. Boston (cf. Blüthgen 1964).
Rogić V (1957) Coastal slopes of Mount Velebit,
I. Natural elements (in Croat), Geogr. Glasnik
Univ. Zagreb 19, 61-102.
Schubert L (1942) Praktische Orkankunde. 156 p.,
Berlin (cf. Blüthgen 1964).
Sloane E (1956) The book of storms, 109 p., New
York (cf. Blüthgen 1964).
Stipaničić V (1955) Bora storms with ice at
Port Senj (in Croat), Hidrogr. Godišnjak 2, 162-
163.
Stipaničić V (1956) The icing in Adriatic (in
Croat). Vijesti Pom. Met. Službe 2(2), pp.7.

Tannehill IR (1959) Hurricanes, their nature and history. Princeton, 308 p.

Yoshino MM, ed. (1972-1973) Studies on Bora, I-III, Climat. Notes Univ. Tokyo, 10, 1-78; 14, 1-40 and 15, 1-48.

Yoshino MM (1973) Studies on wind-shaped trees, Climat. Notes Univ. Tokyo 12, 1-52.

Yoshino MM, ed. (1976) Local wind Bora, Univ. Tokyo Press, 290 p.

Weischet W and Barsch D (1963) Studien der Deformation von Baumkronen durch Wind, Freiburg Geogr. Hefte 1, 1-130.

Zgur V (1954) Bora hurricanes at Port Senj (in Croat), Pomorstvo 9(2), 74-76.

Zgur V (1956) Effects of Bora hurricanes in Velebit channel (in Croat), Pomorstvo 11(4), 131-133.

LEAF FORM OF THE WOODY PLANTS OF INDIANA AS RELATED TO ENVIRONMENT*

GARY E. DOLPH (Indiana University at Kokomo, Kokomo, Indiana 46902 USA)

1. INTRODUCTION

Paleobotanists commonly use one of two methods to determine the paleoclimate under which a fossil angiosperm flora existed: the nearest living relative method (Axelrod, Bailey, 1969) or the foliar physiognomy method (Dilcher, 1973; Wolfe, 1969, 1971, 1978, 1981; Wolfe, Hopkins, 1967). The nearest living relative method is more subjective than the foliar physiognomy method. Use of the nearest living relative method requires the correct identification of each genus in a fossil flora. If a leaf form is misidentified, the correct identification of its nearest living relative is impossible, and any estimate of paleoclimate based on the climatic tolerances of the wrong living genus and/or species has a high probability of being inaccurate. In addition, the nearest living relative method assumes that a genus has a single and constant response to climate through time (D.L. Dilcher, personal communication). Few genera can meet this criterion over the millions of years involved. When using the foliar physiognomy method, no dependence is placed on the correct taxonomic identification of the fossil leaf forms. Therefore, the foliar physiognomy method should be easier to apply. As long as leaf form is closely tied with climate, the foliar physiognomy method should give an accurate estimate of paleoclimate. Eight different leaf characteristics are assumed to vary with climate

*This material is based upon work supported by the National Science Foundation under Grant No. SER 8004789.

(Dolph, Dilcher, 1979): 1) leaf size distribution, 2) leaf margin type, 3) presence or absence of drip-tips, 4) organization (compound vs simple), 5) major venation pattern, 6) venation density, 7) leaf texture, and 8) leaf base shape. Only the variation in the percentage of species having leaves with entire margins in a modern or fossil angiosperm flora has been studied in any detail.

The correlation between leaf margin type and climate was first studied by Bailey and Sinnott (1915, 1916). After analyzing 35 published floras and associated herbarium material, they concluded that as mean annual temperature decreases, either in response to changing latitude or increasing altitude, the percentage of woody dicotyledonous species having leaves with entire margins in a flora also decreases. The only exceptions occur in physiologically dry areas, such as tundra, where this percentage is always high. Other researchers have expanded on the results of Bailey and Sinnott (1915, 1916). For example, Dilcher (1973) studied the influence of both mean annual temperature and average annual precipitation on the percentage of dicotyledonous species having leaves with entire margins in the forests of the Western Hemisphere. Dolph (1979) analyzed the influence of mean annual biotemperature, average annual precipitation, and potential evapotranspiration ratio on leaf margin type in 38 forest stands in Costa Rica. Another researcher and proponent of the foliar physiognomy method is Wolfe (1969, 1971, 1978, 1981). When studying the flora of southeast Asia, Wolfe (1981) showed that the percentage of woody dicotyledonous species having leaves with entire margins decreased by 3% for every $1^{\circ}C$ decrease in temperature. Using this correlation, Wolfe (1978) was able to estimate the paleoclimate under which 52 fossil floras from the western United States and Alaska existed and to graph climatic trends from the Paleocene to the Present in the western United States.

Many of these studies have correlated generalized regional floras with broad regional climatic parameters, particularly mean annual temperature (Bailey, Sinnott, 1915, 1916; Wolfe, 1969, 1971, 1978, 1981). Although these studies have led to the development of significant new insights into the relationship between leaf margin type and climate, the successes achieved at the regional level have not been duplicated at the local (or county) level. After studying the correlation between leaf margin type and climate at 38 sample sites in Costa Rica using Holdridge's (1967) life zone chart, Dolph (1979) concluded that the leaf margin variation was not directly correlated with the variation in any of the three climatic parameters studied (mean annual biotemperature, mean annual precipitation, and potential evapotranspiration ratio). After mapping the variation in leaf form in the flora of North and South Carolina, Dolph and Dilcher (1979) found that the variation in leaf margin type was much greater than expected and that this variation was not strongly correlated with any of the five climatic parameters studied (mean annual temperature (equivalent to mean annual biotemperature in the Carolinas), average annual precipitation, potential evapotranspiration ratio, effective temperature, and equability). These studies indicated 1) that extreme caution should be used when applying the foliar physiognomy method in local vegetational analyses and 2) that local

environmental influences (such as soil type) as well as regional climatic parameters (such as mean annual temperature) must be considered. The role of every environmental influence could not be covered in detail in these studies due to a lack of environmental data from Costa Rica (Dolph, 1979) and due to the size and environmental diversity of the Carolinas (Dolph, Dilcher, 1979). Because Indiana is smaller, is less environmentally diverse, and has a well-studied environment, a detailed analysis of the relationship between leaf margin type and environment was undertaken there.

2. MATERIALS AND METHODS

The relationship between leaf margin type and environment was analyzed at the local (county) level in Indiana using two large data bases. The first data base contains climatic data from 91 weather stations in Indiana and the adjacent states of Michigan, Ohio, Kentucky and Illinois. Climatic information was obtained from the National Climatic Center in Asheville, North Carolina (U.S. Dept. of Commerce, 1973a-e). Seven different climatic parameters were studied using this data base (mean annual temperature, mean annual range in temperature, average annual precipitation, mean annual biotemperature, potential evapotranspiration ratio, effective temperature, and equability). The climatic variables could be studied for the whole year or for a selected portion of the year, such as the growing season. The second data base contains information on each of the 227 woody dicots native to Indiana. Floristic information was obtained from the writings of Deam (1932, 1940, 1953) and from the author's twelve years of field experience in Indiana. The data base records the name of each species, its leaf characteristics

(margin type as well as the range in leaf
length and width), its growth form (tree,
shrub, or vine), its growth habit (deciduous or evergreen), and the counties
in Indiana where it grows. The woody
gymnosperms were not included in this
study. The maps presented in this paper
were generated using these two data bases.
When calculating the percentage of species
having leaves with entire margins in a
county, the leaflets of compound leaves
were considered analogous to simple leaves.
Lobed leaves were considered to be non-
entire even if the actual margin lacked
teeth.

3. LEAF FORM OF THE WOODY VEGETATION

Because a strong correlation has
been shown to exist between the variation
in leaf margin type and mean annual
temperature, any analysis of the variation
in leaf margin type must begin by considering the influence of this climatic
parameter. Figure 1 shows the variation
in mean annual temperature across Indiana.
The mean annual temperature increases from
lows of less than 10°C in northern Indiana
to highs in excess of 13°C in southern
Indiana. The average temperature in
Indiana is 11.5$\pm$1.4°C. If approximately
3% of the woody dicotyledonous species
present should have leaves with entire
margins for every degree centigrade of
the mean annual temperature (Wolfe, 1981),
30% of woody dicotyledonous species present in northern Indiana should have
leaves with entire margins. This percentage should increase to about 40% in
southern Indiana. The average percentage
of woody dicotyledonous species having
leaves with entire margins should be
approximately 35%.

An analysis of the data for Indiana

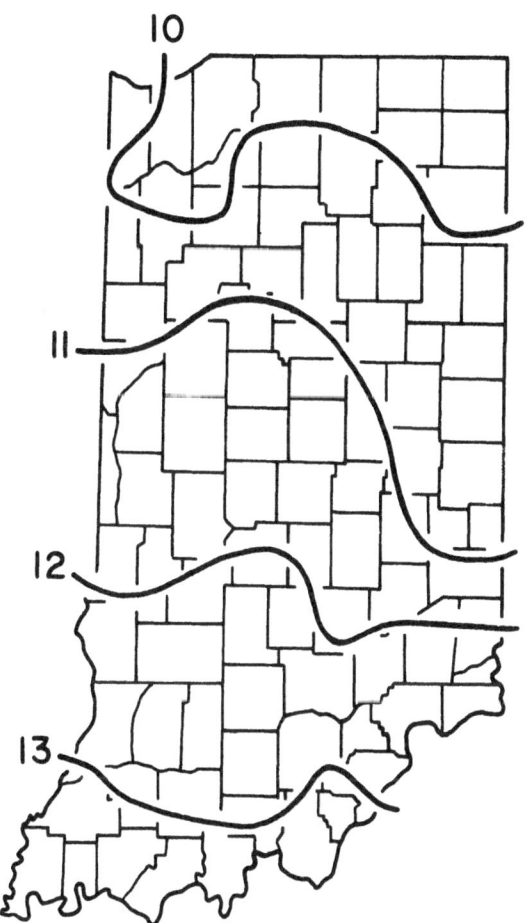

FIGURE 1. Mean annual temperature in $^{\circ}$C.

did not yield these predicted values
(Fig. 2). An average of only 18.9$\pm$2.7% of
the woody dicotyledonous species in Indiana
have leaves with entire margins. Rather
than showing the predicted north to south
increase along with mean annual temperature,
the percentage of woody dicotyledonous
species having leaves with entire margins
is lowest in east-central Indiana (with
lows of less than 15%) and highest in northern and southeastern Indiana (with highs in
excess of 20%). The percentages range from
a high of 22.6% in Brown County in southeastern Indiana to a low of 13.5% in Delaware County in east-central Indiana.
Although these results are quite different

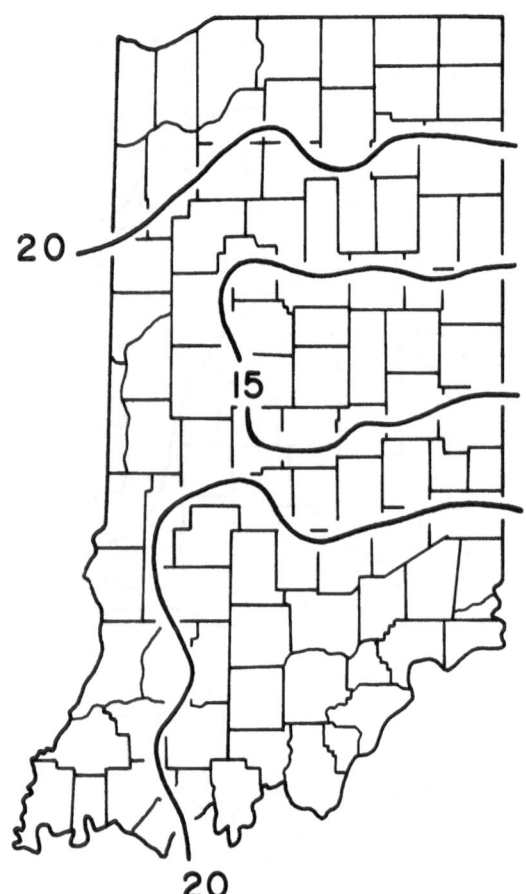 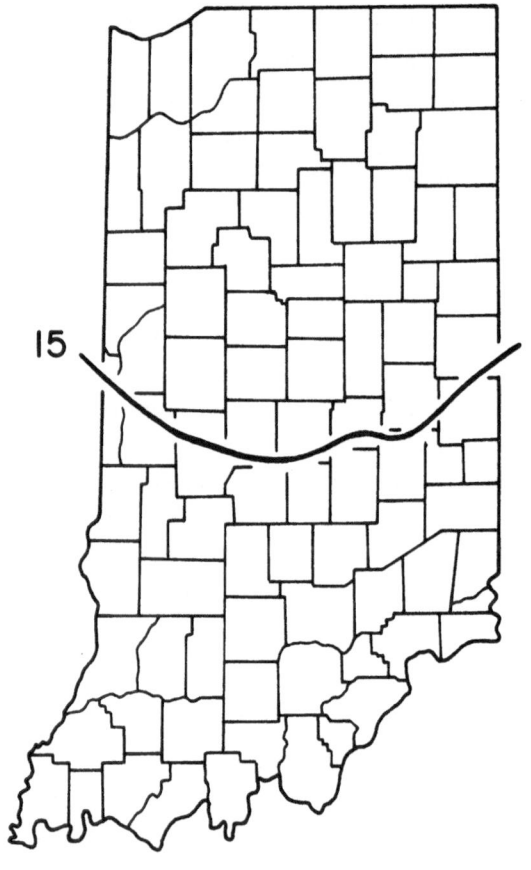

FIGURE 2. Variation in the percentage of woody dicotyledonous species having leaves with entire margins.

FIGURE 3. Variation in the percentage of dicotyledonous tree species having leaves with entire margins.

from the predicted results, a careful analysis of the data will reveal 1) why the initial predictions are inaccurate, 2) what environmental variables are responsible for the observed variation and 3) which plants are responding most strongly to these environmental variables.

4. LEAF FORM OF THE TREE SPECIES

In order to study the influence of environment on the leaf form of the woody dicotyledonous plants of Indiana (227 species) in greater detail, the total woody dicotyledonous vegetation was subdivided into trees (92 species), shrubs (117 species), and vines (18 species). The results of this subdivision are discussed in detail in this paper. To help clarify the situation further, each of these three groups could be subdivided into their deciduous and evergreen components. The results of these analyses are referred to in the text, but a more detailed discussion of them will be found elsewhere (Dolph, manuscript in preparation).

Figure 3 shows the variation in the percentage of deciduous dicotyledonous tree species having leaves with entire margins across Indiana. There are no evergreen dicotyledonous trees native to Indiana.

The percentages increase from north to
south and their values range from a low of
12.7% in LaPorte and St. Joseph Counties
in northern Indiana to a high of 18.8% in
Shelby County in southern Indiana. An
average of 15.0±1.5% of the dicotyledonous
tree species present have leaves with entire margins.

Both the percentage of dicotyledonous
tree species having leaves with entire
margins and the mean annual temperature
increase from north to south. This cor-
respondence is significant, because a
careful analysis of Wolfe's chief data
source (Wang, 1961) indicates that most
of his percentages reflect only the dominant trees in the region. In both Indiana
and southeast Asia, there is a correspon-
dence between the variation in the percentage of dicotyledonous tree species
having leaves with entire margins and
mean annual temperature. Therefore, the
leaf margin variation discussed in the
previous section and based on the total
dicotyledonous vegetation of Indiana
differs from the predicted results, because two different types of vegetation
are being compared.

Although the percentage of deciduous
dicotyledonous tree species having leaves
with entire margins in Indiana appears to
be correlated with mean annual temperature, I do not mean to imply that a
direct relationship exists between these
two variables. It is unlikely that any
one climatic parameter can determine the
variation in any one plant character.
Almost all of the other climatic variables
studied in Indiana vary from north to
south. For example, average annual precipitation varies from lows of less than
90 cm/year in northern Indiana to highs
in excess of 110 cm/year in southern

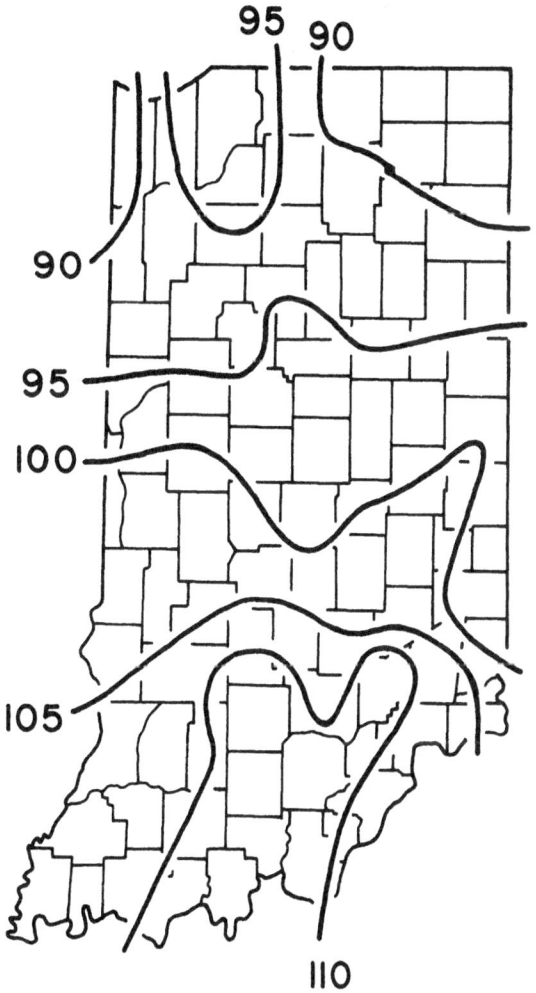

FIGURE 4. Average annual precipitation in cm.

Indiana (Fig. 4). The local high in excess
of 95 cm/year in northwestern Indiana reflects the influence of moisture-laden winds
coming off of Lake Michigan. Undoubtedly,
all the major climatic parameters contribute
to determining the percentage of dicotyledonous species having leaves with entire margins in a flora. However, the major climatic
parameters, such as temperature and precipitation, are more likely to influence leaf
margin variation in the trees, which are
directly exposed to them, than they are in the
shrubs, which are protected from them. As we
shall see in the next two sections, it is

56

edaphic influences on the shrubs and vines which influence their distribution and hence their leaf margin variation. Because the trees vary so slightly in leaf margin percentage, the shrubs and vines exert the major influence over how the leaf margin variation of the total woody dicotyledonous flora looks when mapped.

5. LEAF FORM OF THE SHRUB SPECIES

Figure 5 shows the variation in the percentage of dicotyledonous shrub species, both deciduous and evergreen, having leaves with entire margins. The percentages vary from lows of less than 20% in east-central Indiana to highs in excess of 25% in southeastern and northern Indiana. The average percentage of dicotyledonous shrub species having leaves with entire margins is 22.9±3.9%. The percentages range from a high of 29.0% in Dubois County in south-central Indiana to a low of 15.0% in Delaware County in east-central Indiana. The leaf margin variation of the dicotyledonous shrub species (Fig. 5) is very similar to and in large part determines the variation seen for the total woody dicotyledonous flora of Indiana (Fig. 2).

The dicotyledonous shrubs exert a dominant influence on the total leaf margin variation seen in Indiana for four reasons. First, the range in leaf margin variation for the dicotyledonous trees (12.7% to 18.8%) is less than half that for the shrubs (15.0% to 29.0%). Second, the number of shrubs present (117) is greater than the number of tree species present (92). A more variable and more numerous group (i.e., the dicotyledonous shrubs) will suppress the influence of a less variable and less numerous group (i.e., the dicotyledonous trees) when the influence of the two groups are averaged together. Third, the percentage of dicoty-

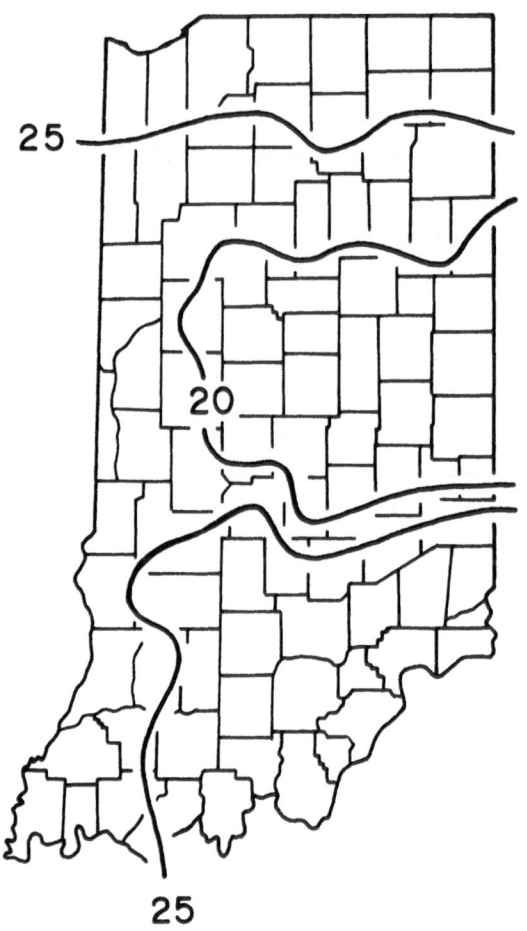

FIGURE 5. Variation in the percentage of dicotyledonous shrub species having leaves with entire margins.

ledonous shrub species (63.2%) showing a restricted distribution within the State is greater than that for the dicotyledonous trees (48.9%). Finally, the vast majority of shrub species, particularly those having leaves with entire margins, are restricted to four very specific habitats: 1) the lake area of northern Indiana, 2) the flats of south-eastern Indiana, 3) the knobs of south-central Indiana and 4) the sandstone ridges (Crawford Upland) and outcrops of the Mansfield Formation which connect the knobs in the south with the lake area in the north (Fig. 6). The lake area found in northern Indiana has a high density of bogs, swamps, and ponds (Deam, 1932, 1940,

1953; Taylor, 1907) as well as peat
deposits dating back to the retreat of
Wisconsin ice (Schneider, Moore, 1978;
Taylor, 1907). The flats consist of a
swampy area (Deam, 1932, 1940, 1953) without peat deposits (Schneider, Moore, 1978).
The knobs consist of two deeply dissected
uplands, the Norman Upland to the east and
the Crawford Upland to the west, separated
by an area of low relief called the
Mitchell Plain (Schneider, 1966). The high
ridges found in the Norman Upland are
capped by the resistant siltstones of the
Edwardsville and Locust Point Formations
(Carr, Hatfield, 1975). The high ridges
found in the Crawford Upland are capped
by the resistant sandstones of the Big
Clifty and Mansfield Formations (Carr,
Hatfield, 1975; Gray, 1962). Sandstones
of the Mansfield Formation continue out
of the knobs area as an extension of the
Crawford Upland (Schneider, 1966) and
finally as localized outcrops (Murray,
Patton, 1953; Gray, 1962; Greenberg, 1960)
which connect the lake area with the knobs.

The aerial distribution of these four
physiographic regions corresponds nicely
to where the highest percentages of dicotyledonous shrub species having leaves with
entire margins are found. The correspon-
dence does not mean that the same shrub
species occur throughout these physiographic regions, however. In northern
Indiana, a large number of the dicotyledonous shrubs are swamp or marsh species
having a more northern center of distribution (e.g., _Betula_ _pumila_, _Cornus_ _rugosa_,
and _Sambucus_ _pubens_). In southern Indiana,
a large number of the shrub species are
reaching the northern limit of their distribution in either the Lower Wabash and
Ohio River Valleys (e.g., _Forestiera_
acuminata), the knobs (e.g., _Rubus_

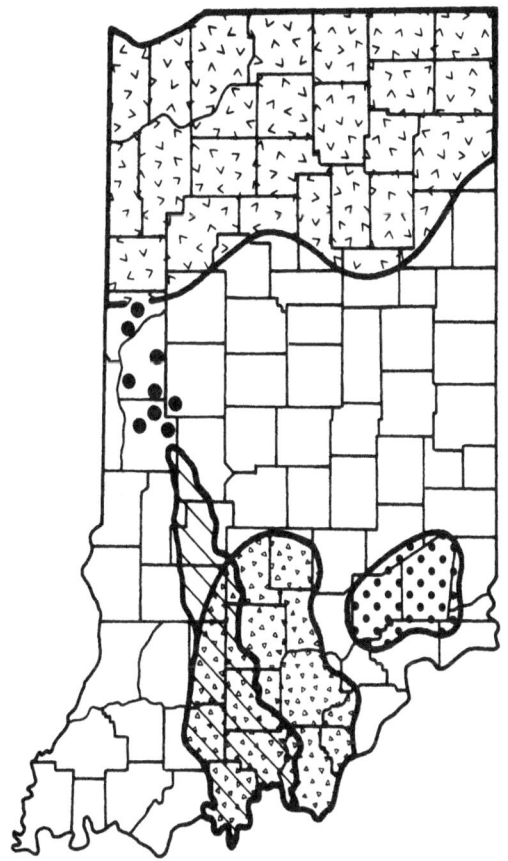

☐ Lake Area

☐ Knobs

☐ Flats

☐ Crawford Upland

● Mansfield Outcrops

FIGURE 6. Physiographic regions in Indiana
which influence the distribution of the
dicotyledonous shrubs and vines.

odoratus), or the flats (e.g., _Viburnum_
molle). The sandstone outcrops of the
Mansfield Formation which connect northern
and southeastern Indiana serve as a bridge
for species inhabiting both the lake area
and the knobs and flats (e.g., _Gaylussacia_
baccata or _Vaccinium_ _vacillans_).

In each of these physiographic regions,
the environmental parameter that limits
dicotyledonous shrub distribution is different. In northern Indiana, high soil
moisture levels exert the greatest influence. In southern Indiana, dicotyledonous
shrub distribution is influenced by the protected nature of the valleys in the knobs
and the longer growing season (170 days or
more in southern Indiana as compared to 160
days or less in northern Indiana; Schaal,
Newman, 1981). Dicotyledonous shrubs found
associated with the sandstone outcrops of
the Mansfield Formation thrive there because the outcrops are found in deeply
dissected stream beds. Therefore, the
distribution of the dicotyledonous shrub
species in Indiana is not directly determined by major climatic factors, such as
mean annual temperature or average annual
precipitation. The distribution of the
dicotyledonous shrubs is determined by more
subtle secondary environmental factors,
such as degree of protection and the availability of soil moisture.

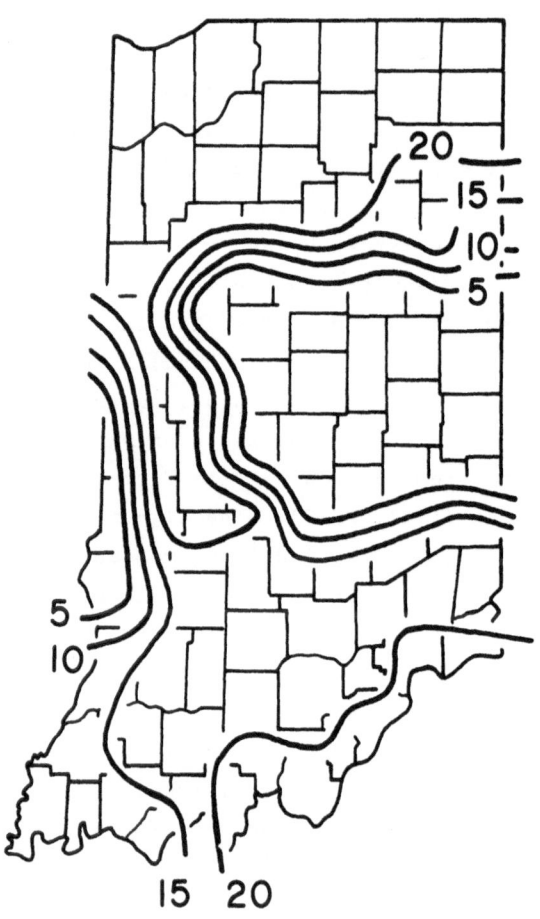

FIGURE 7. Variation in the percentage of
dicotyledonous vine species having leaves
with entire margins.

6. LEAF FORM OF THE VINE SPECIES

The variation in the percentage of
dicotyledonous vine species, both deciduous and evergreen, having leaves with
entire margins is shown in Figure 7. The
average percentage of dicotyledonous vine
species having leaves with entire margins
is 13.5+8.7%. The percentages range from
a high of 21.4%, found all along the Ohio
River in south-central and southeastern
Indiana, to a low of 0.0%, found in east-
central and west-central Indiana. The
extreme variation encountered in leaf
margin variation for the dicotyledonous
vines is a reflection of the small number
of species present (18). No county contained more than 14 or less than 8 vine

species, but slight changes in species distribution caused radical changes in the
percentage of dicotyledonous vine species
having leaves with entire margins.

By comparing the variation in the percentage of dicotyledonous vine species having
leaves with entire margins (Fig. 7) with the
physiographic regions shown in Figure 6, it
is obvious that the same environmental factors which influence leaf margin variation in
the dicotyledonous shrubs influence leaf margin variation in the dicotyledonous vines.
Some vine species are restricted to the lake
area (e.g., _Parthenocissus_ _vitacea_). Others

are restricted to the Lower Wabash and
Ohio River Valleys (e.g., <u>Anisostichus</u>
<u>capreolata</u> or <u>Calycocarpum Lyoni</u>). Two
species occur in the lake area, southeastern Indiana (knobs and flats), and
in the counties of west-central Indiana
where the Mansfield Formation outcrops
(<u>Lonicera dioica</u> and <u>L. prolifera</u>). No
species is restricted solely to either
the knobs or flats region.

7. CONCLUSIONS

A number of important conclusions
can be reached as a result of this study.
First, as in the Carolinas (Dolph,
Dilcher, 1979) and Costa Rica (Dolph,
1979), the variation in the percentage
of woody dicotyledonous species in the
total flora having leaves with entire
margins is not closely correlated with
any single climatic parameter in Indiana.
That this result is not due to faulty
data collection or small sample size
(Wolfe, 1981) is shown by the close correlation between the variation in the
leaf margin type of the trees, shrubs,
and vines and the variation in a number
of environmental parameters. Second, the
variation in the percentage of dicotyledonous tree species having leaves with
entire margins is strongly influenced
by the major climatic parameters, such
as mean annual temperature and average
annual precipitation. All the major
climatic variables probably influence
the leaf margin variation of the dicotyledonous trees, because these trees are
directly exposed to their action. Third,
the variation in the percentage of dicotyledonous shrub or vine species having
leaves with entire margins is not influenced by the variation in the major
climatic parameters, such as mean annual

temperature. Leaf margin variation in the
dicotyledonous shrubs and vines is most
strongly influenced by secondary environmental factors, the most obvious of which
are soil moisture level and degree of protection. Fourth, at least in Indiana, the
variation in the number of woody dicotyledonous species in the total flora having
leaves with entire margins is most strongly
influenced by the presence of the shrubs and
vines, because they make up more than 50%
of the total flora. If paleoclimatic estimates
made using fossil angiosperm floras are accurate, it is because they contain a higher
representation of dicotyledonous trees than
of dicotyledonous shrubs and vines. Finally,
the extreme variation encountered when analyzing the leaf margin variation of the dicotyledonous vines underscores the danger of
working with small samples.

Although problems exist with current
theories relating foliar physiognomy with
climate, the work that has been carried out
to date has led to a notable increase in our
understanding of the variation of leaf form
with climate. If past work has been somewhat restricted in terms of the types of
plants and geographic areas covered, this
work did point out that correlations between
leaf form and climate do exist. Further work
will use larger data sets and more environmental variables. Total floras and not simply
the dominant trees will have to be studied.
The total environment, including such neglect-
ed parameters as soil type and soil moisture
levels, and not just the major climatic parameters, such as mean annual temperature and
total annual precipitation, will have to be
studied. If past work has emphasized the
freedom of the foliar physiognomy method from
taxonomic constraints, future research might
indicate that certain subsets of the total
vegetation (e.g., the dicotyledonous trees)

are more useful than others in determining paleoclimate. If this is true, then a renewed emphasis on the taxonomy of fossil angiosperm leaves will be required to allow this subset to be identified in a fossil flora. If past work has been based primarily on published floras, future work will have to emphasize field work. Field work will be particularly necessary to find out how deposition alters the composition of the modern forest, if the theory is to be used to estimate paleoclimate (Roth, Dilcher, 1978).

REFERENCES

Axelrod DI and Bailey HP (1969) Paleotemperature analysis of Tertiary floras, Palaeogeogr., Palaeoclimatol., Palaeoecol. 6, 163-195.

Bailey IW and Sinnott EW (1915) A botanical index of Cretaceous and Tertiary climates, Science 41, 831-834.

Bailey IW and Sinnott EW (1916) The climatic distribution of certain types of angiosperm leaves, Amer. J. Bot. 3, 24-39.

Carr DD and Hatfield CB (1975) Dimension sandstone resources of Indiana, Indiana Dept. Natur. Resources, Geol. Surv. Bull., 42-M.

Deam CC (1932) Shrubs of Indiana, Indianapolis, Dept. of Conserv.

Deam CC (1940) Flora of Indiana, Indianapolis, Dept. of Conserv.

Deam CC (1953) Trees of Indiana, Indianapolis, Dept. of Conserv.

Dilcher DL (1973) The Eocene floras of southeastern North America. In Graham A, ed. Vegetation and vegetational history of northern Latin America, pp. 39-59. New York, Elsevier.

Dolph GE (1979) Variation in leaf margin with respect to climate in Costa Rica, Bull. Torrey Bot. Club 106, 104-109.

Dolph GE and Dilcher DL (1979) Foliar physiognomy as an aid in determining paleoclimate, Palaeontographica, 170B, 151-172.

Gray HH (1962) Outcrop features of the Mansfield Formation in southwestern Indiana, Indiana Dept. Conserv., Geol. Surv. Rep. Progress 26.

Greenberg SS (1960) Petrography of Indiana sandstones collected for high-silica evaluation, Indiana Dept. Conserv., Geol. Surv. Bull. 17.

Holdridge LR (1967) Life zone ecology. Tropical Science Center, San Jose, Costa Rica.

Murray HH and Patton JB (1953) Preliminary report on high-silica sand in Indiana, Indiana Dept. Conserv., Geol. Surv. Rep. Progress 5.

Roth JL and Dilcher DL (1978) Some considerations in leaf size and leaf margin analysis of fossil leaves, Cour. Forsch.-Inst. Senckenberg 30, 165-171.

Schaal LA and Newman JE (1981) Determining spring and fall frost-freeze risks in Indiana, Coop. Ext. Service, Purdue Univ., Agronomy Guide (Weather) AY-231.

Schneider AF (1966) Physiography. In Lindsey AA, ed. Natural features of Indiana, pp. 40-56. Indianapolis, Indiana Academy of Science.

Schneider AF and Moore MC (1978) Peat resources of Indiana, Indiana Dept. Natur. Resources, Geol. Surv. Bull., 42-0.

Taylor AE (1907) The peat deposits of northern Indiana, Indiana Dept. Geol. Natur. Resources, Annu. Rep. 31.

U.S. Dept. of Commerce Environmental Data Service (1973a) Climatology of the United States, No. 81 (Illinois). Monthly normals of temperature, precipitation, and heating and cooling degree days 1941-1970. Asheville, North Carolina.

U.S. Dept. of Commerce Environmental Data Service (1973b) Climatology of the United States, No. 81 (Indiana). Monthly normals of temperature, precipitation, and heating and cooling degree days 1941-1970. Asheville, North Carolina.

U.S. Dept. of Commerce Environmental Data Service (1973c) Climatology of the United States, No. 81 (Kentucky). Monthly normals of temperature, precipitation, and heating and cooling degree days 1941-1970. Asheville, North Carolina.

U.S. Dept. of Commerce Environmental Data Service (1973d) Climatology of the United States, No. 81 (Michigan). Monthly normals of temperature, precipitation, and heating and cooling degree days 1941-1970. Asheville, North Carolina.

U.S. Dept. of Commerce Environmental Data Service (1973e) Climatology of the United States, No. 81 (Ohio). Monthly normals of temperature, precipitation, and heating and cooling degree days 1941-1970. Asheville, North Carolina.

Wang CW (1961) The forests of China, Harvard Univ., Publ. Maria Moors Cabot. Found. 5.

Wolfe JA (1969) Paleogene floras from the Gulf of Alaska region, U.S. Geol. Surv. Open-file Rep.

Wolfe JA (1971) Tertiary climatic fluctuations and methods of analysis of Tertiary floras, Palaeogeogr., Palaeoclimatol., Palaeoecol. 9, 27-57.

Wolfe JA (1978) A paleobotanical interpretation of Tertiary climates in the Northern

Hemisphere, Amer. Sci. 66, 694-703.

Wolfe JA (1981) Paleoclimatic significance of the Oligocene and Neogene
floras of the northwestern United States.
In Niklas KJ, ed. Paleobotany, paleoeco-
logy, and evolution, Vol. 2, pp. 79-101.
New York, Praeger.

Wolfe JA and Hopkins DM (1967) Climatic
changes recorded by land floras in northwestern North America. In Hatai K, ed.
Tertiary correlations and climatic changes
in the Pacific, pp. 67-76. Tokyo, Symp.
Pacific Sci. Congr.

FIRE ADAPTATION STRATEGIES OF PLANTS IN THE FRENCH MEDITERRANEAN AREA

L. TRABAUD (Centre d' Etudes Phytosociologiques et Ecologiques, route de Mende, B.P. 5051, 34033-Montpellier, France)

1. INTRODUCTION

Due to its frequent occurrence in the past and the present, fire is an important factor influencing the dynamics of the plant communities in the French mediterranean area. Former studies (Braun-Blanquet, 1936; Kuhnholtz-Lordat, 1938, 1958; Kornas, 1958; Barry, 1960) considered vegetation changes through a succession of several stages in the regressive sequence. More recently, Trabaud (1970, 1974, 1977, 1980),and Trabaud and Lepart (1980, 1981) showed that burned phytocoenoses both floristically and structurally recovered very quickly to the state similar to the previous ones which existed before fire. Thus after fire there is also a return, more or less rapid according to the different communities, toward a stable (at least temporarily) metastable equilibrium condition, similar to that existing before fire. The plants which appeared immediately after fire were the same ones which already existed before fire. This result is due to the types of re-generation they use: either principally by vege-tative survival organs, or by residual seeds buried in the soil or seeds coming from unburned areas of the same communities.

Thus plants have been obliged to acquire adaptive traits allowing them to withstand and survive fires.

2. EXPERIMENTAL APPROACH OF THE SURVIVAL STRATEGIES

Fire frequency is extremely important to under-stand the relative stability of phytocoenoses. As a matter of fact when fires occur every two or three years, a frequency which is not encountered in wildlands, there are noticeable changes in the plant dominance (some species can even dis-appear) and consequently in the structure of the communities. Therefore a factorial type experi-mentation has been set up in a *Quercus coccifera*

L. garrigue located 10 km north of Montpellier (Trabaud, 1974, 1977, 1980). This experimentation was intended to analyse:

a. vegetation changes as compared to the state previous to fire in order to determine the plant's degree of resistance to fire;

b. the impact of repeated burns with different frequencies on vegetation: a burn every six years, a burn every three years, a burn every two years. This would allow determination of the effect of frequent burns on the vegetation. Control plots were never burned;

c. the effect on the vegetation of the burning season (either in late spring or early autumn), in order to determine if the seasonal conditions influencing the changes of the phenology and physiology of species are correlated with the effects of fire, and to determine if they modify the species behavior and the vegetation equilibrium. For example, when prescribed burns are repeated within frequent intervals there is a change in phytomass. During the first six years of the experimentation the phytomass progressively de-creased, then reached a rather stable level (Trabaud, 1977, 1980). This decrease is due to a reduction in the woody plant phytomass. There is no time for the stems and twigs located well above ground to develop in frequently burned plots. On the other hand there is an increase of the herbaceous plant phytomass. These tendencies are more markedly striking when the burning season is early in autumn (Fig. 1).

Nevertheless, in spite of the repeated frequency of the burns there is no noticeable change in the flora of the *Quercus coccifera* garrigue (Trabaud, 1980; Trabaud and Lepart, 1981). This result comes from the behavioral differences of the species with regards to fire action. The perennial species possessing vegetative regenerative means,

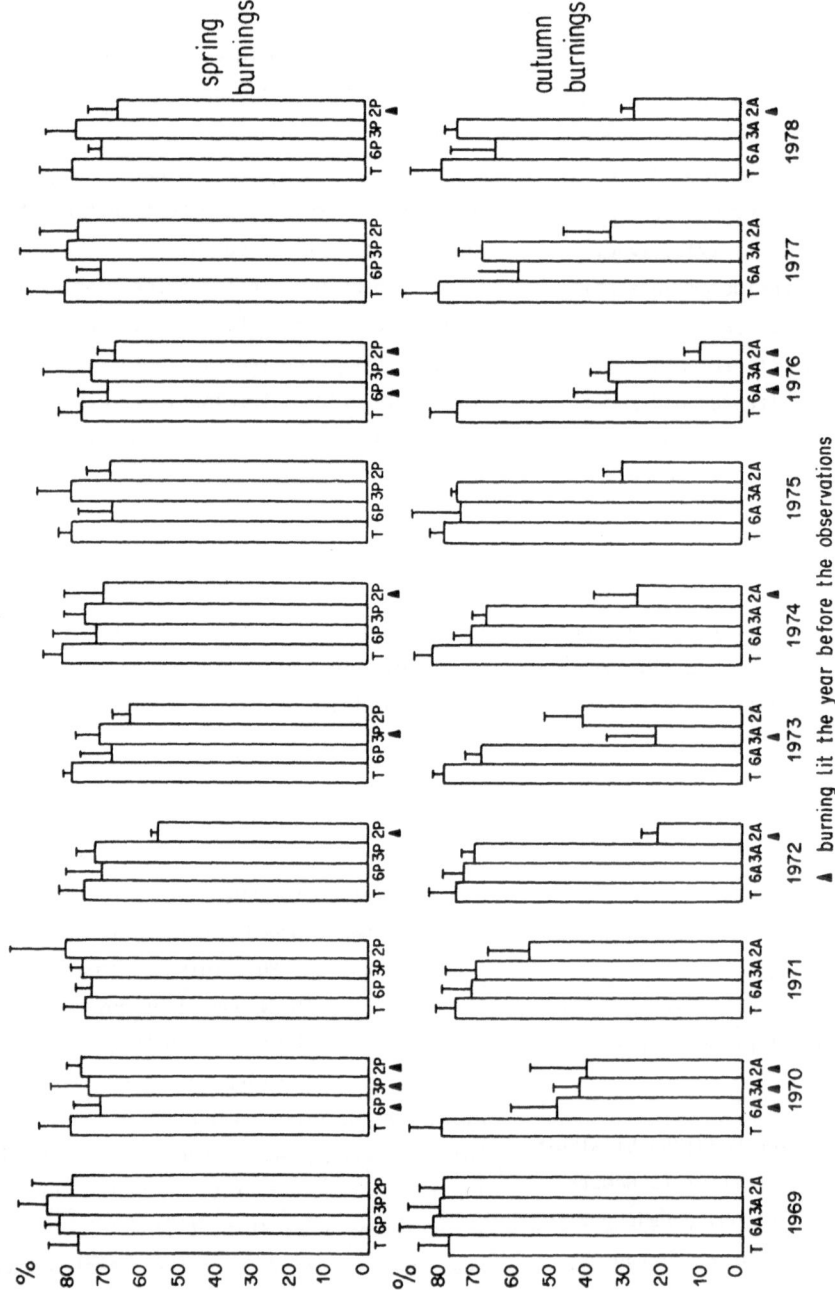

FIGURE 1. Changes in the ratio of the hits made by woody plants to the total hits for the prescribed burned *Quercus coccifera* garrigue.

which were present before the burns, always massively occupy the plots. However, the frequency and the season of burns can lead to some very little changes in the floristic composition. Thus, the frequency of fire occurrence is particularly important for species which can only reproduce by sexual means (i.e., by seeds; e.g. *Cistus*). They can be eliminated if the time interval between two successive burns is shorter than the period needed by these plants to reach sexual maturity. This has been also established by Specht et al. (1958), Hanes (1971), Barney and Frischknecht (1974), and Gill (1977).

In fact, the general model for the increase of the phytomass after each burn is similar to a sigmoid type curve (Trabaud, 1977, 1980). Frequent burns lead to a progressive decrease in this curve down to a stable level. But plants do not have the same behavior toward fire action: some are more susceptible than others (Trabaud, 1974, 1980). The survival strategies used by species play an important role in their success after fire, and the study of the strategies allows to understand and confirm what has previously been written. Three pairs of perennial species (one woody and one herbaceous) corresponding to the three survival strategies usually used to regenerate after fire have been chosen in order to study plant behavior after fire regularly set every two years in late spring (2P) or in early autumn (2A):

- Species which regenerate only by sprouts (*Quercus coccifera, Brachypodium ramosum*);
- Species which regenerate both by sprouts and seeds (*Dorycnium suffruticosum, Arrhenatherum elatius*);
- Species which reproduce only by seeds (*Cistus monspeliensis, Sanguisorba minor*).

The occurrences of only perennial species on 10 cm segments along a line of 10 m have been compared in such a manner that biases due to the importance of the phytomass or the cover introduced by dominant species are avoided.

The occurrences of species regenerating only by sprouts remain relatively constant althrough the period of the experimentation (Table 1); however, the frequency of fires has an effect on *Q. coccifera*; there is a little change in its occurrence. The species which reproduce only by seeds are very rare in this type of garrigue, and appear only erratically after fire (*Sanguisorba minor*); there is even a disappearance of some woody species (*Cistus monspeliensis*). On the other hand, species which regenerate both by sprouts and seeds become more and more important (Table 1). This is due to the survival means which are utilized. These allow the latter to have quick establishment and good dissemination.

Therefore the survival strategies that are the most widespread among the plants of the *Q. coccifera* garrigue are sprouting and sprouting plus seeding. Is this inertia (Holling, 1973; Boesch, 1974) or resilience (Orians, 1974, 1975) of the vegetation? It is not important, the fact is that the dominant species of this type of garrigue have acquired adaptations which allow them to survive fires.

3. WILDFIRE EFFECTS AND SURVIVAL ADAPTIVE SYSTEMS

To have a better understanding of wildfire action on vegetation a set of 47 observation plots was established in eight burned phytocoenoses representative of the most frequently encountered plant communities in the region of Montpellier (Trabaud, 1980). The direct method, i.e. diachronic method (Pavillard, 1935) was used to follow the stages of vegetation recovery. The aim was to follow the modifications of the vegetation through time on permanent plots established after wildfires. This method allowed to precisely record relatively small floristic and structural changes. The observation technique consisted of a permanent transect line 20 m long. Observations were made every 10 cm. At each observation point the occurrence made by species at every layer was recorded as the number of hits along a sight needle. These two kinds of measures permit an estimate of the layering of the vegetation and its quantity. In addition to these measures, a floristic list of

TABLE 1. Changes in the segment occurrence of six species with different survival strategies with regard to fire (a burn every two years in spring or in autumn): perscribed burned *Quercus coccifera* garrigue (means and standard error). (▲ : burning period).

		1969	1970	1971	1972	1973	1974	1975	1976	1977	1978
Quercus coccifera	2P	99,3 ±0,7	89,7±0,3	97,3±1,7	75,3±5,5	90,7±2,7	82,7±2,3	81,7±2,3	73,7 ±3,4	88,3±5,5	73,3±2,1
	2A	96,0±2,1	30,3±9,8	86,0±3,0	27,0±12,3	77,3±7,0	51,0±8,7	67,3±8,3	26,3± 2,2	58,3±6,3	34,0±2,5
Brachypodium ramosum	2P	49,7±7,6	28,3±2,8	41,7±7,8	34,7± 5,0	45,0±7,0	43,3±3,7	47,0±5,5	42,0±4,0	49,7±2,4	43,0±3,2
	2A	60,0±8,1	32,3±3,8	53,3±9,9	48,3±8,1	63,0±7,0	64,7±13,4	79,3±6,7	65,7±9,2	75,3±8,1	68,7±11,8
Dorycnium suffruticosum	2P	17,3±6,8	23,3±5,9	35,7±6,7	28,3±5,4	36,0±6,6	42,0±8,5	36,7±6,6	35,3±5,5	39,7±7,5	39,3± 5,8
	2A	25,7±9,9	12,7±1,3	32,3±0,5	27,3±8,7	41,3±5,3	45,3±14,1	39,7±7,3	26,0±11,7	45,7±9,0	35,3±6,9
Arrhenatherum elatius	2P	6	1	2,0±0,0	20,0±10,6	38,7±18,4	32,0±15,7	43,5±17,0	19,0±5,2	43,0 ±9,4	20,3±8,4
	2A	6	10,5±0,5	15,7±6,4	38,3±14,8	41,3±12,7	37,3±17,1	28,5±14,7	10,5±4,9	23,0±8,1	31,3±14,9
Cistus monspeliensis	2P	0	1	1	0	0	0	0	0	0	0
	2A	34	1	3	0	12	0	5	0	0	0
Sanguisorba minor	2P	0	0	0	3	1	2	0	0	0	0
	2A	1	1	0	0	0	0	0	0	1	1

all the species which were present at the time of the observations of the transect line was noted in a plot of 100 m^2. All these records were regularly carried out every year in spring for the five first years following fire. Later, as the communities seemed to physiognomically stabilize, the observations were carried out only every two years up to the tenth, or twelfth,year after fire. All the species of the studied communities were able to survive more or less frequent fires. Therefore, their resistance depends on the survival strategies they utilize.

The species observed in the field during the studies of the plots burned by wildfires, have been grouped according to the adaptive systems they use to survive and regenerate after fire:
- Species reproducing only by seeds;
- Species reproducing preferentially by seeds;
- Species reproducing both by seeds and sprouts, without a clear tendency one way or another;
- Species reproducing preferentially by sprouts;
- Species reproducing only by sprouts.

Table 2 shows the percentage, in number of hits, with regards to the total vegetation sampled at each period, of the species which regenerate preferentially or exclusively by sprouts, for the whole observation periods (10 to 12 years). In all the considered communities this percentage is always higher than 50. Only the pine woodlands of *Pinus halepensis* and the *Rosmarinus officinalis* garrigue have a much lower value (31 and 21% respectively). This is due to the fact that these two species, which dominate these communities, reproduce only by seeds after fire.

The most frequent adaptive trait of survival to fire corresponds to sprouts coming from root crowns, followed by sprouts from rhizomes, finally plants possessing organs deeply buried in the

TABLE 2. Percentage of the hits made by species preferentially or exclusively regenerating by sprouts to the total vegetation hits during recovery after wildfire.

	Range of hits in percent of total vegetation	Percentage of cases for the cited ranges
Quercus ilex dense woodlands	75 - 99	79,5 for the ranges 80 to 100%
Quercus ilex open woodlands	51 - 80	71,0 for the ranges 60 to 80%
Quercus coccifera dense garrigues	70 - 96	77,9 for the ranges 80 to 100%
Quercus coccifera open garrigues	62 - 97	66,7 for the ranges 80 to 100%
Pinus halepensis forests	31 - 79	52,2 for the ranges 40 to 60%
Rosmarinus officinalis garrigues	21 - 90	64,7 for the ranges 70 to 90%
Brachypodium ramosum swards	63 - 90	94,4 for the ranges 70 to 90%
Brachypodium phoenicoides swards	64 - 86	76,5 for the ranges 70 to 80%

soil such as bulbs or tubers have the lowest frequency in regenerating by vegetative means. Plants which reproduce only by seeds belong to a less important group even though they possess a long spectrum of strategies, which include everything from fortuitous introduction to an extreme adaptation.

Long-lived perennial species having a k strategy type (Mac Arthur and Wilson, 1967; Pianka, 1970) prevail throughout the period of the vegetation recovery as compared to species which have a r strategy type. The latter are only numerous during the second and third year after wildfire, but they never constitute an important phytomass (Trabaud, 1980; Trabaud and Lepart, 1980).

4. DISCUSSION AND CONCLUSION

With frequent fire cycles, only plants or phyto-coenoses able to survive the passage of the flame owing to special means have been able to perpetuate. Grime (1977) in his study about the strategies utilized by plants to survive disturbances or environmental stresses considered four types of habitats according to the importance of the disturbance or the stress. This author only recognized three possibilities for survival or adaptive strategies: competitive plants, stress-tolerant plants and ruderal plants. He did not consider a fourth type because he thought plants could not survive in an environment with both strong disturbances and severe stress. However, plants which survive fire (Grime considered fire as a disturbance) seem to fit this fourth case since they withstand fire and they experience severe stresses; as a matter of fact most of the species that vegetatively sprout after fire grow during the worst conditions of the dry summer and early autumn period. The only biological characteristics which can be applied to the plants withstanding fire belong to the group of plants defined by Grime as stress-tolerant; unless fire is a weak disturbance! According to the results obtained during the study it seems that a fourth type of strategy (plants both tolerant to strong

disturbances and severe stresses) does exist: they are plants belonging to the mediterranean vegetation which were burned for millenia, Thus they have acquired unusual adaptive traits. Naveh (1975) compared these adaptations to a cybernetic system possessing feedback homeostatic responses with regards to change due to fire. He thought the information from the "fire event and the ecological after-effects" are largely transformed and recoded by more and more numerous "fire-resistant" genes, whereas genes leading to a fire susceptibility were progressively eliminated. This author added that genotypes in which the feedback is large enough to allow them to endure the fire effects have better and greater chances of survival.

Positive feedback responses are those which help overcome the ecological effects of fire by increasing the physiological activity. This activity is showed by a high production of vegetative shoots (as for Quercus coccifera, Brachypodium ramosum) or by an important germination rate stimulated by fire or by a high production of flowers and seeds (as for Cistus spp, Pinus spp). On the other hand, the survival mechanisms allowing plants to withstand fire, either by passive direct tolerance of the organs or seeds to fire, or by a reduction of the physiological activity during the fire periods, can be considered as negative feedback responses: Quercus ruber with its nonflammable and heat-insulating cork bark shows such mechanisms.

Thus, a typology begins to stand out that permits the distinction of plants resistant to fire, or pyrophytes (Kuhnholtz-Lordat, 1938, 1958): active pyrophytes possessing positive responses to fire action, whereas passive pyrophytes would present negative responses.

Among the active pyrophytes, plants regenerating by sprouts, or both by sprouts and seeds, can start to occupy the area immediately after the passage of fire. In Southern France they, however, must have a strong resistance to drought. On the other hand plants reproducing only by seeds will appear later on: seeds remain in the soil and germinate only after the first rains of autumn or most of the time in spring the following year. Therefore plants which sprout vigourously (or those producing sprouts and seeds) are generally among the most abundant and dominant in the communities of the French mediterranean region.

REFERENCES

Barney MA and Frischknecht NC (1974) Vegetation changes following fire in thepinyon-juniper type of west-central Utah, J. Range Manage. 27, 91-96.

Barry JP (1960) Contribution à l' étude de la végétation de la région de Nimes, Année Biologique, 36, 311-550.

Boesch DF (1974) Diversity, stability and response to human disturbance in estuarine ecosystems, Proc. 1st Int. Congress Ecol., The Hague, 109-114.

Braun-Blanquet J (1936) La forêt d'Yeuse languedocienne (*Quercion ilicis*). Monographie phytosociologique, Mem. Soc, Etudes Sci. Nat. Nimes, 5, 147 p.

Gill AM (1977) Plant traits adaptive to fire in mediterranean land ecosystems, Proc. Symp. Environm. Consequences Fire and Fuel Manage. Medit. Ecosystems, US For. Serv. Gen. Tech. Rep. WO-3, 17-26.

Grime JP (1977) Evidence for the existence of three primary strategies in plants and its relevance to ecological and evolutionary theory, Am. Natur. 111, 1169-1194.

Hanes TL (1971) Succession after fire in the chaparral of Southern California, Ecol. Monogr. 41, 27-52.

Holling CS (1973) Resilience and stability of ecological systems, Ann. Rev. Ecol. Systematics, 4, 1-23.

Kornas J (1958) Succession régressive de la végétation de garrigue sur calcaires compacts dans la montagne de la Gardiole près de Montpellier, Acta Soc. Bot. Poloniae, 27, 563-596.

Kuhnholtz-Lordat G (1938) La terre incendiée. Essai d'agronomie comparée, 361p., Nimes, La Maison Carrée.

Kuhnholtz-Lordat G (1958) L'écran vert. Mem. Mus. Natl Hist. Natur. 9, 276p.

Mac Arthur RH and Wilson ED (1967) The theory of island biogeography, 203p., Princeton Univ. Press, Princeton.

Naveh Z (1975) The evolutionary significance of fire in the Mediterranean region, Vegetatio 29, 199-208.

Pavillard J (1935) Eléments de sociologie végétale. Act. Sci. Industrie, 102 p., Paris, Herman et Cie.

Pianka ER (1970) On r and K selection, Amer. Natur. 104, 592-597.

Specht RL, Rayson P and Jackman ME (1958) Dark Island heath (Ninety Miles Plain, South Australia), VI Pyric succession: Changes in composition, coverage, dry weight and mineral nutrient status, Aust. J. Bot. 6, 59-88.

Trabaud L (1970) Quelques valeurs et observations sur la phyto-dynamique des surfaces incendiées dans le Bas-Languedoc, Natural. Monspel. 21, 231-242.

Trabaud L (1974) Experimental study of the effects of prescribed burning on a *Quercus coccifera* L. garrigue, Proc. Tall Timbers Fire Ecol. Conf. 13, 97-129.

Trabaud L (1977) Comparison between the effects of prescribed fires and wildfires on the global quantitative development of the kermes scrub oak (*Quercus coccifera* L.) garrigues, Proc. Symp. Environm. Consequences Fire and Fuel Manage. Medit. Ecosystems. US For. Serv. Gen. Tech. Rep. WO-3, 271-282.

Trabaud L (1980) Impact biologique et écologique des feux de végétation sur l'organisation la structure et l'évolution de la végétation des garrigues du Bas-Languedoc, 288 p., Thèse Etat, Univ. Sci. Tech Languedoc, Montpellier.

Trabaud L and Lepart J (1980) Diversity and stability in garrigue ecosystems after fire, Vegetatio 43, 49-57.

Trabaud L and Lepart J (1981) Changes in the floristic composition of a *Quercus coccifera* L. garrigue in relation to different fire regimes, Vegetatio 46, 105-116.

PART II

ADAPTATIONS TO WATER (AVAILABILITY OR DEFICIENCY)

ROOT SHEATHS AS AN ADAPTATION TO SOIL MOISTURE STRESS IN ARID ZONE GRASSES

R.C. BUCKLEY (AMDEL, P.O. Box 114, Eastwood, SA 5063, Australia)

1. INTRODUCTION

Plant adaptations to drought are many and varied. One adaptation which is apparently peculiar to grasses of arid zone sands is the possession of cylindrical sand sheaths surrounding the shallow lateral roots. These roots also have an aerenchymatous outer cortex which disintegrates at maturity. Various functions have been proposed for the sand sheaths: evidence reviewed here suggests that they are primarily an adaptation to soil moisture stress.

2. DISTRIBUTION

Root sheaths were first described by Volkens (1887), Price (1911), Thomas (1922) and Warming (1925) in North Africa; Henrici (1929) in South Africa; and Weaver (1919) and Cannon (1925) in North America. They are most prevalent on perennial grasses growing in loose desert sands, being reported, for example, on all perennial sandridge grasses in central Australia (Latz, pers. comm.) and the southern Kalahari (Leistner 1967); on Aristida pungens and other grasses from the Sahara (Price 1911); and on a variety of grasses from the North American arid zone. Thin or partial sheaths have also been reported from some annual grasses, such as Aristida browniana and A. stipitata, and several other monocotyledons, such as Corynetheca lateriflora, Dipcadi marlothii and D. vaginatum. Within arid regions sheaths are thicker and more common, under given conditions of soil and rainfall, on species characteristic of drier and sandier habitats than on those habitually growing in damper and less sandy areas; on given species they are better developed in the more arid and sandy parts of the overall species range (Leistner 1967). They are particularly abundant in sandy habitats with marked fluctuations in the soil moisture content (Henrici 1929). The sandhill grasses, in particular, have very

extensive root systems, a length of 20 m being reported for the horizontal roots of Ammodendron connollyi (Vassiliev 1931, Oppenheimer 1960) and over 20 m for Aristida pungens (Price 1911). The horizontal laterals generally run 10-30 cm beneath the soil surface (Volkens 1887, Massart 1898, Price 1911, Buckley 1982a).

3. MORPHOLOGY

Sheaths may be up to 1 mm thick, with total diameter up to 5 mm. They extend to various depths, becoming thinner and sparser in moister soil strata. A zone of 10-20 mm immediately behind the root tip is free of grains, and roots may have long sheath-free zones at depth. Leistner (1967) reported that sheaths rarely extend below 20 cm depth in the Kalahari. Sheaths are only found on roots with persistent root hairs, and sheath thickness and density are related to root hair length and density (Leistner 1967). Plants possessing root sheaths generally also have an expanded cortex which breaks down, in the mature root, to isolated strands of cortical cells or lamellae which hold the stele centrally in the sheath. In sheath-free zones the cortex remains intact, though spongy. The fragile or fragmented cortex often enables the steles to be pulled from the ground, leaving the sheaths embedded. High soil aeration and fluctuating soil moisture content may promote the development and enlargement of cortical aerenchyma independently of root sheath production. In some genera root-hair development and cortical disintegration are also enhanced in well-aerated soils (Kutschera 1960, Beckel 1956): this may provide a clue to the evolution of root sheaths. The classic studies of such cortical morphology are those of Henrici (1929); Goossens (1935), who figured a section of the enlarged cortex of Eragrostis plana; and Beckel (1956) who described the developmental sequence of cortical formation and disintegration in Bouteloua gracilis. More recently Robards et al.

(1979) used light, fluorescence and electron microscopy to analyse the structure and development of the cortical aerenchyma and epidermal/hypodermal sleeve of Carex arenaria roots, and Buckley (1982a) used petrological techniques to section the root and intact sand sheath of Zygochloa paradoxa, the 'sandhill canegrass' dominant on the dunecrests of Australia's Simpson Desert. Mature roots of this species possess a polyarch stele with sclerenchymatous conjunctive tissue, a heavily sclerotised inner cortex and an aerenchymatous and partially-disintegrated outer cortex. The hypodermis is also sclerenchymatous, and together with the exodermis and epidermis forms an outer cylinder bearing a dense mat of short root hairs which invest the sand grains. The grains themselves are approximately isodiametric and even-sized, as would be expected from their aeolian origin.

What holds the sand grains in place? The most likely candidate appears to be a mucilaginous cement. Goossens (1935) showed histochemically that mucilage produced by the root cap and root-tip epidermis extends further back from the root tip in more xerophytic grasses. He found soil particles adhering to the mucilage in species such as Anthephora pubescens and Themeda triandra. Price (1911) reported mucilage in the root sheaths of Aristida pungens and Lygeum spartum but not A.obtusa or Schismus calycinus. Oades (1978), in a very complete review of plant root mucilages, noted that they are produced more abundantly in moist soils, appear to be composed of a glucan and a complex of polysaccharides including a pectic or pectic-like component, and produce uronic acids and various sugars on hydrolysis.

Buckley (1982a) found that dilute perchloric acid or pectinases softened the sheaths of Zygochloa paradoxa whereas water, ethanol or carbon tetrachloride did not, indicating that any cement is mucilaginous rather than resinous. Even with the pectinase treatment, however, the grains were not dislodged unless brushed gently,

indicating that the root hairs are probably the primary means of attachment.

4. FUNCTION

Four possible functions have been suggested for the root sheaths and expanded cortex. Thomas (1922) and Oppenheimer (1960) suggested that they protect the stele from high temperatures, mechanical abrasion or soil compression; and Warming (1925), Goossens (1935) and Walter (1939) that they insulate it against moisture loss. Henrici (1928), Killian (1937), and Kutschera (1960), on the other hand, thought that the sheaths' function was to increase moisture absorption. Most recently a role in nutrient uptake has been suggested (Wullstein 1980).

Protection against mechanical abrasion of the stele can be dismissed: the sheaths are considerably more fragile than the central conducting strands, and disintegrate if exposed by aeolian sand excavation. This does not eliminate a mechanical protective function for the sheath entirely, however, as noted later. A moisture-absorptive function can also be dismissed, since the sheath is not connected to the stele save by isolated strands of disintegrated cortical cells: moisture uptake is presumably restricted to an apical zone a few centimetres long, where the cortex is still intact, the sand sheath absent and the hypodermis still permeable. The root sheaths of Oryzopsis hymenoides contain bacteria, possibly Bacillus polymyxa, which can reduce acetylene at rates equivalent to approximately 20 micrograms N/plant/day or 50-100 nm acetylene/10 gm isolated rhizosheath material/day (Wullstein et al 1979, Wullstein 1980). Nutrient transfer to the plant, however, has not yet been demonstrated conclusively. The sand-filled mounds of Zygochloa paradoxa contain significantly higher concentrations of extractable phosphorus than control soils (Buckley 1982b), suggesting that the Zygochloa rhizosheaths might be mycorrhizal, but anatomical studies (Buckley 1982a) revealed only sparse hyphae, and the soil P enrichment is probably due to animals living in the perennial

Zygochloa mounds.

Effective insulation of the stele could be provided by an impermeable endodermis, an insulative cortex or an impermeable hypodermis and sheath. It was once believed that the endodermis was of primary importance in controlling the passage of water (and ions) to the stele (e.g. Tanton and Crowdy 1972), but it now appears that in some species the endodermis may be considerably more permeable than was thought previously, and the hypodermis considerably less so (Clarkson and Hanson 1980). The hypodermal sleeve of _Carex arenaria_ has a very low permeability, 200 times lower than that of onion roots (Robards et al 1979). Hypodermal permeability has not been measured for any of the arid-zone sandhill grasses, but Buckley (1982a) found that water could be passed through lengths of the hypodermal/exodermal cylinder and associated sand sheath of _Zygochloa_ paradoxa (with the stele and inner cortex removed) for 24 h without leakage. Such sections could also withstand a head of 50 cm water, and the outer cylinder is therefore relatively impermeable.

CONCLUSIONS

Overall, it appears that the main function of the expanded cortex, hypodermis and rhizosheath in these arid-zone grasses is to minimise moisture loss from the stele as water is translocated along the lateral roots through many metres of air-dry sand. Robards et al (1979) concluded that 'the development of the root of _Carex arenaria_ seems to afford the central conducting strand strong protection against adverse conditions at the expense of sacrificing any role in water in ion uptake'. In _Zygochloa paradoxa_, the overall structure of root and sheath is analogous to a central water-conducting pipe held axially in a tube of insulating foam plastic sheathed by an impermeable outer cylinder, the whole being sufficiently flexible to accommodate minor slumping of the dune sands. The main role of the sand sheath is perhaps to provide

mechanical protection and limited insulation for the fragile hypodermis and cortex. It may play different roles in other species, and indeed the nutrient-uptake hypothesis seems most likely at present for _Oryzopis_ _hymenoides_, but in the perennial grasses of arid-zone sands it appears that root sheaths are primarily an adaptation to minimise water loss under soil moisture stress. As such they must be a significant part of the plants' overall adaptive armoury, and deserve further attention.

ACKNOWLEDGEMENTS

This study was commenced in 1975 under an Australian National University Ph.D scholarship and completed in 1981 under a Rothmans Research Fellowship at the Department of Biogeography and Geomorphology, ANU. Thanks are due to Dr R.J. Wasson and Peter Latz for providing some of the samples, to Max Campion for preparing the thin sections, to David Moser and Dr G.A. Chilvers for histochemical assistance and advice, to Jim Caldwell for soil nutrient determinations and advice on cement chemistry, to Dr D.L. Bill of Pfizer Pty Ltd for a sample of the liquid pectinase 'Clariphase GH', and to Dr R. Lange and Dr J. Jessop for criticism of an earlier draft.

REFERENCES

Beckel DKB (1956) Cortical disintegration in the roots of Bouteloua gracilis (HBK) Lag. New Phytol. 55, 183-190.

Buckley RC (1982a) Sand rhizosheath of an arid-zone grass. Plant and Soil, in press.

Buckley RC (1982b) Soil requirements of seven central Australian sandridge plants in relation to the dune-swale soil catena. Aust. J. Ecol. 7, 309-313.

Cannon WA (1911) The root habits of desert plants. Carnegie Inst. Publ. 131, Washington.

Cannon WA (1925) Physiological features of roots with especial reference to the relation of roots to aeration of the soil. Carnegie Inst. Publ. 368, Washington.

Clarkson DT and Hanson JB (1980) The mineral nutrition of higher plants. Ann. Rev. Plant. Physiol. 31, 239-298.

Goossens AP (1935) Notes on the anatomy of grass roots. Trans. Roy. Soc. S. Afr. 23, 1-21.

Henrici M (1929) Structure of the cortex of grass roots in the more arid regions of South Africa. S. Afr. Dep. Agric. Sci. Bull. 85.

76

Killian C (1937) Contribution a l'étude
écologique des végétaux du Sahara et du Soudan
tropical. Bull. Soc. Hist. Nat. Afr. Nord. 28,
12-18.

Leistner OA (1967) The plant ecology of the
Southern Kalahari. Bot. Surv. South Afr. Mem. 28.

Ling-Lee M, Ashford AE and Chilvers GA (1977)
A histochemical study of polysaccharide distribution in eucalypt mycorrhizas. New Phytol. 78,
329-335.

Oades JM (1978) Mucilages at the root surface.
J. Soil Sci. 29, 1-16.

Oppenheimer HR (1960) Adaptation to drought:
xerophytism. Arid Zone Res. 15, 105-138.

Price SR (1911) The roots of some North
African desert-grasses. New Phytol. 10, 328-329.

Robards AW, Clarkson DT and Sanderson J (1979).
Structure and permeability of the epidermal/
hypodermal layers of the sand sedge (Carex
arenaria, L). Protoplasma 101, 331-347.

Tanton TW and Crowdy SH (1972) Water pathways
in higher plants. II. Water pathways in roots.
J. exp. Bot. 23, 600-618.

Thomas HH (1922) Some observations on plants
in the Libyan Desert. J. Ecol. 9, 75-89.

Vassiliev JM (1931) Uber den Wasserhaushalt
von Pflanzen der Sandwuste im sudostlichen
Kara-Kum. Planta 14, 225-309.

Walter H (1939) Grasland, Savane und Busch
der ariden Teile Afrikas in ihrer okologischen
Bedingheit. Jahr. Wiss. Bot. 87, 750-860.

Warming E (1925) Ecology of plants. London.

Weaver JE (1919) The ecological relations of
roots. Carnegie Inst. Publ. 286, Washington.

Wullstein LH (1980) Nitrogen fixation
(acetylene reduction) associated with rhizosheaths of Indian ricegrass used in stabiliza-
tion of the Slick Rock, Colorado, tailings pile.
J. Range Manage 33, 204-206.

Wullstein LH, Bruening ML and Bollen WB (1979)
Nitrogen fixation associated with sand grain
root sheaths (rhizosheaths) of certain xeric
grasses. Physiol. Plant. 46, 1-4.

OSMOTIC AND TURGOR RELATIONS
IN SELECTED CHAPARRAL SHRUB SPECIES.

Stephen W. Roberts
William D. Bowman

Systems Ecology Research Group
San Diego State University
San Diego, California 92182 USA

ABSTRACT

Internal water relations were studied in
three chaparral shrub species in southern
California. Osmotic changes were detected in
leaf tissue both throughout the course of the
summer drought season, and additionally
throughout the course of a single day. The
diurnal osmotic adjustments were comparable to
the adjustments over the season. These
adjustments are viewed as adaptive mechanisms
promoting turgor maintenance and turgor-related
processes such as growth and stomatal opening
during periods of diurnal and seasonal water
stress.

INTRODUCTION

Water, with all its attendant significance
for plant life, is often highly variable in
natural plant communities. Strong selection is
often presumed for mechanisms which might
minimize plant tissue stress and damage, and
permit plant growth and development to continue
under conditions of often wide variation in
environmental water availability. Such
mechanisms ought to be particularly significant
for plants such as chaparral shrubs, and shrubs
of other mediterranean-type regions, which are
relatively long-lived and must therefore endure
repeated periods of water limitation of varying
severity during recurring summer drought
(Mooney and Dunn, 1970; Walter, 1973).
Chaparral shrubs experience wide variation in
water stress on both a seasonal and a diurnal
basis (Figures 1 and 2, respectively).

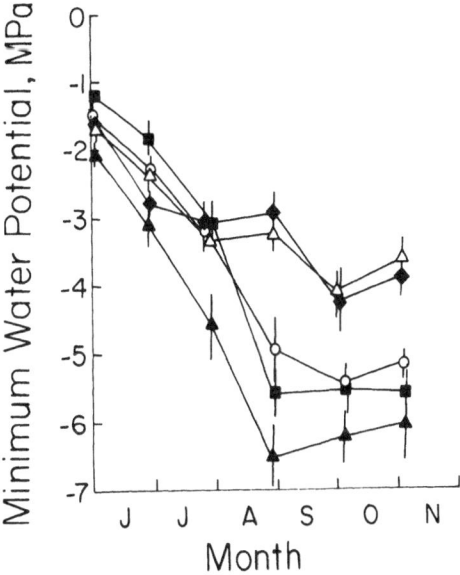

Figure 1. Seasonal course of daily minimum leaf
water potentials, with standard errors,
measured in the southern California chaparral
shrubs Adenostoma fasciculatum on a pole- (Δ)
and equator-facing slope (▲), and Quercus
dumosa (◆), Ceanothus greggii (■), and
Arctostaphylous glauca (O), on a pole-facing
slope. (from Roberts et al., 1981).

Chaparral plants have been shown to
possess adaptations which promote survival
during drought such as strong stomatal control
of transpiration and the capacity to develope
and endure very negative water potentials
(Poole and Miller, 1975; Burk, 1978; Gigon,
1979; Miller and Poole, 1979; Roberts et al.,
1981; and others). Previous chaparral water
relations studies have generally focussed on

78

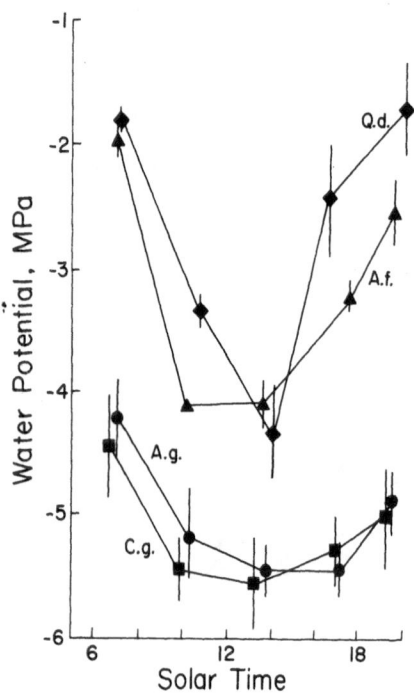

Figure 2. Diurnal course of leaf water potentials, with standard errors, measured late in the southern California drought period. Quercus dumosa (◆), Adenostoma fasciculatum (▲) Arctostaphylous glauca (●), and Ceanothus greggii.■ (from Roberts et al., 1981).

stomatal opening. Studies of these properties have been performed using the pressure-volume approach (Tyree and Hammel, 1972) in forest tree species (Hellkvist et al., 1974; Cheung et al., 1975; Huzulak, 1976; Roberts and Knoerr, 1977; Tyree et al., 1978; Hinckley et al., 1978; Roberts et al., 1980). These studies have provided insight into the ways in which leaf tissue properties (such as tissue elasticity, solute content, and water content) affect leaf water status, and additionally how these properties may change in response to tissue developmental stage and environmental conditions.

The pressure-volume approach has been applied largely to studies of forest tree species and has not been widely employed in studying the water relations of plants which typically occupy water stress habitats. Hinckley et al. (1980) have studied water relations of several plants occurring in the Mediterranean basin using the pressure-volume method. The objective of the present study was to obtain estimates of leaf tissue parameters which govern tissue responses to seasonal water stress conditions in several southern California chaparral shrub species, and to focus particularly on seasonal and diurnal trends in osmotic and turgor relations.

daily and seasonal patterns of stomatal conductances and leaf water potentials. Increasingly, the importance of characterizing the components of water potential has been recognized (Hsiao et al., 1976; Fereres et al., 1978) as well as the importance of understanding osmotic and structural properties of leaves, and other plant organs, which determine the component potentials. This is required for proper understanding of the dependencies of tissue biochemical processes on the osmotic/solute milieu in which they occur, and for understanding the links between osmotic and turgor components and their effects on turgor-requiring processes such as growth and

METHODS

Internal water relations were studied in
Ceanothus greggii, Arctostaphylous glandulosa,
and Quercus dumosa, three common and widespread
shrub members of the southern California
chaparral vegetation (Munz and Keck, 1959).
These studies were carried out at the Sky Oaks
Biological Field Station, which is operated by
the College of Sciences, San Diego State
University. Pressure-volume measurements were
made on the study species using terminal leafy
stems of similar exposure from the crowns of
mature shrubs in mixed stands. The
pressure-volume measurements (Tyree and
Hammel, 1972) were made following the method
described by Roberts and Knoerr (1977), with
the exception that the pressure-volume curves
were measured without the usual rehydration of
the tissue. In these measurements, the PV
curves were measured immediately following
collection of the leaf material from the field.
PV measurements were made at dawn and again at
midday for the same species. At each
measurement time four replicates were measured
simultaneously using a multiple pressure bomb
system. The dawn and midday sampling was
designed to detect possible diurnal changes in
the PV relationship between a period of minimal
(dawn) and maximal (midday) daily water stress.

In addition to the pressure-volume
experiments, field measurements of xylem
pressure potential were obtained using a
portable pressure bomb (Scholander et al.,
1965). Climate above the canopy and soil
moisture below the canopy were continuously
recorded using a Campbell CR-21 battery
operated data logger.

RESULTS

The seasonal trends of soil moisture at -7
cm, air temperature and relative humidity above
the canopy are shown in Figure 3. There was a
general trend toward drying soils, increasing
air temperatures, and decreasing relative
humidities from mid-May to mid-July. In late
July there were 30mm of precipitation which
temporarily recharged the soil moisture at the
-7 cm depth, and which resulted in decreased
air temperatures and increased relative
humidities. Following this precipitation event
the soil resumed a drying trend, air
temperatures increased, and humidities
decreased.

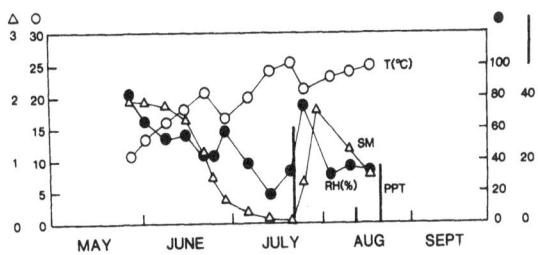

Figure 3. Seasonal course of climatic data at
the Sky Oaks research site, showing daily mean
temperature (T○) and relative humidity (RH●)
at two meters height, precipitation (PPT) in
mm, and soil moisture (SM △) at -7 cm depth.
Soil moisture values are uncalibrated, relative
values.

The results of the dawn and midday pressure-volume experiments indicated shifts in the relationship between total water potential and the non-pressure (osmotic and matric) component of the total potential. These relationships were analysed as plots of the sum of the osmotic and matric components against the total water potential. The diagonal line in such plots is the locus of zero turgor; deviations from the diagonal indicate regions where there is positive turgor. The point where the relationship just departs from the diagonal indicates the point where the bulk of the cells in the tissue just become flaccid, and may be called the zero turgor point or the turgor loss point.

Early in the drought period, in late May, none of the species showed a diurnal change in the relationship between water potential and the sum of the non-pressure potentials. (Figure 4). Later in the season, in early August, during the course of a day all three species showed a diurnal shift in the water potential corresponding to the zero turgor point. In general the diurnal changes were greatest in Quercus dumosa and Ceanothus greggii. Smaller diurnal changes were measured in Arctostaphylous glandulosa. In early August Q. dumosa and C. greggii underwent shifts of approximately 0.42 and 0.40 MPa, respectively. A. glandulosa showed a shift of approximately 0.19 MPa (Figure 5). One month later, in early

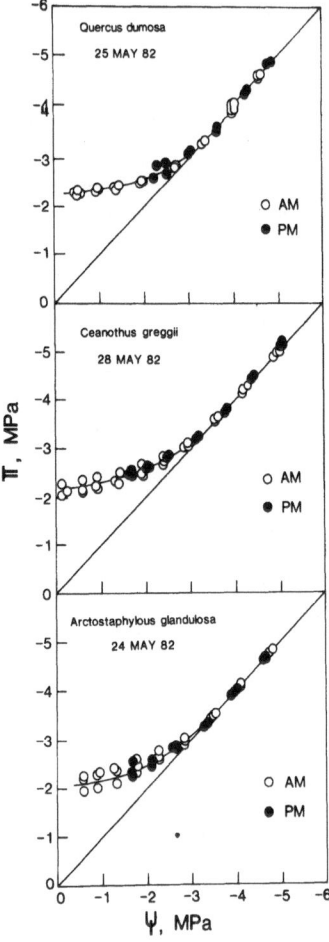

Figure 4. Plots of the relationship betweeen the total water potential and the non-pressure components early in the drought season at the Sky Oaks research site. Relationships are shown for leaf tissue analysed at dawn (AM) and later at midday (PM) from the same individual shrubs.

September, Q. dumosa showed a diurnal shift of the zero turgor point of about 0.48 MPa, C. greggii showed a 0.58 MPa shift, and A. glandulosa showed a 0.33 MPa shift (Figure 6).

The pressure-volume measurements may be related to the field setting of the plants by comparing the pressure-volume results with concurrent field measurements of water potentials on the same plants used for the PV

analyses. This was done by plotting, for each
of the species, the seasonal course of the mean
midday minimum water potential measured in the
field with the pressure bomb; the dawn PV
estimate of the water potential corresponding
to the turgor loss point; the midday PV
estimate of the water potential corresponding
to the turgor loss point; and the dawn estimate
of the osmotic potential at full turgidity
(Figure 7).

These data indicate a seasonal progression
toward increasingly negative midday water
potentials, coupled with a corresponding
decrease in the dawn and midday estimates of
the turgor loss point (Figure 7). The dawn
estimates of the osmotic potential at full
turgidity also showed a seasonal decrease. The
difference between the dawn and midday
estimates of the turgor loss point indicates
the magnitude of the shift in this parameter
from dawn to midday.

DISCUSSION

The data presented indicate that these
chaparral shrubs accomplish both seasonal and
diurnal adjustments in their internal water
relations. Early in the season, when soil
moisture is available and temperatures are
moderate, the adjustments were relatively
small. As soils dried and the evaporative

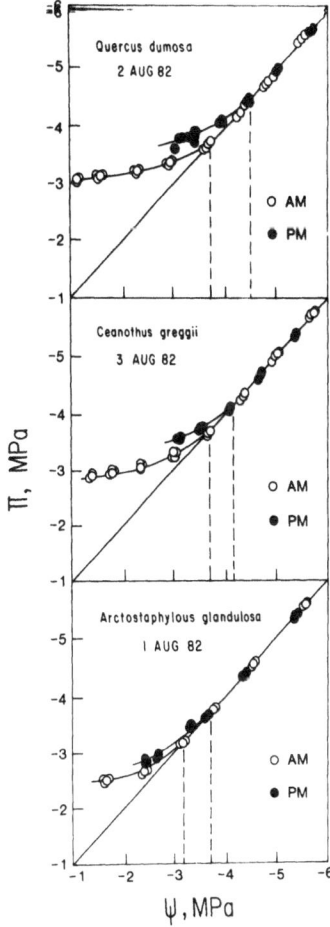

Figure 5. Plots of the relationship between the
total water potential and the non-pressure
components in August at the Sky Oaks research
site.

demand of the atmosphere increased, there was
both a seasonal decrease in the midday turgor
loss point, and in general a greater difference
between the dawn and midday estimates of the
turgor loss point. The significance of the
capacity for a diurnal adjustment in the turgor
loss point may be seen in the September
measurements for Quercus dumosa and Ceanothus
greggii. In these two shrubs, if the turgor
loss point were not adjusted from dawn to
midday, the tissue would essentially be

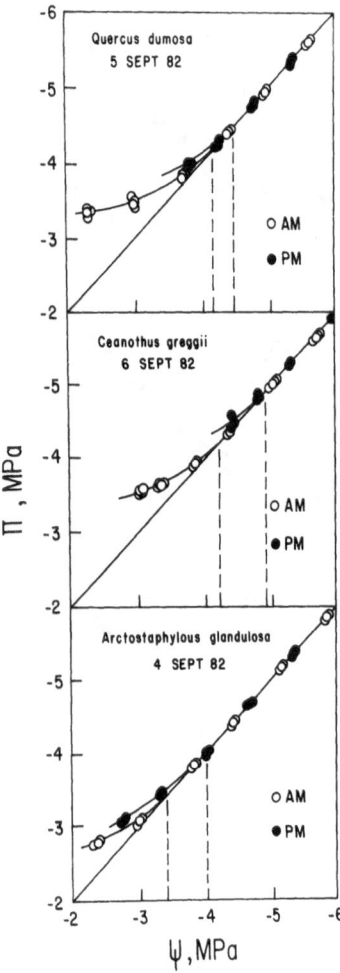

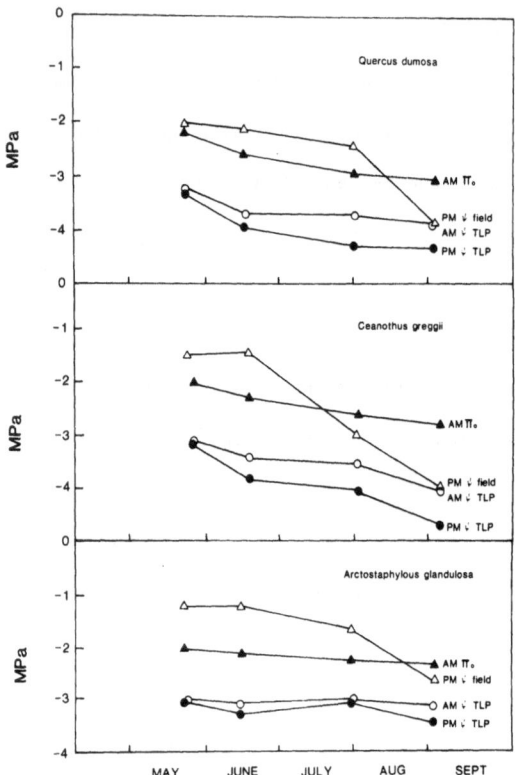

Figure 6. Plots of the relationship between the total water potential and the non-pressure components in September at the Sky Oaks research site.

Figure 7. Seasonal course of the initial osmotic potential at full turgidity (▲), the midday leaf water potential measured in the field on the same days as when the PV analyses were performed (△), the water potential corresponding to the turgor loss point (TLP) from the PV analyses performed on leaf material sampled at dawn (○), and the TLP from the leaf material sampled at midday (●).

operating at zero turgor during midday. By making a diurnal adjustment, the tissue is able to maintain approximately 0.4 to 0.6 MPa of turgor during the midday period. Presumably as the drought season progresses the result of these daily adjustments will be even more significant in terms of turgor maintenance and for turgor-requiring processes in the tissue. Arctostaphylous glandulosa did not reach water potentials as negative as those of the other

two species, and thus could have maintained turgor throughout the midday period without any adjustment. However, a midday adjustment was measured of approximately 0.33 MPa. We expect that as drought stress increases the diurnal adjustments will be of increasing magnitude.

The mechanism of the diurnal adjustments measured in this study could be an osmotic adjustment, a tissue elasticity adjustment, or a combination of adjustments in both parameters. Since the midday PV curves were measured in unrehydrated tissue, we cannot easily compare dawn and midday samples on a water deficit basis. Thus it is difficult to compare osmotic properties such as the osmotic potential at full turgidity, which is a standard comparative measure in pressure-volume studies. We can, however, compare osmotic potentials at the turgor loss point, and there were clear differences here both diurnally and seasonally. If we assume that in the plots of non-pressure potentials and water potentials (Figures 4, 5, and 6) the curves would continue smoothly to intersect the y-axis, as they do in the dawn data and in similar plots of rehydrated tissue (Roberts, unpublished data), then the bulk of the diurnal adjustment would appear to be an osmotic adjustment. An elasticity adjustment is not precluded, but we do not see evidence of such adjustment in data plots such as those illustrated in Figures 4 through 6.

The tissue water relations adjustments reported here appear adaptive under developing drought and may be interpreted as a mechanism promoting turgor maintenance during midday periods of low water potential. Short-term osmotic adjustments over a few hours have been measured in field-grown maize (Acevedo Hinojosa, 1975; Hsiao et al., 1976) and in apple (Davies and Lakso, 1979). The apparent osmotic shifts indicated in the present study

might result from accumulation of photosynthate during the day, from a more direct response of the tissue to water deficits by interconversion of solute molecules to more osmotically active species, or by importation and sequestering of extracellular solute. Clarification of this mechanism must await studies of the solute complement of these tissues, and its seasonal and diurnal variation.

Nearly all pressure-volume studies have proceeded using rehydrated tissue. In view of the kinds of short-term adjustments reported here, results of PV studies on rehydrated tissue may substantially overestimate osmotic potentials in the in situ field material. While measurements of short-term osmotic adjustments in two forest tree species showed only small (<0.1 MPa) changes from morning to afternoon (Roberts et al., 1980), the results of the present study show that species in differing environments may differ substantially in their ability to undergo changes in tissue osmolality. We hypothesize that tissue osmotic adjustment should be particularly important in plants growing in mediterranean-type environments where recurrent water shortage, both diurnally and seasonally, is a recurring theme throughout the lifetime of the plant.

ACKNOWLEDGEMENTS

This study was supported under National Science Foundation grant DEB-8025997. We wish to thank Penny Roberts for preparation of the figures.

LITERATURE CITED

Acevedo Hinojosa E (1975) The growth of Zea mays L. as affected by its water relations. Ph.D. Diss. University of California, Davis, U.S.A.

Burk JH (1978) Seasonal and diurnal water potentials in selected chaparral species. Amer. Midland Naturalist. 99:244-248.

Cheung YNS, Tyree MT and Dainty J (1975) Water relations parameters on single leaves obtained in a pressure bomb and some ecological implications. Can. J. Bot. 53:1342-1346.

Davies FS and Lakso AN (1979) Diurnal and seasonal changes in leaf water potential components and elastic properties in response to water stress in apple trees. Physiol. Plant. 46:109-114.

Fereres E, Acevedo E, Henderson DW and Hsiao TC (1978) Seasonal changes in water potential and turgor maintenance in sorghum and maize under water stress. Physiol. Plant. 44:261-267.

Gigon A (1979) CO_2-gas exchange, water relations, and convergence of mediterranean shrub-types from California and Chile. Oecol. Plant. 14:129-150.

Hellkvist J, Richards GP and Jarvis PG (1974) Vertical gradients of water potential and tissue water relations in Sitka spruce trees measured with the pressure chamber. J. Appl. Ecol. 11:637-667.

Hinckley TM, Duhme F, Hinckley AR and Richter H (1980) Water relations of drought hardy shrubs: Osmotic potential and stomatal reactivity. Plant, Cell, and Environment. 3:131-140.

Hsiao TC, Acevedo E, Fereres E, and Henderson DW (1976) Stress metabolism. Phil. Trans. R. Soc. Lond. B. 273:479-500.

Huzulak J (1976) Relationship between water saturation deficit and water potential in the leaves of three forest tree species. Biologia (Bratislava) 31:25-32.

Miller PC and Poole DK (1979) Patterns of water use by shrubs in southern California. For. Sci. 25:84-98.

Mooney HA and Dunn EL (1970) Convergent evolution of mediterranean-climate evergreen sclerophyll shrubs. Evolution 24:292-303.

Munz PA and Keck DD (1959) A California Flora. University of California Press, Los Angeles.

Poole DK and Miller PC (1975) Water relations of selected species of chaparral and coastal sage communities. Ecology. 56:1118-1128.

Roberts SW and Knoerr KR (1977) Components of water potential estimated from xylem pressure measurements in five tree species. Oecologia (Berl.) 28:191-202.

Roberts SW, Miller PC and Valamanesh A (1981) Comparative field water relations of four co-occurring chaparral shrub species. Oecologia (Berl.) 48:360-363.

Roberts SW, Strain BR, and Knoerr KR (1980) Seasonal patterns of leaf water relations in four co-occurring forest tree species: Parameters from pressure-volume curves. Oecologia (Berl.) 46:330-337.

Scholander PF, Hammel HT, Bradstreet ED and Hemmingsen EA (1965) Sap pressure in vascular plants. Science 148:339-346.

Tyree MT and Hammel HT (1972) The measurement of the turgor pressure and the water relations of plants by the pressure-bomb technique. J. Exp. Bot. 23:267-282.

Tyree MT, Cheung YNS, Macgregor MT and Talbot AJB (1978) The characteristics of seasonal and ontogenetic changes in the tissue-water relations of Acer, Populus, Tsuga, and Picea. Can. J. Bot. 56:635-647.

Walter H (1973) Vegetation of the earth in relation to climate and the eco-physiological conditions. Springer-Verlag, New York-Heidelberg-Berlin. 237 pp.

ADAPTATIONS OF THE PHOTOSYNTHETIC PIGMENT SYSTEM TO ECOLOGICAL CONDITIONS WITH RESPECT TO WATER IN DIFFERENT TERRESTRIAL PLANTS*

Z. TUBA (Research Institute for Botany of the Hungarian Academy of Sciences, H-2163 Vácrátót, Hungary)

1. INTRODUCTION

Although many studies have been done on water stress, relatively few of these papers deal with the photosynthetic pigment system. Of the pigment system components, primarily chlorophylls have been studied (Virgin, 1965; Duysen, Freeman, 1974, 1976; Alberte et al., 1975; Alberte, Thornber, 1977). Only the total amount of carotenoids has been recorded (Duysen, Freeman, 1974, 1976; Spyropoulos, Mavrommatis, 1978) and there is no information available on the response of the total photosynthetic pigment system to water deficit in either drought adapted or in non-drought adapted plants. Because of the high sensitivity and variability of the photosynthetic pigment components, especially carotenoids, with varying ecological and internal factors (Claes, 1967; Maróti, Szajkó, 1972; Tuba, 1977; Seslák, 1978; Krinsky, 1978; Goedheer, 1979; Ida, 1981) it is apparent that, among extreme environmental conditions, such as water availability, the adaptation of photosynthetic pigments to drought should be one of the first steps in studying plant adaptation to drought. It is though possible that, in the plant species adapted to water deficit, the adaptation of pigment system is controlled by a well established mechanism. This mechanism supposedly is lacking or less important in species not adapted to water deficit.

In the investigations described here the pigment systems of three plant species, each adapted to xeric conditions by different kinds of water economy, were related. These species were studied during dry and wet periods. The study reported here contains data on the changes of the photosynthetic pigment system of three different species in a dry perennial sandy grassland community.

* "Tece Studies" No. 31.

2. MATERIALS AND METHODS

The studies were carried out in a calciphylous dry perennial sandy grassland community (Festucetum vaginatae danubiale) at Vácrátót, Hungary, in the experimental area of the Institute for Botany of the Hungarian Academy of Sciences, in the summer of 1981.

Three different drought adapted species of the community were chosen for the investigations: the sclerophyllic *Festuca vaginata* W. et K., the semi-succulent *Achillea ochroleuca* Ehrh. and the succulent *Sedum sexangulare* L. The investigations were carried out in 10x10 m sampling plots.

2.1. The course of investigations

The investigations were carried out, after a two weeks drought period, on the 28th July 1982 (dry period), the 29th July (wet period), on the 3rd of August (dry period), the 4th of August (wet period) and on the 6th of August (dry period). The wet periods during the extraordinary dry summer of 1981 were produced by rainlike irrigation, with an amount of 25 $mm.m^{-2}$ of water supplied 5-6 hours before the day of wet periods.

2.2. Microclimate measurements

Air temperature, air humidity and light intensity were recorded continuously at 10 cm height above ground level with auto-recording thermohygrographs and photoelectric light meter. Soil temperatures were measured at 5 cm depth during the investigations with soil mercury thermometers.

2.3. Water relations determinations

The soil water content at 0-10 cm soil depth was determined at 5 replicates after drying at 105°C for 24 hrs. Leaf water content and water deficit of the examined species were measured according to Weatherley (1950), each time with 10 replicates per species.

2.4. Photosynthetic pigment determination

The extraction and the separation of the pigments were carried out by thin-layer chromatography

according to Maróti, Gabnai (1971), with 5 replicates per sample period. The "total pigment
extract", obtained by alternating acetonic and
petroleum ether procedures, contains the chlorophylls, carotenes and xanthophylls in full. The
pigment quantities as mg.g^{-1} dry wt., were
calculated according to the formulas of Hager,
Meyer-Bertenrath (1966). In the course of the
work the following pigment categories were determined: combined α and β carotenes, chlorophyll
α and β, combined lutein and zeaxanthin, antheraxanthin, violaxanthin, neoxanthin.

2.5. Sampling

For the study, soil samples and leaves were collected at 07.00 h and extracted immediately. Since
the distribution of the pigments changes in the
direction from the peak towards the base, the
leaf samples were taken from middle parts of the
leaves.

The leaf samples were chosen from leaves of
similar age since the quantity and the ratio of
the pigments change considerable during the ontogenesis (Šesták, 1977, 1978).

3. RESULTS AND DISCUSSION

3.1. Changes of the soil water content

During the wet period, the soil water content of
the plant communities is 11-12% dry wt., while
during dry periods is only 2.5-5.5% (Table 1).
The water holding capacity of light sandy soil is
low, but the changes of water content are considerable and the drying process is rapid. Therefore by irrigation and natural drying two extremes
of the soil water conditions were produced.

TABLE 1. Changes of the soil water content (in
dry weight %) during the investigation.

Period	dry	wet	dry	wet	dry
%	2.56	11.66	3.48	12.1	5.54

3.2. Changes of water relations in the studied species

The sclerophyllic, half-succulent and succulent
designation appears reasonable considering the

relative water content values of the leaves belonging to the different species (Table 2). The change
in water content between wet and dry periods is
the lowest in the sclerophyllic *Festuca* (10-15%)
and medium in the semisucculent *Achillea* (40-50%).
During the first wet period, the leaf water content
of the succulent *Sedum* increased by 100%, while
after the later one the amount of increase is much
smaller (8-65%); Rychnovská (1966, 1972) measured
similar values on sclerophyllic sandy and rocky
grassland species. The absolute value and change of
the water deficit (Table 2) is the greatest by the
sclerophyllic *Festuca* and lowest by the succulent
Sedum.

3.3. Microclimatic conditions

There was no significant difference between the
microclimatic parameters of dry and wet periods.
The daily mean air temperature at 10 cm was 21.6-
23°C and 23-26°C in soil at 5 cm depth. The daily
averages of the relative air humidity were between
50 and 60%. The daily maximum light intensity at
13.00 hrs was between 1140 and 1226 $\mu E.m^{-2}.sec^{-1}$.
Therefore the changes in pigment systems during the
subsequent dry and wet periods should be assigned
to the changes in moisture status of the plants.

3.4. Changes in the photosynthetic pigment system

3.4.1. Chlorophyll concentrations. The chlorophyll
α and β concentrations in the *Festuca* and *Achillea*
are lower during the dry periods than in the wet
periods (Table 3), whereas the chlorophyll α and β
content of *Sedum* gradually increases from the first
wet period. During the dry period, chlorophyll α
increases more than chlorophyll β in *Festuca* and
Achillea (Table 4). Among wet conditions the chlorophyll α/β ratio decreases through time. This is
caused by the more rapid increase of chlorophyll β
content compared to that of chlorophyll α. The
above pattern is more pronounced in *Achillea* than
in *Festuca*. The general decrease in the chlorophyll
α/β ratio through drought cycles coincide with
former records on chlorophyll changes in relation
to the changes of water conditions (Duysen, Freeman,
1974, 1976; Alberte, Thornber, 1977; Spyropoulos,
Mavrommatis, 1978).

TABLE 2. Water relations

Period	Festuca vaginata		Achillea ochro		Sedum sexangulare	
	water content	water deficit	water content	water deficit	water content	water deficit
Dry	92.60	32.40	228.70	30.60	851.10	16.90
Wet	103.50	22.30	272.80	26.50	957.80	7.65
Dry	93.80	31.60	233.80	29.10	949.30	8.47
Wet	108.20	18.80	286.80	22.70	1015.50	3.09
Dry	97.90	28.50	242.60	26.40	980.00	5.51

TABLE 3. Changes of the chlorophyll α and β concentrations ($mg.g^{-1}$ dry weight) in the three species during the dry and wet periods.

Period	Festuca vaginata		Achillea ochroleuca		Sedum sexangulare	
	chlorophyll α	chlorophyll β	chlorophyll α	chlorophyll β	chlorophyll α	chlorophyll β
Dry	0.567	0.138	1.216	0.327	0.514	0.140
Wet	0.672	0.194	1.432	0.434	0.552	0.153
Dry	0.627	0.160	1.279	0.355	0.567	0.162
Wet	0.725	0.206	1.525	0.476	0.593	0.169
Dry	0.658	0.177	1.483	0.449	0.594	0.183

TABLE 4. Values of the photosynthetic pigment ratio in the examination periods. Abbreviations are: chl α :chlorophyll α; chl β:chlorophyll β; car : carotene; xant:xanthophylls.

Period	chl α/chl β	car/xant	car/chl β
	Festuca vaginata		
Dry	4.10	1.60	0.60
Wet	3.40	1.40	0.30
Dry	3.90	1.60	0.70
Wet	3.50	1.30	0.30
Dry	3.70	1.50	0.60
	Achillea ochroleuca		
Dry	3.70	1.20	0.13
Wet	3.30	1.10	0.08
Dry	3.60	1.30	0.21
Wet	3.20	1.10	0.10
Dry	3.30	1.20	0.16
	Sedum sexangulare		
Dry	3.60	1.30	0.40
Wet	3.60	1.20	0.42
Dry	3.50	1.50	0.59
Wet	3.50	1.20	0.49
Dry	3.40	1.20	0.44

3.4.2. Carotene and lutein content. The carotene content of Festuca and Achillea increases during dry and decreases during the wet periods (Table 5). The lutein (and the zeaxanthin) contents decrease during dry periods and increase during wet periods (Table 5 and unpublished data). The carotene content of Sedum changes similarly to the preceding two species, while its lutein content increases continuously with time in all conditions (Table 5).

3.4.3. The catorene-lutein hydroxylation-dehydro-xylation cycle. In Festuca and Achillea the decrease of carotene content is always followed by the increase of lutein content, while the increase of carotene content is followed by the decrease of lutein. It is known that as carotenes are hydroxylated lutein and zeaxanthin develop (Donochue et al., 1967). Carotenes formation due to lutein and zeaxanthin dehydroxylation has also beeen verified (Maróti, Szemenkei, 1972; Maróti, Szajkó, 1972). On this basis, the author proposes that during dry conditions the increase of carotene content origi-nates from the dehydroxylation of lutein content. During wet periods, as the effect of water surplus,

TABLE 5. Changes of the carotene and lutein concentrations (mg.g^{-1} dry weight) in the three species examined during the dry and wet periods.

Period	Festuca vaginata		Achillea ochroleuca		Sedum sexangulare	
	carotene	lutein	carotene	lutein	carotene	lutein
Dry	0.084	0.064	0.042	0.090	0.057	0.107
Wet	0.066	0.079	0.035	0.147	0.064	0.165
Dry	0.107	0.057	0.076	0.119	0.096	0.179
Wet	0.050	0.086	0.048	0.197	0.085	0.196
Dry	0.113	0.061	0.073	0.169	0.079	0.231

part of the carotene is hydroxylated; so the carotene decreases and the lutein increases.
In the case of *Sedum* both the increase and the decrease of carotene content is accompanied by the increase of lutein content from the first wet period onwards. Probably, lutein synthesis starts via *de novo* carotene formation from the first wet period. During extreme wet conditions, the carotenes which were *de novo* formed undergo a further transformation towards hydroxylation. Then, in the succulent, the lutein formation via carotene continues even during the shorter drought following a wet period. However, the intensity of the process is slower, the lutein content increases at a lower rate and also the carotenes accumulate. The author attaches major potential importance to this proposed carotene-lutein, hydroxylation - dehydroxylation, cycle.

3.4.4. Pigment ratios and pigment system ratios.
The distribution of the chlorophylls and carotenoids within the two pigment systems is different and characteristic of each (Gross et al., 1966; Briantais, 1967; Boardman, Anderson, 1967; Lichtenthaler, 1969). Based on the different pigment distributions of the two pigment systems, pigment ratios can be established, which indicate the ratio of the two pigment systems. A ratio of this kind is the generally known chlorophyll α/β one. Obviously, the ratio of the two pigment systems can be given with greater certainty if one has the knowledge of more than one set of pigment ratios. Relying on this for characterizing the ratio between two pigment systems, the author set

up new ratios: the ratio of carotene/chlorophyll β and carotene/xanthophylls. Higher values of the ratios carotene/chlorophyll β, chlorophyll α/β and carotene/xanthophyll indicate a higher ratio of the pigment system I as against system II. On the other hand, a decrease in the ratios means an increase of the share of the second pigment system (Tuba, 1981). As a consequence of the above, in dry conditions the ratio of pigment system I/II is higher, while among wet conditions this ratio decreases. The same change can be observed on the base of the ratio of chlorophyll α/β, carotene/chlorophyll β and carotene/ xanthophylls (Table 4). The above pigment system ratio changes were the highest in the sclerophyllic *Festuca*, lower but well defined in *Achillea* and only barely noticable in *Sedum*.
Based on the above results it can be concluded, that, amond dry conditions the ratio of pigment system I is highest while in wet conditions the ratio of the pigment system II is the highest by the *Festuca*, moderate but definitive in *Achillea* and only at the beginning, after longer drought, is observable in *Sedum*.

4. CONCLUDING REMARKS
During shorter dry periods even the drought resistant species show typical pigment reactions due to the water shortage. Based on these results it is concluded, that in the examined species, these symptoms are apparent at different rates. The sclerophyllic *Festuca vaginata* reacts most sharply to the changes in the soil water content, while the response of semisucculent *Achillea ochroleuca*

is less expressed and the succulent *Sedum sexangulare* has a similar response only after longer drought. The degree of the responses decreases towards the succulency. In the adaptation to drought the author assigns basic importance to the amount of carotenes and β-carotene is able to restore the photosystem I activity and cyclic photophosphorylation (Tukendorf et al., 1980). Its increase is due to lutein and zeaxanthin dehydroxylation. These processes should be accompanied by the release of extra water on the thylakoid membrane surfaces (Chua, 1970). This mechanism can coincide with the highest degree of the violaxanthin desepoxidation via antheraxanthin favouring the zeaxanthin and lutein accumulation, which again helps the above production of metabolic water. The changes of the epoxyxanthophylls are also important. In wet conditions, a better supply of water results in more water consuming carotenoid synthesis. This increases the lutein content in the plant and the amount of epoxyxanthophylls as well. The changes of the chlorophylls and carotenoids among dry and wet conditions also modify the ratio of pigment systems. These changes in pigment system ratios can highly influence the ratio among functions connected to pigment systems. During drought there appears to be a higher ratio of pigment system I and, presumably the reactions connected to pigment system I are more active. Actually, the lower water potential reduces the activity of pigment system II (Vieira-Da-Silva, Veltkamp, 1970; Fry, 1972; Potter, Boyer, 1973; Pospišilová et al., 1976) and electron transfer is reduced (Keck, Boyer, 1974). Among these conditions the cyclic photophosphorylation connected to the photosystem I should play an important role in the ATP supply for energy consuming processes. This view is supported by Keck and Boyer (1974), who considered that cyclic photophosphorylation continuous at full activity in the range of -10 bars water deficit. Accordingly, the changes of the ratio of the two pigment systems are considered approximate compositional proof of the changes

ensued in the ratio of the functions associated with the two pigment systems. It is very probable, that the controlled transformations of the components of the pigment system constitute one of the bases of the photosynthetic system's adaptation to different ecological and internal conditions. To obtain better knowledge about the details, it seems necessary to investigate the pigment system together with the activities of photosystems, photophosphorylation and chloroplast ultrastructure.

ACKNOWLEDGEMENTS

The author expresses his sincere thanks to Dr. T. Pócs for translating the manuscript and to Ms. I. Divald, Mrs M. Dinka and Mrs E. Petényi for their technical assistance.

REFERENCES

Alberte RS, Fiscus EL and Maylor AW (1975) The effects of water stress on the development of the photosynthetic apparatus in greening leaves, Plant Physiol. 55, 317-321.

Alberte RS and Thornber JP (1977) Water stress effects on the formation and organization of chlorophyll in mesophyll and bundle sheats chloroplasts of maize, Plant Physiol. 59, 351-353.

Boardman NK and Anderson JM (1967) Fractionation of the photochemical system of photosynthesis II. Cytochrome and carotenoid content of particles isolated from spinach chloroplasts, Biochim. Biophys. Acta 31, 187-203.

Briantais JM (1967) Spectroscopie de la chlorophylle dans les chloroplastes entiers et des fragments chloroplastiques, Photochem. Photobiol. 6, 155-162.

Chua KS (1970) Structure of anomalous water and its mechanism, Nature 227, 834-836.

Claes H (1967) Action spectrum of light-depended carotenoid synthesis in *Chlorella vulgaris*. In Goodwin TW, ed. Biochemistry of chloroplasts, pp. 341-344. London, New York, Acad. Press.

Donochue HV, Nakayama TOM, Chichester CO (1967) Oxygen reactions of xanthophylls. In Goodwin TW, ed. Biochemistry of chloroplasts, pp. 431-440. London, New York, Acad. Press.

Duysen ME and Freeman TP (1974) Effect of moderate water deficit (stress) on wheat seedling growth and plastid pigment development, Physiol. Plant. 3 , 31 262-266.

Duysen ME and Freeman TP (1976) Promotion of plastid pigment accumulation in water stressed wheat leaf sections by hormone treatment, Amer. J. Bot. 63, 1134-1139.

Fry KE (1972) Inhibition of ferricyanide reduction in chloroplasts prepared from water-stressed cotton leaves, Crop. Sci. 12, 698-701.

Goedheer JC (1979) Carotenoids in the photosynthetic apparatus, Ber. Deutsch. Bot. Ges. 92, 427-436.

Gross JA, Shefner AM and Becker MJ (1966) Distribution of chlorophylls in a chloroplast fragment, Nature 209, 615.

Hager A and Meyer-Bertenrath T (1966) Die Isolierung und quantitative Bestimmung der Carotenoide und Chlorophylle von Blättern, Algen und isolierten Chloroplasten mit Hilfe dünnschichtchromatographischer Methoden, Planta 69, 198-217.

Ida K (1981) Eco-physiological studies on the response of Taxodiaceous conifers to shading with special reference to the behavior of leaf pigments II. Chlorophyll and carotenoid contents in green leaves grown under different grades of shading, Bot. Mag. Tokyo 94, 181-196.

Keck RW and Boyer JS (1974) Chloroplast response to low water potential III. Differing inhibition of electron transport and photophosphorylation, Plant Physiol. 53, 474-479.

Krinsky NI (1978) Non-photosynthetic functions of carotenoids, Phil. Trans. R. Soc. Lond. 284, 581-590.

Lichtenthaler H (1969) Localization and functional concentrations of lipoquinones in chloroplasts. In Metzner H, ed. Progress in photosynthesis research, pp. 304-314. Tübingen.

Maróti I and Gabnai É (1971) Separation of chlorophylls and carotenoids by thin-layer chromatography, Acta Biol. Szeged 17, 67-77.

Maróti I and Szemenkei (1972) Light-induced transformations of pigments I. Transformations of carotenoids under aerobic and anaerobic conditions, Acta Biol. Szeged 18, 71-80.

Maróti I and Szajkó I (1972) Light-induced transformations of pigments II. The role of water in the transformation of carotenoids, Acta Biol. Szeged 18, 81-91.

Potter JR and Boyer JS (1973) Chloroplast response to low water potentials II. Role of osmotic potential, Plant Physiol. 51, 993-997.

Pospišilová J, Zima J and Šesták Z (1976) Effect of hydration level in primary bean leaves on the activity of photosystems 1 and 2 in isolated chloroplasts, Biologia Plantarum (Praha) 18, 473-479.

Rychnovská M (1966) Wasserhaushalt einiger Stipa-Arten am natürlichen Standort, Ropravy CSAV 76, 1-32.

Rychnovská M, Květ J, Gloser J and Jarklová J (1972) Plant water relations in three zones of grassland, Acta Sc. Nat. Brno 6(5), 1-38.

Šesták Z (1977) Photosynthetic characteristics during ontogenesis of leaves. 1. Chlorophylls, Photosynthetica 11, 367-448.

Šesták Z (1978) Photosynthetic characteristics during ontogenesis of leaves. 3. Carotenoids, Photosynthetica 12, 89-109.

Spyropoulos CG and Mavrommatis M (1978) Effect of water stress on pigment formation in Quercus species, J. Exp. Bot. 29, 473-477.

Tukendorf A, Krupa Z and Baszynski T (1980) β-carotene as a factor in the reconstitution of cyclic phosphorylation in damaged chloroplast.

membranes, Acta Soc. Bot. Pol. 49, 435-443.

Tuba Z (1977) Examination of the vertical pigment structure in an oak forest (Quercetum petraeae-cerris), Acta Soc. Bot. Pol. 49, 435-443.

Tuba Z (1981) The changes of the photosynthetic pigment systems of two paprika (red pepper) varieties from the fully developed vegetative stage to the ripening of the fruit, Bot. Közl. 68, 123-131.

Vieira-Da-Silva JB and Veltkamp J (1970) Action du potentiel osmotique de la solution nutritive sur la réaction de Hill et la phosphorylation de chloroplastes de Cottonier, Compt. rend. Acad. Sci. Paris Sér D 271, 1367-1379.

Virgin HI (1965) Chlorophyll formation and water deficit, Physiol. Plant. 18, 994-1000.

Weatherley PE (1950) Studies in the water relations of the cotton plant I. The field measurement of water deficits in leaves, New Phytol. 49, 81-97.

WAYS OF DETECTING ADAPTIVE RESPONSES OF CULTIVATED PLANTS TO DROUGHT. AN AGRONOMIC APPROACH.

A.J. KARAMANOS (Laboratory of Crop Production, Athens College of Agriculture, Athens 301, Greece).

1. INTRODUCTION

Plants are exposed to water deficits whenever the rate of water loss exceeds that of water absorption. The rate of water loss is determined by the atmospheric evaporative demand and the stomatal control of transpiration while the rate of water absorption depends on both the availability of soil water and the extent of the root system. By definition, cells and tissues are regarded as water-deficient or water-stressed when their water potential falls below zero (Crafts, 1968; Kramer, 1969). Given that the state of full turgor is hardly achieved (for a review, see Karamanos, 1980), plants are usually water-stressed during the greatest part of the day showing the most negative values of their water potential around midday (Namken et al., 1969). Under conditions of high evaporative demand plants are water-stressed to a considerable extent even when soil water is adequate (Denmead, Shaw, 1962). It follows that water stress is particularly intense in plants growing in places of low rainfall and high evaporative demand.

Water stress affects adversely many physiological processes, thus retarding plant growth and reducing the overall productivity of the stands. There is a variety of adaptive mechanisms through which plants are able to withstand water shortage. Levitt (1972) classified these mechanisms into two main categories: those which enable plants to avoid drought to a certain extent('drought avoidance') and those which refer to plants already water-stressed to a given degree and enable them to endure satisfactorily this adverse situation ('drought tolerance mechanisms').Both categories are detectable in cultivated plants as well as in the natural vegetation. However, the adaptive strategies are as a rule more intense in the natural vegetation than in the cultivated species. The necessity to meet the World food demands by increasing the productivity of the cultivated plants led to the production of new highly productive genotypes at the expense of their adaptability. The high demands for water and nutrients were the main cause of the failure of the highly productive grain varieties in the dry areas during the Green Revolution (Dahlberg, 1979). As a result, a greater emphasis on adaptability is given in the new breeding programs.

From the agronomic point of view, the adaptive responses of cultivated plants cannot be evaluated without assessing their impact on crop productivity. To give some examples, a reduction in the transpiring area in the way that phrygana do (Orshan, 1964) is not a desirable response for cultivated plants. Furthermore, sclerophyll-like modifications which are bound with low rates of photosynthesis (Larcher, 1975) are not desirable as well. Taking into account that the maintainance of the productivity potential to a satisfactory level is the main priority, it is interesting to examine how plants withstand the degrees of water stress usually encountered in the field. In the present contribution, an attempt is made to detect and evaluate the drought-induced adaptive mechanisms of the cultivated plants by means of some physiological parameters. These parameters, which refer to mechanisms of drought tolerance, drought avoidance, as well as to water stress-induced concentration of some metabolites may also be valuable as screening criteria to plant breeders for the identification and production of new genotypes able to withstand more effectively water stress.

2. DROUGHT TOLERANCE

An aspect of drought tolerance is the 'dehydration avoidance' (Levitt, 1972) which refers to the ability of the water-stressed plants to maintain their turgor above zero at high degrees of water stress. This is important, because the significance of cell turgor to many physiological processes is well-established (Hsiao, 1973). The manner in which plants maintain their turgor above zero at low values of their water potential can be visualized if one considers the total plant water potential (Ψ) as the algebraic sum of the pressure (ψ_p) and solute (ψ_s) potentials:

$$\Psi = \psi_s + \psi_p \qquad (1)$$

The matric potential is no more regarded as playing a significant role in the water exchanges of the tissues (Tyree, Karamanos, 1981) and, therefore, can be omitted. Tissues have two main ways of preventing their turgor from rapidly falling to zero. First, by lowering the osmotic potential in the vacuoles of their component cells, and, secondly, by changing the elasticity of their cell walls. Both ways do not appear to influence adversely plant productivity (Turner, 1979).

2.1. Lowering of the solute potential

A lowering of ψ_s arises from a simple cell dehydration. According to the law of van't Hoff, the osmotic pressure (π) is inversely related to the volume of the symplast (V):

$$\pi = k \, \frac{1}{V} \qquad (2)$$

where k is a constant. Thus, the passive lowering of ψ_s is expected to be represented by a linear fall of ψ_s in relation to cell volume or, to a close approximation, to the cell relative water content (R). Since the cells experience small volume changes, between 10 to 30 %, the passive changes in ψ_s are also small and, therefore, of little ecological

advantage (Tyree, Karamanos, 1981).

In addition to dehydration, cells are also able to lower their ψ_s by accumulating osmotically active substances such as sugars, inorganic ions, organic acids, etc. inside their vacuoles. This response, known as osmoregulation or osmotic adjustment, brings about a decrease in the solute potentials at maximum (ψ_{sm}) and zero turgor (ψ_{so}) (Fig. 1a).

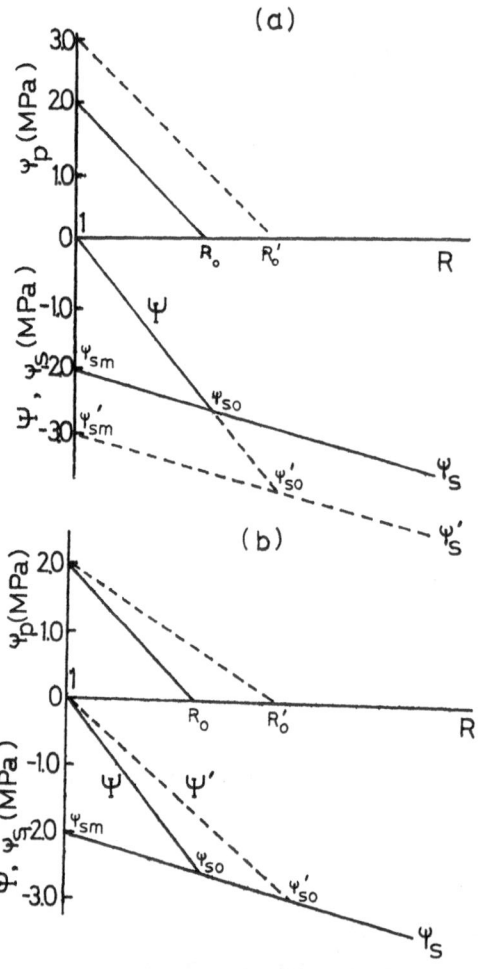

FIGURE 1. Water and component potential isotherms in plant tissues exhibiting 'dehydration avoidance'. a) Osmotic adjustment by 50 %. The continuous lines represent the isotherms in non-adjusted ($\psi_{sm} = -2.0$ and $\psi_{so} = -2.5$ MPa) while the dotted ones in osmotically-adjusted tissue ($\psi'_{sm} = -3.0$ and $\psi'_{so} = -3.75$ MPa). b) Increase in elasticity by 50 %. The continuous lines represent the isotherms in the less elastic ($\psi_{sm} = -2.0$ and $\psi_{so} = -2.5$ MPa) while the dotted ones in the more elastic tissue ($\psi'_{so} = -2.8$ MPa).

The benefits for the cell are both a decrease in the relative water content at zero turgor (R_O) which allows larger amounts of water to be lost before the turgor loss point and considerably higher values of cell turgor during the dehydration cycle. A 50 % fall in ψ_S by means of osmotic adjustment increases ψ_p and decreases both ψ_{sm} and ψ_{so} in an equal proportion.

There is ample evidence that osmotic adjustment is quite a usual adaptive response to drought in cultivated plants. It arises as a result of progressively increasing soil dryness (Karamanos, 1978a) or diurnally, following variations in light or water stress (Hsiao et al., 1976; Wenkert et al., 1978). Moreover, the ability for drought-induced osmotic adjustment can be an inheritable property: intraspecific differences have been reported in wheat (Morgan, 1977) and sorghum (Stout, Simpson, 1978). Finally, it is not so easy to distinguish between drought-induced and ontogenetically-controlled osmotic adjustment which is also known to occur (Knipling, 1967; Millar et al., 1968).

To detect osmotic adjustment, it is necessary to separate the passive changes in ψ_S. This can be done by examining the values of ψ_S at standard reference levels of tissue hydration, such as at maximum (ψ_{sm}) or zero turgor (ψ_{so}). The seasonal or diurnal courses of these parameters reveal osmotic adjustment caused by drought or plant development (Fig. 2). However, it is important to note that a lowering of ψ_{so} is not caused only by osmotic adjustment, but also, to a minor extent, from an increase in cell wall elasticity (Turner, 1979; see also Fig. 1b). From this point of view, it appears that ψ_{sm} is a better criterion for detecting exclusively osmotic adjustment. ψ_{so} and ψ_{sm} can be determined from pressure-volume curves (Tyree, Hammel, 1972; Richter et al., 1981; see also Fig. 3). ψ_{so} can also be determined from the change in length of leaf strips floating

in osmotica of varying ψ_S (Kassam, 1972; Fig. 4).

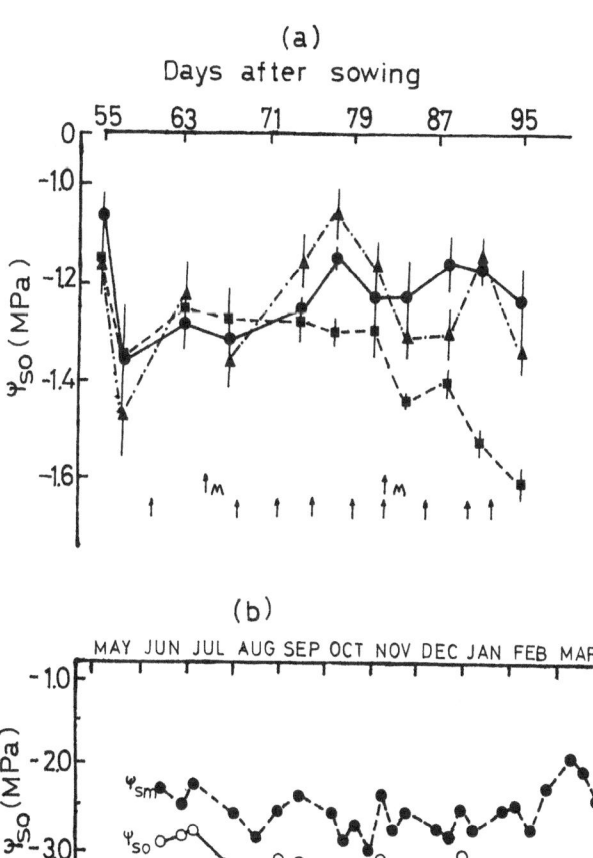

FIGURE 2. a) The time course of the solute potential at zero turgor (ψ_{so}) in crops of field beans growing under different irrigation regimes (●——●: wet, ▲–·–·▲: medium, and ■–––■: dry). The arrows indicate the irrigation timing in the wet and medium (M) treatments (Karamanos, 1978a). b) The seasonal courses of the solute potentials at maximum (ψ_{sm}, dashed line) and zero turgor (ψ_{so}, continuous line) in a non-irrigated olive-tree. Both ψ_{sm} and ψ_{so} were determined from pressure-volume curves (Karamanos, unpublished).

A psychrometric determination of ψ_{sm} is also possible on leaf tissue previously saturated with water (Jones, Turner, 1978). However, psychrometric determinations of ψ_{sm} which presuppose the destruction of cell membranes are subjected to errors arising from the dilution of cell sap by the water of the cell walls (Tyrée, 1976).

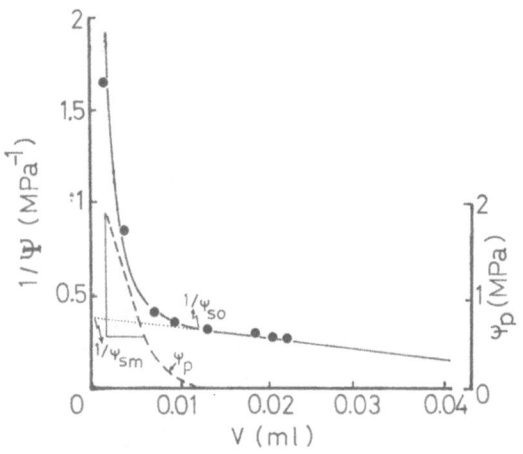

FIGURE 3. Pressure-volume curve taken from an olive-tree leaf. The arrows indicate how ψ_{sm} and ψ_{so} are determined. The enclosure with the dashed line shows the curvilinear relationship between ψ_p and V. The bulk modulus of elasticity was determined only in the straight-line portion of the relationship (Karamanos, unpublished).

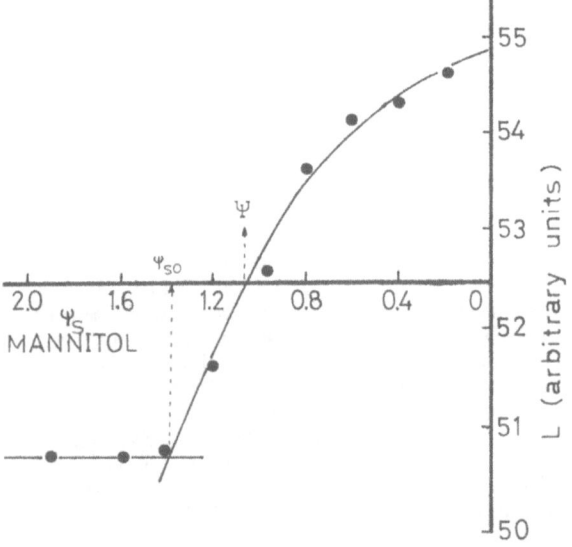

FIGURE 4. The determination of ψ_{so} in field beans from the length change of strips floated in mannitol solutions. ψ_{so} is determined as the point beyond which no change in the overall length of strips is detectable (Karamanos, 1976).

2.2. Cell wall elasticity

The extent to which ψ_p is reduced for a given amount of water lost is determined from the elastic properties of the cell walls. An increase in elasticity induces a smaller drop in ψ_p for a given degree of dehydration, thus maintaining cell turgor at lower water contents (Fig. 1b). In contrast with osmotic adjustment, the changes in the elasticity affect the relation between water potential and relative water content.

A suitable parameter to express cell wall elasticity is the bulk modulus of elasticity (ε) (Tyree, Hammel, 1972) which can be defined from the equation:

$$\varepsilon = V \frac{d\psi_p}{dV} \qquad (3)$$

where $d\psi_p$ is the change in cell turgor for an infinitesimal change in volume dV relative to the initial volume V. Thus, the higher the value of ε, the quicker the drop in ψ_p for a given amount of water lost, i.e. the lower the elasticity of the walls, and vice versa. The value of ε can be estimated by plotting ψ_p versus water volume (V) or relative water content (R). The relationship between ψ_p and R is non-linear (Haines, 1950; Gardner, Ehlig, 1965; Hellkvist et al., 1974; Cheung et al., 1976). This creates problems both in the calculation and comparability of ε among plant species. A way of dealing with this problem is to consider only high values of cell turgor where the relationship between ψ_p and R is linear (Cheung et al., 1976; Tyree et al., 1978; see also Fig. 3).

The experimental evidence concerning the association of ε with drought is not so clear. In most cases, higher values of ε were associated with drier conditions of growth. For example, ε was higher in the drought-resistant sorghum than in maize (Sanchez-Diaz, Kramer, 1971, 1973). Furthermore, previous experience in water stress increased the values of ε in cotton (Brown et al., 1976) and sorghum (Jones, Turner, 1978). In olive trees, ε

exhibited higher values in the dry (summer) and lower values in the wet period (autumn-winter) (Fig. 5b). On the other hand, water shortage induced a lowering of ε in field beans in comparison with well-watered controls (Fig. 5a).

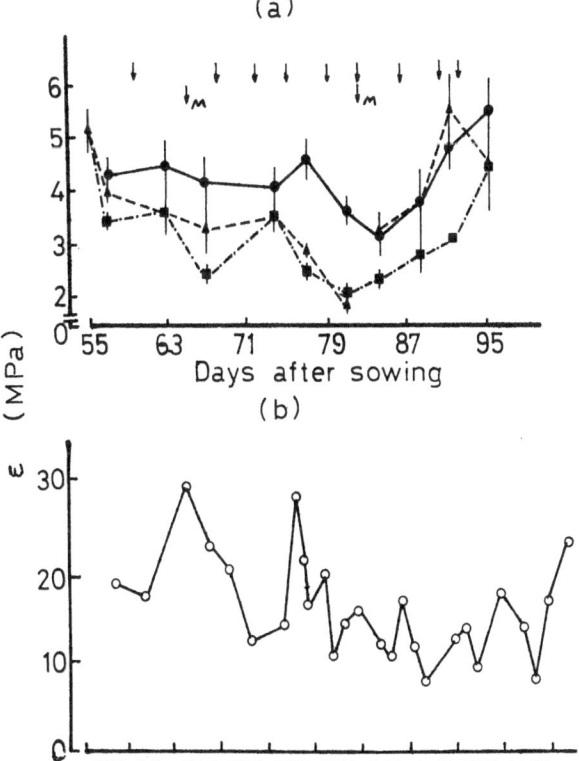

FIGURE 5. (a) The time course of the bulk modulus of elasticity (ε) in field bean crops growing under three different irrigation regimes (●——●: wet, ▲---▲ medium, ■-·-·-■: non-irrigated). The arrows show the irrigation timing in the wet and medium (M) treatments (Karamanos, 1978a). (b) The seasonal course of ε for leaves of a non-irrigated olive-tree calculated from pressure-volume curves. Note the higher values during the summer and the lower ones in the winter (Karamanos, unpublished).

An increase in ε (decrease in elasticity) is normally expected when walls become more rigid or thicker because of the deposition of materials. Since the rigidity and thickness of the cell walls are regarded as characteristics of xerophytism (Slatyer, 1967), the higher values of ε usually encountered under dry conditions are not surprising. Nevertheless, even the

increase in elasticity observed in field beans can be explained in view of the smaller size of leaf cells in the water-stressed plants (Farah, 1979); it has been shown (Steudle et al., 1977) that small cells are more elastic, thereby leading to turgor maintainance.

There are different ways of interpreting the ecological importance of ε. Higher values of ε bring the tissues to the turgor loss point faster than lower values for a given degree of dehydration (Fig. 1b). Such a response favours drought avoidance by stomatal control of transpiration for relatively small amounts of water lost. Alternatively, lower values of ε enable the tissues to endure higher degrees of dehydration (they reach a lower R_o; see Fig. 1b) before their turgor drops to zero. Thus, stomata remain open and assimilation is maintained at lower water contents than in the first case. In addition, the maintainance of turgor enables many physiological processes to proceed unimpaired. Finally, a lower ε brings about also a decrease in ψ_{so}. However, both this lowering of ψ_{so} and the maintainance of ψ_p are less effective for a decrease in ε by a given amount in comparison with the corresponding effects brought about by osmotic adjustment to an equal degree (Fig. 1).

In conclusion, it appears that osmotic adjustment, estimated by means of ψ_{sm} and, in second place, ψ_{so}, offers possibilities for more drastic adaptive responses to drought than the changes in cell wall elasticity expressed in terms of ε. The ecological importance of the latter as an adaptive mechanism to drought still remains vague and merely speculative (Tyree, Karamanos, 1981).

3. DROUGHT AVOIDANCE

Plants can avoid drought by increasing water supply and/or by reducing water loss. An increase in water supply by means of a deeper root system has been occasionally observed in water-stressed crop plants, such as flax (Newman, 1966), cotton (Klepper et al., 1973), barley (Ellis et al.,1977), and field beans (Karamanos, 1981), provided that a water table exists in a depth accessible to the roots. If not, a greater root density is observed close to the soil surface in plants growing under dryland conditions in comparison with irrigated ones (Weaver, 1926). This is due to the small amounts of current rain which do not penetrate deeply into the soil. In addition, the extent of the root system in the soil is also drastically affected by other factors such as nutrient availability and soil structure. It is, therefore, difficult to regard the extent of the root system as a reliable evidence for drought avoidance. In this section the attention will be focused on the adaptive responses which enable plants to save water.

A reduction in water loss can be accomplished either by restricting the available transpirational area or by controlling the rate of transpiration.

3.1. Restriction of transpirational area

The transpirational area can be reduced irreversibly in severe cases of water shortage by means of an acceleration of leaf senescence and death. Such a response is very common both to the natural vegetation, as in phryganic formations (Orshan, 1964, 1972), and in crop plants such as flax (Milthorpe, 1945), cereals (Boyer, McPherson, 1975), and field beans (Karamanos, 1978b; Finch-Savage, Elston, 1982). In cotton, drought-induced leaf shedding is accomplished only after rewatering, provided that the water potential of the stressed plants was below -0.8 to -1.0 MPa (McMichael et al., 1972). However, in most cases it is difficult to search for a threshold value of Ψ below which leaf shedding is triggered. It is not known whether the effects of water stress arise instantaneously whenever a critical value of Ψ is reached or they are cumulative for a given period. Leaf shedding can be regarded as the ultimate means of saving water for the stressed plants. It is a response for survival

when all other means have been exhausted. Consequently, from the agronomic point of view leaf shedding is of little interest as an adaptive response. An irreversible loss in leaf area brings about a significant reduction in the productive potential of the cultivated plants. A reversible reduction in the transpiring area would be more desirable.

Leaf rolling, a response exhibited in water-stressed cereals and grasses, provides a reversible reduction in the exposed transpiring area. In addition, it reduces the radiative load intercepted by leaves and interferes with the dynamics of gas exchange (Blum, 1979). The extent to which the transpiration of some Mediterranean grasses is reduced by leaf rolling amounts 46 to 63 % (Parker, 1968). Similar estimates are not available for cultivated plants. Hurd (1976) stated that the more drought-resistant durum wheat cultivars rolled their leaves more easily than the less resistant ones. Moreover, the evidence (Hurd, 1976) that photosynthesis and grain filling proceed satisfactorily with rolled leaves is extremely interesting and requires further attention. In general, the effects of leaf rolling on plant productivity as a function of water stress have not been determined yet, although rolling is considered as a screening criterion for drought resistance in breeding programs of rice (O'Toole, Chang, 1979) and wheat (Austin, Jones, 1975).

The loss of turgor of the bulliform or motor cells of the leaf upper epidermis in cereals and grasses is thought to be the cause of rolling (Parker, 1968), although other factors cannot be excluded (Esau, 1965). It follows that rolling provides a visual estimate of the turgor loss point in *Gramineae*. Results from an experiment with two wheat cultivars revealed a clear difference between them in the value of the leaf water potential (Ψ_1) at which rolling was observed (Table 1). This implies a differentiation between the cultivars in the value of Ψ_1 at which turgor drops to zero, with possible effects on gas exchange, grain filling etc.

TABLE 1. The average Ψ_l (MPa) and the corresponding proportions of the sampled leaves that exhibited rolling or had a normal appearance in the wheat cultivars Yecora and Generoso on day 172 after sowing (adapted from Karamanos et al., 1982).

	Yecora		Generoso	
	Ψ_l	%	Ψ_l	%
Rolled leaves	-2.39	33	-	0
Normal leaves	-2.09	67	-2.47	100

In conclusion, the value of Ψ_l at which leaf rolling occurs may be a very useful parameter to assess adaptive responses to water stress in cereals. It gives information about the ability of the plants to avoid drought by reducing reversibly their exposed leaf area, and, at the same time, provides evidence for the degree of water stress that plants are able to withstand before their turgor falls to zero.

3.2. Regulation of the transpiration rate

The regulation of the transpiration rate according to the leaf water status is achieved by means of the stomata in a complex manner (Raschke, 1975). In general, stomata remain open as long as Ψ_l is maintained above a threshold value. When the threshold value is reached, stomata close rapidly within a narrow range of Ψ_l. At this threshold value of Ψ_l, the values of ψ_p are between 0 to 0.3 MPa (Fig. 6).

Two useful parameters can be produced from the general relations between Ψ_l and stomatal resistance (r_s): the threshold value of Ψ_l and the minimal value of r_s.

The threshold value of Ψ_l for stomatal closure varies considerably among the cultivated species: from -0.8 MPa in field beans (Kassam, 1973) to -2.8 MPa in cotton (Ackerson,

1980) and olive trees (Karamanos, unpublished). It also varies with leaf position, leaf age, and growth conditions (for a review, see Turner, 1979). For leaves of the same age and position, the threshold Ψ_l is adjusted according to the previous stress history of the plants. It has been found that sorghum and cotton plants preconditioned to drought exhibited more negative values of threshold Ψ_l than non-stressed plants (McCree, 1974; Brown et. al., 1976; Stout, Simpson, 1978; Ackerson, 1980). This is a very important adaptive response, because the 'hardened' plants are able to photosynthesize at lower water potentials, thus increasing their water use efficiency, provided that the enzymatic systems of photosynthesis are unaffected by the increased stress and the water supply is not restricted. From this point of view, the lowering of the threshold Ψ_l can be regarded as a desirable adaptive response. A similar explanation could be given to the fact that the threshold Ψ_l is usually more negative in species regarded as more 'drought resistant' than others (e.g. sorghum, olive trees, cotton). The lowering of the threshold Ψ_l in pre-stressed plants has been ascribed to an osmotic adjustment which has been induced during the previous stress cycles (Begg, Turner, 1976; Ackerson, Krieg, 1977). Thus, the evidence that sorghum is superior to maize in the ability to adjust the threshold Ψ_l under similar conditions of growth, can be ascribed to a greater osmotic adjustment achieved by sorghum (Ackerson, Krieg, 1977).

The minimal value of stomatal resistance (r_{so}) is indicative of the intensity of the gas exchange in leaves under conditions of low water stress when irradiance and CO_2-concentration are not limiting. Thus, the magnitude of r_{so} is related to the potential transpiration of the plants and hence, it might be of some ecological interest. However, an examination of the values of r_{so} in a number of crop species (Table 2) shows that their magnitude cannot be related straightforward to their overall drought resistance. Furthermore,

98

it appears that the previous stress history does not affect the magnitude of r_{so} (McCree, 1974; Ackerson, 1980).

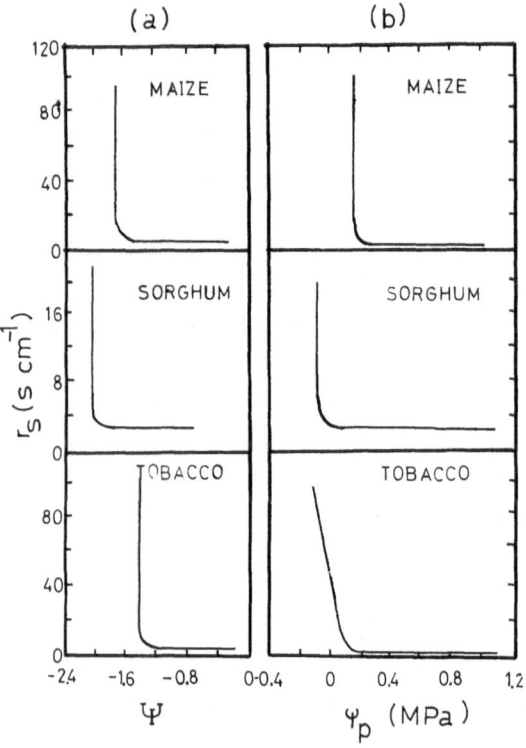

FIGURE 6. The relation between stomatal resistance (r_s) and leaf water potential (Ψ) (a) or pressure potential (ψ_p) (b) for maize, sorghum and tobacco. The leaves were irradiated at greater than 0.6 cal cm^2 min^{-1} (redrawn after Turner, 1974).

We conclude that the threshold value of Ψ_1 for stomatal closure is an important ecological parameter which can be adjusted according to the previous stress history of the plants. In contrast, the minimal value of stomatal resistance does not appear to play a role in the adaptive responses of plants.

4. PROLINE ACCUMULATION

The accumulation of free-proline in water-stressed tissues of several plant species is a well-established phenomenon. It arises mainly from an enhancement of proline synthesis from glutamic acid (Boggess et al., 1976; Stewart, 1980) and, in second place, from an inhibition

TABLE 2. The minimal value of stomatal resistance (r_{so}) (s.cm^{-1}) reported by various authors for some cultivated species.

Species		r_{so}	Source
Sorghum	(C_4)	2.4	Turner, Begg (1973)
Maize	(C_4)	2.8	" " "
Tobacco	(C_3)	1.3	" " "
Cotton	(C_3)	2.5	Ackerson (1980)
Wheat	(C_3)	1.0-1.5	Biscoe et al. (1976)
Sunflower	(C_3)	1.1	Berger (1971)
Sugarbeet	(C_3)	1.5	Brown (1976)
Soybean	(C_3)	0.7	Wein et al. (1979)
Cowpea	(C_3)	1.0	" " " "
Olive-tree	(C_3)	4.5	Karamanos (unpublished)

of both protein synthesis (Stewart, 1973) and proline oxidation to other soluble compounds (Stewart et al., 1977). There is a controversy as regards the physiological significance of proline. Most investigators ascribe to the iminoacid a positive role associated with some sort of adaptive response. Stewart, Lee (1974) found that proline was involved in the osmotic adjustment of some halophytes. Singh et al. (1973) considered proline as a source of energy, carbon, and nitrogen for the recovering tissues. Recently (Paleg et al., 1981), it was reported that proline protected several enzymatic systems against the inactivating effects of heat and postulated a similar role in water-stressed tissues where stomatal closure induces a rise in leaf temperature. Yet, there are some authors who consider the accumulation of proline as a result of injury to plant tissues by the imposed stress (Hanson et al., 1977).

Free proline is detected in all organs of the stressed plants (Karamanos et al., 1982). In a given organ, its concentration increases as Ψ_1 becomes more negative in a curvilinear fashion (Fig. 7). There is evidence for a threshold Ψ_1 (ranging between -1.0 to -2.0 MPa for several plant species) beyond which the amounts of proline increase abruptly for a further decline in Ψ_1 (Waldren et al., 1974; Hanson et al., 1977).

Nevertheless, plant development appears to affect drastically the relation between free proline and Ψ_l (Karamanos et al., 1982)(Fig.7).

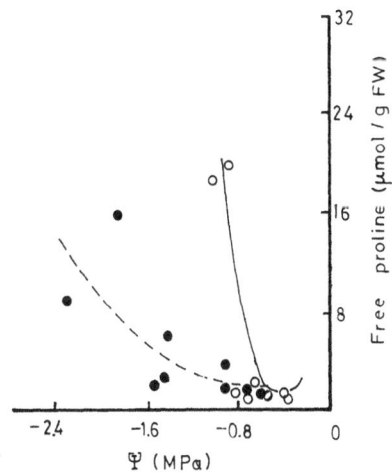

FIGURE 7. The relations between the amounts of free proline and the water potential (Ψ) in wheat leaves (cv. Yecora) during the vegetative (o——o) and reproductive stages (●----●) (adapted from Karamanos et al., 1982).

Many attempts have been made to associate the accumulation of proline with the drought resistance of crop species and genotypes within species. In most cases, positive correlations between the amounts of proline and some parameters related to drought resistance were found. Here, it is important to emphasize that drought resistance is a complex of many morphological, physiological, and biochemical characteristics. As a result, one of the major problems of dealing with drought resistance is to define the most appropriate factors to express it adequately. In the present case, different criteria were used by the various investigators. Singh et al. (1972) used as an expression of the drought resistance in barley the stability of yield over a range of environments. The ability of leaves to resume growth after the alleviation of stress was used by Singh et al. (1973) in barley and Blum, Ebercon (1976) in sorghum. Karamanos et al. (1982) used parameters of plant water status, such as the

Ψ_l at which leaf rolling occurred and the relation between Ψ_l and r_s, to assess drought resistance in two wheat cultivars which showed different patterns of proline accumulation with water stress. On the other hand, proline was not correlated with drought resistance in other works (Waldren et al., 1974; Hanson et al., 1977). Thus, there are still some doubts as to whether the levels of the accumulated free proline reflect a real adaptive response or they are simply a result of the imposed stress. If further evidence supports the former hypothesis, then the levels of proline at a given value of plant or soil water potential can be used as a measure of the adaptability of genotypes and, accordingly, as a screening criterion for plant breeding.

REFERENCES

Ackerson RC (1980) Stomatal response of cotton to water stress and abscissic acid as affected by water stress history, Pl. Physiol. 65, 455-459.

Ackerson RC and Krieg DR (1977) Stomatal and non-stomatal regulation of water use in cotton, corn and sorghum, Pl. Physiol. 60, 850-853.

Austin RB and Jones HG (1975) The physiology of wheat, Rep. Plant Breed. Inst. 1974, pp 20-73.

Begg JE and Turner NC (1976) Crop water deficits, Adv. Agron. 28, 161-217.

Berger A (1971) La circulation de l'eau dans le système sol-plante. Etude de quelques résistances, en relation avec certains facteurs du millieu, Thèse de doctorat d'état, Montpellier.

Biscoe, PV, Cohen Y and Wallace JS (1976) Daily and seasonal changes of water potential in cereals, Phil. Trans. R. Soc. Lond. B 273, 565-580.

Blum A (1979) Genetic improvement of drought resistance in crop plants: A case for sorghum, In Mussell H and Staples RC eds, Stress Physiology in Crop Plants, pp 429-445, New York, Academic Press.

Blum A and Ebercon A (1976) Genotypic responses in sorghum to drought stress. III. Free proline accumulation and drought resistance, Crop Sci. 16, 428-431.

Boggess SF, Stewart CR, Aspinall D and Paleg LG (1976) Effect of water stress on proline synthesis from radioactive precursors, Pl. Physiol. 58, 398-401.

Boyer JS and McPherson HG (1975) Physiology of water deficits in cereal crops, Adv. Agron. 27, 1-23.

Brown KW (1976) Sugar beet and potatoes, In Monteith JL ed., Vegetation and the Atmosphere, vol. 2, pp 65-86, London, Academic Press.

Brown KW, Jordan WR and Thomas JC (1976) Water stress induced alterations of the stomatal response to decreases in leaf water potential, Physiologia Pl. 37, 1-5.

Cheung YNS, Tyree MT and Dainty J (1976) Some possible sources of error in determining bulk elastic moduli and other parameters from pressure-volume curves of shoots and leaves, Can. J. Bot. 54, 758-765.

Crafts AS (1968) Water structure and water in the plant body, In Kozlowski TT ed., Water Deficits and Plant Growth, pp 23-47, New York, Academic Press.

Dahlberg KA (1979) Beyond the Green Revolution, New York, Plenum.

Denmead OT and Shaw RH (1962) Availability of soil water to plants as affected by soil moisture content and meteorological conditions, Agron. J. 54, 385-389.

Ellis FB, Elliott JG, Barnes BT and Howse KR (1977) Comparison of direct drilling, reduced cultivation and ploughing on the growth of cereals. II Spring barley on a sandy loam soil: Soil physical conditions and root growth, J. agric. Sci. 89, 631-642.

Esau K (1965) Plant Anatomy, New York, John Wiley and Sons.

Farah SM (1979)An examination of the effects of water stress on leaf growth of crops of field beans *Vicia faba* L. Ph.D. Thesis University of Reading.

Finch-Savage WE and Elston J (1982) The effect of temperature and water stress on the timing of leaf death in *Vicia faba*, Ann. appl. Biol. 100, 567-579.

Gardner WR and Ehlig CF (1965) Physical aspects of the internal water relations of plant leaves, Pl. Physiol. 40, 705-710.

Haines FM (1950) The relations between cell dimensions, osmotic pressure and turgor pressure, Ann. Bot. 14, 385-394.

Hanson AD, Nelsen CE and Everson EA (1977) Evaluation of free proline accumulation as an index of drought resistance using two contrasting barley cultivars, Crop Sci. 17, 720-726.

Hellkvist J, Richards GP and Jarvis PG (1974) Vertical gradients of water potential and tissue water relations in sitka spruce trees measured with the pressure chamber, J appl. Ecol. 11, 637-667.

Hsiao TC (1973) Plant responses to water stress, A. Rev. Plant Physiol. 24, 519-570.

Hsiao TC, Acevedo E, Fereres E and Henderson DW (1976) Water stress, growth, and osmotic adjustment, Phil. Trans. R. Soc. Lond. B 273, 479-500.

Hurd EA (1976) Plant breeding for drought resistance,In Kozlowski TT ed. v. IV pp 317-353, New York, Academic Press.

Jones MM and Turner NC (1978) Osmotic adjustment in leaves of sorghum in response to water deficits, Pl. Physiol. 61, 122-126.

Karamanos AJ (1976) An analysis of the effect of water stress on leaf area growth in *Vicia faba* L. in the field, Ph.D. Thesis, University of Reading.

Karamanos AJ (1978a) Understanding the origin of the response of plants to water stress by means of an equilibrium model, Praktika Acad. Athens, 53, 308-341.

Karamanos AJ (1978b) Water stress and leaf growth of field beans (*Vicia faba* L.) in the field: Leaf number and total leaf area, Ann. Bot. 42, 1393-1402.

Karamanos AJ (1980) Response in plant water status to integrated values of soil matric potential calculated from soil water depletion by a field bean crop, Aust. J. Pl. Physiol. 7, 51-66.

Karamanos AJ (1981) Case examples of research progress in drought-stress physiology. *Vicia faba*. In Simpson GM ed., Water Stress on Plants, pp. 199-234, New York, Praeger.

Karamanos AJ, Drossopoulos JB and Niavis CA (1982) Free proline accumulation during development of two wheat cultivars with water stress, J. agr. Sci., in press.

Kassam AH (1973) Influence of light and water deficit upon diffusive resistance of leaves of *Vicia faba*, New Phytol. 72, 557-570.

Klepper B, Taylor HM, Huck MG and Fiscus EL (1973) Water relations and growth of cotton in drying soil, Agron. J. 65, 307-310.

Knipling EB (1967) Effect of leaf aging on water deficit-water potential relationships of dogwood leaves growing in two environments, Physiologia Pl. 20, 65-72.

Kramer PJ (1969) Plant and Soil-Water Relationships. A Modern Synthesis, New York, Mc Graw-Hill.

Larcher W (1975) Physiological Plant Ecology, Berlin, Springer-Verlag.

Levitt J (1972) Responses of Plants to Environmental Stresses, London, Academic Press.

Mc Cree KJ (1974) Changes in stomatal response characteristics of grain sorghum produced by water stress during growth, Crop Sci. 14, 273-278.

Mc Michael BL, Jordan WR and Powell RD (1972) An effect of water stress on ethylene production by intact cotton petioles, Pl. Physiol. 49, 658-660.

Millar AA, Duysen ME, and Wilkinson GE (1968) Internal water balance of barley under soil moisture stress, Pl. Physiol. 43, 968-972.

Milthorpe FL (1945) Fibre development of flax in relation to water supply and light intensity, Ann. Bot. 9, 31-53.

Morgan JM (1977) Changes in diffusive conductance and water potential of wheat plants before and after anthesis, Aust. J. Pl. Physiol. 4, 75-86.

Namken LN, Bartholic JF and Runkles JR (1969) Monitoring cotton plant stem radius as an indication of water stress, Agron. J. 61, 891-893.

Newman, EI (1966) Relationship between root growth of flax (*Linum usitatissimum*) and soil water potential, New Phytol. 65, 273-283.

O'Toole JC and Chang TT (1979) Drought resistance in cereals. Rice: A case study. In Mussell H and Staples RC eds, Stress Physiology in Crop Plants, pp 373-405, New York, John Wiley and Sons.

Orshan G (1964) Seasonal dimorphism of desert Mediterranean chamaephytes and its significance as a factor in their water economy, In Rutter AJ and Whitehead FH eds, The Water Relations of

Plants, pp 206-222, Edinburgh, Blackwell.

Orshan G (1972) Morphological and physiological plasticity in relation to drought, Proc. Int. Symp. Wildland Shrub Biol. and Utilization at Utah State Univ. pp 245-254.

Paleg LG, Douglas TJ, van Daal A and Keech DB (1981) Proline, betaine and other organic solutes protect enzymes against heat inactivation, Aust. J. Pl. Physiol. 8, 107-114.

Parker J (1968) Drought resistance mechanisms In Kozlowski TT ed. Water Deficits and Plant Growth, v. I, pp 195-234, New York, Academic Press.

Raschke K (1975) Stomatal action, A Rev. Plant Physiol. 26, 309-340.

Richter H, Duhme F, Glatzel G, Hinckley TM and Karlic H (1980) Some limitations and applications of the pressure-volume curve technique in ecophysiological research, In Grace J, Ford ED and Jarvis PG eds, Plants and their Atmospheric Environment, pp 263-272, Oxford, Blachwell.

Sanchez-Diaz MF and Kramer PJ (1971) Behaviour of corn and sorghum under water stress and during recovery, Pl. Physiol. 48, 613-616.

Sanchez-Diaz MF and Kramer PJ (1973) Turgor differences and water stress in maize and sorghum leaves during drought and recovery, J. Exp. Bot. 24, 511-515.

Singh TN, Aspinall D and Paleg LG (1972) Proline accumulation and varietal adaptability to drought in barley: a potential metabolic measure of drought resistance, Nature New Biol. 236, 188-190.

Singh TN, Aspinall D, Paleg LG and Boggess SF (1973) Stress metabolism. II Changes in proline concentration in excised plant tissues, Aust. J. biol. Sci. 26, 57-63.

Stayer RO (1967) Plant-Water Relationships, London, Academic Press.

Steudle E, Zimmermann U and Lüttge U (1977) Effects of turgor pressure and cell size on the wall elasticity of plant cells, Pl. Physiol. 59, 285-289.

Stewart CR (1973) The effect of wilting on proline metabolism in excised bean leaves in the dark, Pl. Physiol. 51, 508-511.

Stewart CR (1980) The mechanism of ABA-induced proline accumulation in barley leaves, Pl. Physiol. 66, 230-233.

Stewart, CR, Boggess SF, Aspinall D and Paleg LG (1977) Inhibition of proline oxidation by water stress, Pl. Physiol. 59, 930-932.

Stewart GR and Lee JA (1974) The role of proline accumulation in halophytes, Planta 120, 279-289.

Stout DG and Simpson GM (1978) Drought resistance of *Sorghum bicolor*. 1. Drought avoidance mechanisms related to leaf water status, Can. J. Pl. Sci. 58, 213-224.

Turner NC (1974) Stomatal behaviour and water status of maize, sorghum and tobacco under field conditions. II At low soil water potential, Pl. Physiol. 53, 360-365.

Turner NC (1979) Drought resistance and adaptation to water deficits in crop plants, In Mussel H and Staples RC eds, Stress Physiology in Crop Plants, pp 343-372, New York, John Wiley and Sons.

Turner NC and Begg JE (1973) Stomatal behaviour and water status of maize, sorghum and tobacco under field conditions. I. At high soil water potential, Pl. Physiol. 51, 31-36.

Tyree MT (1976) Negative turgor in plant cells: fact or fallacy ? Can. J. Bot. 54, 2738-2746.

Tyree MT and Hammel HT (1972) The measurement of the turgor pressure and the water relations of plants by the pressure-bomb technique, J. exp. Bot. 23, 267-282.

Tyree MT and Karamanos AJ (1981) Water stress as an ecological factor, In Grace J, Ford ED and Jarvis PG eds, Plants and their Atmospheric Environment, pp 237-261, Oxford, Blackwell.

Tyree MT, Cheung YNS, McGregor ME and Talbot AJB (1978) The characteristics of seasonal and ontogenetic changes in the tissue-water relations of *Acer*, *Populus*, *Tsuga* and *Picea*, Can. J. Bot. 56, 635-647.

Waldren, RP, Teare RP and Ehrler SW (1974) Changes in free proline concentration in sorghum and soybean plants under field conditions, Crop Sci. 14, 447-450.

Weaver JE (1926) Root Development of Field Crops New York, McGraw-Hill.

Wein HC, Littleton EJ, and Ayanaba A (1979) Drought stress of cowpea and soybean under tropical conditions, In Mussel H and Staples RC eds, Stress Physiology in Crop Plants, pp 283-302, New York, John Wiley and Sons.

Wenkert W, Lemon ER and Sinclair TR (1978) Water content-potential relationship in soya bean: Changes in component potentials for mature and immature leaves under field conditions, Ann. Bot. 42, 295-307.-

STRUCTURAL ASPECTS OF THE ADAPTATION OF SOME
BLUE-GREEN ALGAE AND DIATOMS TO DESICCATION

by

A. ECONOMOU-AMILLI, K. ANAGNOSTIDIS & M. ROUSSOMOUSTAKAKI

Institute of Systematic Botany, University of Athens, Panepistimiopolis, ATHENS 621, Greece.

1. INTRODUCTION

Terrestrial algae consisting mainly of Cyanophyta, Chlorophyta, Bacillariophyta and Xanthophyta show a great capacity to withstand desiccation, dominating in the microbial flora of extreme dry biotopes such as hot and cold deserts, rocks and buildings (Bristol 1920; Bristol-Roach 1927; Petersen 1935; Cameron 1963; Gollerbakh, Shtina 1969; Fogg 1969; Fogg et al. 1973; Brock 1975; Friedmann, Ocampo 1976; Novichkova-Ivanova 1980; Friedmann 1980; Broady 1981a, 1981b; Potts, Friedmann 1981; Starks et al. 1981; Anagnostidis et al. 1982).

Within terrestrial blue-green algae are included: filamentous forms with intertwined trichomes and coalescent sheaths forming close-knit mats (e.g. strands of *Microcoleus,* sheets of *Phormidium, Plectonema, Lyngbya,* bundles of *Oscillatoria, Schizothrix*) as well as unicellular or filamentous forms aggregated into expanded or globose thalli (e.g. *Nostoc, Chroococcus, Gloeocapsa*). The development of gelatinous sheaths or the formation of mats in blue-green algae must be regarded as a means of protection against desiccation and accordingly as an adaptive phase, since this type of organization assists in retaining water during droughts and also in absorbtion of water when moisture reappears (Fritsch 1922, 1945; Petersen 1935; Shtina 1961; Fogg 1969; Starks et al. 1981). In general, the vegetative cells of Oscillatoriaceae withstand desiccation better than do the vegetative cells of species with perennating stages (e.g. *Nostoc, Calothrix, Scytonema*), the resting cells and hormocysts of which enable the algae to survive (Petersen 1935, Anagnostidis 1961, Fogg 1969, Mac Entee et al. 1972, Fogg et al. 1973).

The diatoms which live in the soil are mostly small or minute forms belonging to the Pennales and consist of widely distributed species, though a few endemics have been found (Petersen 1935; Patrick, Reimer 1966). The reduction in size, increase in oil storage and the building of inner plates in the thecae of diatoms may well be special adaptations to withstand drying out and increased salt concentrations (Kolbe 1932; Hustedt 1938; Anagnostidis, Economou-Amilli 1978).

An extensive revision of the taxonomy of blue-green algae on a purely taxonomic basis might appreciably reduce the number of species recognized. On the other hand many authors (Jaag 1945; Golubic 1965, 1967a, 1967b; Anagnostidis 1961, 1968; Anagnostidis, Golubic 1966) have called attention to the necessity of building up ecological systems, adding to the taxonomic designation the word "status" to characterize the variations determined by the ecological conditions at the place of growth. Similarly, the taxonomy of diatoms suffers from several superfluous nominations of systematically insignificant forms. Thus, material from different biotopes described as separate taxa were proved to be completely identical. Also, taxa which have erroneously been accepted as homogenous reveal significant morphological deviations, differing also in ecology (Lange-Bertalot, Bonik 1978). On the other hand, alteration of the sculptural elements in centric and pennate diatoms can also be triggered by a change in environmental conditions (Drum 1964; Geissler 1970a, 1970b; Schultz 1971; Paasche et al. 1975; Schmid 1976, 1979).

In this paper, sheaths and resting cells of blue-green algae (species of *Nostoc, Gloeocapsa, Oscillatoria, Scytonema, Cylindrospermum*) and the thecal structure of diatoms (species of *Hantzschia* and *Navicula*) are studied in material from dry and moist soil. The value of these morphological features in allowing the organisms to exist under the conditions of the terrestrial habitat, as well as the taxonomic problems created are discussed.

2. MATERIAL AND METHODS

Algal growths were obtained from different types of soil in Greece. All samples collected were either observed alive or fixed with formaldehyde solution 4%. The species isolated were obtained from agar plates or liquid cultures (Staub 1961; Stanier et al. 1971) and preserved in inclined tubes of agar.

The treated samples of blue green algae were obtained from the following biotopes:

Cyanothece cedrorum (Sauv.) Komarek: a) Terra rossa with limestone rocks ($CaCO_3$=22%, pH=7.5 - 7.6) in disturbed - uncultivated area; Chalkida to Eretria (Euboea) and b) Layer of dry mud, 15 m far from the sea; Thermaikos gulf (Anagnostidis 1968).

Gloeocapsa montana Kütz.: Brown-gray stony soil free of $CaCO_3$ and saturated with bases (pH=4.5) under *Pteridium aquilinum* (L.)Kuhn in undisturbed - uncultivated forest area of *Castanea sativa* Miller; Ano Steni (Euboea).

Scytonema crustaceum Ag.: a) Loamy stony soil with limestone fragments (pH=7.5, $CaCO_3$=22-54%) in disturbed - cultivated area (*Olea europaea* L.); Rovies (Euboea), b) Alluvial soils with alluvion pebbles (pH=7.3-8, $CaCO_3$=2%) in disturbed uncultivated area; Pissonas (Euboea) and c) Dry or lightly showered and hardly crusted calcareous tuff (pH=7,3) near waterfalls; Edessa (Anagnostidis 1968).

Cylindrospermum licheniforme (Bory)Kütz.: Brown gray forest soil mixed with terra rossa, free of $CaCO_3$ and saturated with bases (pH=7.7), in disturbed - cultivated area (*Zea mays* L.); Steni (Euboea).

Nostoc commune Vauch.: Different types of moist and dry terrestrial habitats in many parts of Greece.

Oscillatoria limosa Ag. ex Gom.: Shallow, temporary pool (height approximately 5 cm, pH=7.9, t= 12°C) in disturbed - uncultivated area of red soil (terra rossa) with limestones rocks ($CaCO_3$=2%); Kato Steni (Euboea).

Oscillatoria margaritifera Kütz. ex Gom.: Shallow saline pool filled with marine algae washed ashore (*Ulva, Enteromorpha, Cystoseira, Sargassum, Padina*); Thermaikos gulf (Anagnostidis 1968).

Collections of diatoms were made from different parts of Greece and from various natural substrates including dry and moist soil, walls of buildings and calcareous rocks. Permanent diatom slides were prepared following standard oxidation techniques (Hasle, Fryxell 1970) or simple burning on a cover slide and using Aroclor as a mounting medium (dn=1.66). The species *Synedra crystallina* (Ag.) Kr. and *Achnanthes brevipes* Ag., used for comparison purposes, were found in rock pools of the supralittoral zone of the Saronikos gulf (Economou-Amilli 1980).

For SEM, the treated material was air-dried onto a circular coverslip attached to a specimen holder and coated with ca 15 μm of gold/palladium using JFC - 1100 Ion Sputter. Electron micrographs were taken on a Jeol - JSM-35 operated at 25 KV.

3. OBSERVATIONS - DISCUSSION

3.1. Sheath formation

3.1.1. *Oscillatoria limosa* Ag. ex Gom. is a collective species, widely distributed in fresh waters. Cell dimensions and other morphological features, such as constriction of cross walls and presence of granules, have been used as diagnostic criteria in the various infraspecific taxa described. Fjerdingstad (1971), for the purpose of variation in trichomes diameter, classified several species under *O. limosa*. In this revision, the presence of sheath is not regarded as a feature of taxonomic usefulness, and f. *phormidioides*

(Rabh.)Elenk. - a form characterized by the presence of sheath - is included within the species variability (Fjerdingstad et a 1976).

The filaments of *Oscillatoria limosa* observed in our material (Figs 1-2) - coming from a shallow water depression - form sheaths in a proportion of 4%. Going away from the liquid phase the proportion keeps in increasing, reaching 4o % on dry soil at a distance 2om from the depression; meanwhile the sheaths become slimy to firm (Fig. 2). Moreover, the trichomes change in colour from dark blue-green to olive-green or brown.

Considering sheath formation as a feature of taxonomic usefulness, the form of *O. limosa* observed on dry soil could be characterized as infraspecific taxon and identified as *O. limosa* f. *phormidioides*. In our opinion this form, being an ecological adaption to the terrestrial environment, represents an ecoform, supporting the view that it should be characterized as "status *phormidioides*". It is noted that in other species of the genus (*O. okenii* Ag. ex Gom., *O. acuminata* Gom.) similar sheathed forms from thermal springs have been described as f. *phormidioides* (*O. okenii* f. *phormidioides* (Hansg.) Poljansk. and *O. acuminata* f. *phormidioides* Anagn.; Elenkin 1949, Anagnostidis 1961).

3.1.2. Forms of *Gloeocapsa* with wide and stratified sheaths corresponding to *G. polydermatica* Kütz. and sometimes to *G. polydermatica* f. *polyzonica* Perty were found in soil watered by melting snow (Fig. 3). At the same biotope during the dry season, the sheaths of *Gloeocapsa* were found to be narrow without an obvious stratification corresponding to *G. montana* Kütz. (Figs 4-5). On the other hand, in cultures of the winter material the cells of *Gloeocapsa* were greater in diameter, while hardly recognizable sheaths or even absence of stratified sheaths, was noticed; these features constitute diagnostic criteria of *G.*

montana.

These observations support the view that the forms studied should be established as status *polydermatica* since they are ecophenes of *G. montana* as Lemmermann (1910), Gollerbakh (1938) and Gollerbakh et al.(1953) stated rather than separate taxa as presented in most identification books. Encouraging opinions have been expressed by Jaag (1945); Golubič (1965; 1967a, 1967b); Anagnostidis (1968); Anagnostidis, Economou-Amilli (1980); Economou-Amilli, Anagnostidis (1981); Anagnostidis et al.(1982), who viewing the various aspects on the subject of the separation of the infraspecific taxa of genus *Gloeocapsa*, considered it necessary to use the word "status" to characterize the ecological variations.

3.1.3. Many problems have arisen in the systematics of the genera *Scytonema* and *Tolypothrix*. The polymorphic species *S. myochrous* (Dillw.) Ag. and *S. crustaceum* Ag. are mainly distinguished by the sheath organization; their difference in sheath thickness is mostly attributed to the grade of the substrate saturation. Thus, either they remain as separate taxa (Gollerbakh et al.1953; Desikachary 1959; Sampaio 1959; Starmach 1962, 1966,1975; Skuja 1964; Kondratjeva 1968,1975; Bourrelly 1970; Kann 1978; Komarek 1978) or *S. crustaceum* is considered a "status" of *S. myochrous* (Dillwyn) Ag. emend. Jaag (Jaag 1945, Kann 1963, Golubič 1967a, Anagnostidis 1968, Starmach 1981).

On dry soil during the summer (soil temperature 30-40°C) the filaments of the *Scytonema* observed form fluffy and brown coatings which macroscopically give the impression of dead material due to strong desiccation (Fig. 9). Dehydration of the soil caused the reviving of the filaments, which are characterized by firm and hard lamellated sheaths corresponding to *Scytonema crustaceum* Ag. and *Tolypothrix Elenkinii* Hollerb. (Fig.10). Stages of *Scytonema* with thinner but always gelatinous sheaths were additionally found in winter collections of the same sampling area. Especially after the soil becomes flooded by rain water different stages are observed: a) long trichomes with thick sheaths reminiscent of *S. crustaceum* Ag., b) short trichomes with basal heterocysts and bag-shaped sheaths corresponding to *S. crustaceum* Ag. (stages *Diplocoleopsis* and *Diplocoleon*) and to *Tolypothrix Elenkinii* f. *saccoideo-fruticulosa* Hollerb. (Fig. 13), c) long motile trichomes without sheaths in the form of free hormogones (Fig. 12), reminiscent of *Oscillatoria* (e.g. *O. lacustris* (Kleb.) Geitler, *O. ornata* Kütz.).

The grade of the substrate saturation does not seem to be related to the structure or thickness of the sheath, at least in the long normally developed filaments of *S. crustaceum*. The remaining stages would probably not be able to develop in the adverse conditions of dry soil. The long, motile trichomes - reminiscent of *Oscillatoria* - could be characterized as status *oscillarioideus*. Similar cases are reported in other species of *Tolypothrix* (Kondratjeva 1975). On the other hand, the sheath organization, as well as the remaining morphological characteristics of the stages *Diplocoleopsis* and *Diplocoleon,* show great similarities to *Tolypothrix elenkinii* f. *saccoideo-fruticulosa* Hollerb. In our opinion, the validity of the latter form, being identical to *S. crustaceum,* is debatable.

3.2. Perennating cells

3.2.1. Mucilaginous black coatings of *Cylindrospermum licheniforme* (Bory)Kütz. found on moist soil (Fig. 14a) were rapidly air dried in daylight for two hours. As a result the vegetative cells wrinkled, not germinating in cultures (Fig. 14b), while the resting cells took over the function of asexual reproduction, tiding the species over periods of adverse conditions (see also Glade 1914, Bristol 1919, Petersen 1935, Fogg et al. 1973). On the other

hand, a similar drying procedure on species characterized by absence of resting cells resulted in a different attitude of the vegetative cells: *Cyanothece cedrorum* (Sauv.) Komarek found in terrestrial habitats and dried for 7 years maintained the vitality of the vegetative cells (Fig. 15); while in the marine *Oscillatoria margaritifera* Kütz. ex Gom. hormogonia formation took place through the development of necridia after a drying procedure of 15 min. (Figs. 6-8).

3.2.2. Our observations on *Nostoc commune* Vauch. cover many terrestrial habitats of Greece. During the winter, the thalli are soft, expanded and olive green with mucilaginous, yellowish or almost colourless sheaths. With the passing of winter the thalli become harder and olive green with gelatinous, yellow-brown sheaths. In summer (t^o= 40^oC) the thalli of *N. commune* wrinkle and change to black with brownish-yellowish sheaths (Fig. 16); the vegetative cells of the trichomes progressively degenerate (Fig. 17) while the heterocysts remain (Fig. 18). After rain has soaked through the soil, the remaining living trichomes start developing.

Granting that resting cells rarely appear in *N. commune,* heterocyst germination takes over their function when only heterocysts have survived in a trichome (see also Bristol 1919; Singh, Tiwari 1970; Geitler 1942,1960). It is noted that the view that heterocysts regulate spore formation or take over the function of asexual reproduction has not entirely been accepted (see Fogg et al. 1973). The recent discovery of the specialized biochemical activities of the heterocysts indicates their possible function as that of nitrogen fixation, but there still remains a controversial area of research (Fay 1973, Fogg et al. 1973, Stewart 1973, Venkataraman 1975).

3.3. Decoration pattern of thecae
The majority of the diatoms found in different types of moist and dry soil of Greece are mostly

minute and widely distributed forms, mainly belonging to Pennales. The reduction in size observed in most of them, being in the lower limits of their type species may be special adaptation to withstand the extreme ecological
conditions (see also Kolbe 1932; Hustedt 1938;
Economou-Amilli 1976; Anagnostidis, Economou-
-Amilli 1978). Among the most common soil species were: *Hantzschia amphioxys* (Ehr.)Grun.,
Navicula mutica Kütz., *N. nivalis* Ehr. and *N.
nivaloides* Bock.

3.3.1. On dry soil of different biotopes investigated, individuals of *N. nivaloides* Bock
were abundant (Fig. 21), sometimes in coexistence with *N. nivalis* Ehr.(Fig. 20) and transitional forms. Both species exhibit a rather similar ecology - mostly in aerial biotopes - ,
while morphologically, the former is distinguished from *N. nivalis* by a reduction of the
sculptural details (Bock 1963, Hustedt 1961-66).
It is noted, however, that the loss of similar
sculptural elements in other species of diatoms
has been attributed to the astatic environment
(Drum 1964; Geissler 1970a, 1970b; Paasche et
al. 1975; Schultz 1971; Schmid 1976,1979). Thus,
it is questionable whether such forms, revealing this kind of morphological variation, must
be regarded as separate taxa.

On the other hand, the individuals of *Navicula mutica* Kütz. (Fig. 25), found also on dry
soil in different parts of Greece, correspond
to a form mainly known from terrestrial habitats and previously described as *N. imbricata*
Bock. The latter species is hardly distinguishable from *N. mutica*, the characteristic features of which are reduced in ecologically extreme biotopes (Bock 1963, Hustedt 1961-66).
However, the original type material of *N. mutica* from brackish waters has already proved to
be completely identical with *N. imbricata* from
terrestrial or river bank habitats (Lange-Bertalot, Bonik 1978).

3.3.2. *Hantzschia amphioxys* (Ehr.)Grun. is the
most widespread soil diatom able to survive strong
desiccation (Petersen 1935, Cholnoky 1968, Geissler, Gerloff 1966; Behre, Schwabe 1970; Mann
1977). The irregular arrangement of fibulae,repre-
senting a difference from the type, is the common
characteristic of most individuals of the dry habitats studied (Figs 19, 22). In our opinion
this morphological variation may be due to the
extreme environment. Similar irregular formation
of fibulae in the form of elongate puncta was
also observed in *H. amphioxys* and *H. trunctata*,
mainly from material of thermal biotopes (Oestrup
1928, thermal springs of Iceland; Guermeur 1954,
Senegal; Economou-Amilli 1976, thermal spring of
Greece; Benson, Rushforth 1975, Utah, U.S.A.). It
is noticed that according to Petersen (1935) *H.
amphioxys* comprises marked hydrophytic, as well
as euterrestrial forms, which would be most cor-
rectly regarded as separate taxa.

Concerning the distribution pattern of the
striae, similar teratologies were found in the
marine species *Synedra crystallina* (Ag.)Kr. and
Achnanthes brevipes Ag. (Figs 23-24, 26-29)
from rock pools of the supralittoral zone (Economou-Amilli 1980).

REFERENCES
  Anagnostidis K (1961) Untersuchungen über die
Cyanophyceen einiger Thermen in Griechenland, Inst.
Syst. Bot. & Pflanzengeogr. Univ. Thessaloniki 7,
1-322, 38 pls.
  Anagnostidis K (1968) Untersuchungen über die
Salz- und Süsswasser - Thiobiocönosen (Sulphure-
tum) Griechenlands, Wiss. Jb. Physik. - Math.
Fak. Univ. Thessaloniki 10, 406-866, 80 pls.
  Anagnostidis K and Economou-Amilli A (1978) Mi-
croorganisms from the volcano of Nea Kammeni
Island (Santorini), Second Intern. Sci. Congr.
Santorini, Thera and the Aegean World 1, 707 -
723, London.
  Anagnostidis K and Economou-Amilli A (1980)
Limnological studies on Lake Pamvotis (Ioannina),
Greece. I. Hydroclimatology, phytoplankton -
periphyton with special reference to the valen-
cy of some microorganisms from sulphureta as
bioindicators. With an appendix by J. Overbeck,
Arch. Hydrobiol. 89, 313-342.
  Anagnostidis K, Economou-Amilli A and Rousso-
moustakaki M (1982) Epilithic and Chasmolithic

Microflora (Cyanophyta, Bacillariophyta) from marbles of the Parthenon (Acropolis - Athens, Greece), Nova Hedwigia (in press).

Anagnostidis K and Golubič S (1966) Über die Ökologie einiger *Spirulina* - Arten, Nova Hedwigia 11, 309-335.

Behre K and Schwabe GH (1970) Auf Surtsey/Island im Sommer 1968 nachgewiesene nicht marine Algen, Schr. Naturw. Ver. Schlesw. - Holst, Sonderband Surtsey Island, 31-100, Kiel.

Benson CE and Rushforth SR (1975) The algal flora of Huntington Canyon Utah, U.S.A., Bibliotheca Phycologica 18, 1-177, 38 pls. J. Cramer, Vaduz.

Bock W (1963) Diatomeen extrem trockener Standorte, Nova Hedwigia 5, 199-254, 2 pls.

Bourrelly P (1970) Les algues d' eau douce. III. Les algues bleues et rouges, les Eugleniens, Peridiniens et Cryptomonadines, 512 pp. N. Boubée & Cie, Paris.

Bristol BM (1919) On the retention of vitality by algae from old stored soils, New Phytol. 18, 92-107.

Bristol BM (1920) On the alga-Flora of some desiccated English soils: An important factor in Soil Biology, Ann. Bot. 34, 36-79.

Bristol - Roach BM (1927) On the algae of some normal English soils, J. Agric. Sci. 17, 563-588.

Broady PA (1981a) The ecology of sublithic terrestrial algae at the Westfold Hills, Antarctica, Br. Phycol. J. 16, 231-240.

Broady PA (1981b) Ecological and taxonomic observations on subaerial epilithic algae from Princess Elizabeth Land and Mac. Robertson Land, Antarctica, Br. Phycol. J. 16, 257-266.

Brock TD (1975) Effect of water potential on a *Microcoleus* (Cyanophyceae) from a desert crust, J. Phycol. 2, 316-320.

Cameron RE (1963) Species of *Nostoc* Vaucher occurring in the Sonoran desert in Arizona, Trans. Amer. Microscop. Soc. 81, 379-384.

Cholnoky BJ (1968) Die Ökologie der Diatomeen in Binnengewässern, 699 pp., J. Cramer, Lehre.

Desikachary TV (1959) Cyanophyta, 686 pp., I.C.A.R. Monographs on Algae. New Delhi, Acad. Press, New York, London.

Drum RW (1964) Notes on Iowa diatoms. VI. Frustular aberrations in *Surirella ovalis*, J. Iowa Acad. Sci. 71, 51-55.

Economou-Amilli A (1976) On diatoms from thermal springs of Greece. Taxonomic, ecological, floristic and phytogeographical research, 1-238 p. 1-243 Figs + 19 Tabl. (Gr. with Engl. summ. & disc.), Athens.

Economou-Amilli A (1980) Marine diatoms from Greece I. Diatoms from the Saronikos Gulf, Nova Hedwigia 32, 63-104.

Economou-Amilli A and Anagnostidis K (1981) Periphyton algae and bacteria from springs of Levadia Boeotia, Greece, Hydrobiologia 80, 67-89.

Elenkin A (1949) Sinezelenije vodorosli SSSR (Monographia algarum Cyanophycearum aquidulcium et terrestrium in finibus URSS inventarum, pars specialis), Fasc. II, 985-1908. Akad. Nauk., Moskva, Leningrad.

Fay P (1973) The Heterocyst. In Carr NG and Whitton BA eds. The Biology of Blue-green Algae, Bot. Monogr. 9, 238-259. Blackwell Sci. Publ., Oxford, London, Edinburgh, Melbourne.

Fjerdingstad E (1971) Dimensions and taxonomy of Oscillatoriaceae I. *Oscillatoria limosa* Ag. and *Oscillatoria nitida* Schkorb. emend., Schweiz. Z. Hydrol. 33, 171-199.

Fjerdingstad E, Holma B and Fjerdingstad EJ (1976) The structure of *Oscillatoria limosa* Ag. (Cyanophycea) and the formation of hormogonia and necridia, Rev. Algol. N.S. 11, 261-272.

Fogg GE (1969) Survival of algae under adverse conditions. In: W. Woolhouse HW ed. Dormancy and Survival p. 123-142, 23rd Symp. Soc. exp. Biol. University Press, Cambridge.

Fogg GE, Stewart WDP, Fay P and Walsby AE (1973) The blue-green algae, 459 pp. Academic Press, New York.

Friedmann EI (1980) Endolithic microbial life of hot and cold deserts, Origins of Life 10, 223-235.

Friedmann EI and Ocampo R (1976) Endolithic blue-green algae in the Dry Valleys: Primary producers in the Antarctic desert ecosystem, Science 193, 1247-1249.

Fritsch FE (1922) The terrestrial alga. Journ. of Ecology 10, 220-226 (cit. in Petersen 1935).

Fritsch FE (1945) The structure and reproduction of the algae. Cambridge, Univ. Press, 2, 1-939.

Geissler U (1970a) Die Schalenmerkmale der Diatomeen. Ursachen ihrer Variabilität und Bedeutung für die Taxonomie. Nova Hedwigia, Beih. 31, 511-535.

Geissler U (1970b) Die Variabilität der Schalenmerkmale bei den Diatomeen. Nova Hedwigia, Beih. 19, 623-773.

Geissler U and Gerloff J (1966) Das Vorkommen von Diatomeen in menschlichen Organen und in der Luft, Nova Hedwigia 10, 565-577.

Geitler L (1942) Schizophyceae. In Engler A and Prantl K eds. Natürliche Pflanzenfamilien, 1b, 1-232, 2 Aufl. 1959 (Unveränd. Fotonachdr.). Dunker and Humbolt, Berlin.

Geitler L (1960) Schizophyceen. In Zimmermann W and Ozenda P eds. Encyclopedia of Plant Anatomy: Handbuch der Pflanzenanatomie, 2nd ed. part I, p. 1-131 Bornträger, Berlin, Germany.

Glade R (1914) Zur Kenntnis der Gattung *Cylindrospermum*. C. Beitr. Biol. Pfl. 12, 295-344 (cited in Petersen JB 1935).

Gollerbakh MM (1938) *Gloeocapsaceae* Elenk. et Hollerb. In Elenkin AA, Sineselenije vodorosli SSSR, Monographia algarum Cyanophycearum aquidulcium et terrestrium in finibus URSS inventarum, pars specialis, 165-261, Akad. Nauk, Moskva, Leningrad.

Gollerbakh MM, Kosinskaya EK and Polyanskij VI (1953) Sineselenije wodorosli. In Opredelitelj presnovodnich vodoroslej SSSR (Cyanophyceae. Bestimmungsbuch für die Algen der Binnegewässer der UdSSR), 2, 1-652, Sowjetskaya, Nauka, Moskwa.

Gollerbakh MM and Shtina ZA (1969) Potsvennije vodorosli (Soil Algae) 228 pp. Izdat. "Nauka", Leningrad Acad. Sci. Press.

Golubič S (1965) Zur Revision der Gattung *Gloeocapsa* Kützing (Cyanophyta). Schweiz. z. Hydrol. 27, 218-232.

Golubič S (1967a) Algenvegetation der Felsen, Binnengewässer 23, 183 pp.

Golubič S (1967b) Die litorale Algenvegetation des Titisees. Arch Hydrobiol. Suppl. 33, (Falkau - Arbeiten 6) 2, 172-205.

Guermeur P (1954) Diatomées de l' A.O.F. (Première liste. Sénégal). Institut Francais d' Afrique Noire. Catalogues XII. Ifan-Dakak, 1-137 p., 24 pls.

Hasle GR and Fryxell GA (1970) Diatoms: Cleaning and mounting for light and electron microscopy, Trans. Amer. Microsc. Soc. 89,4, 469-474.

Hustedt F (1938) Systematische und ökologische Untersuchungen über die Diatomeenflora von Java, Bali und Sumatra. II. Die Diatomeenflora der untersuchten Gewässertypen, Archiv für Hydrobiologie Supplement 16, 1-155.

Hustedt F (1961-1966) Die Kieselalgen. In Rabenhorst L Kryptogamenflora Deutschlands, Österreichs und der Schweiz, 7, 3, 816 pp., Akad. Verlagsgesellschaft, Leipzig. Repr. 1971, Johnson Reprint Corp., New York, London.

Jaag O (1945) Untersuchungen über die Vegetation und Biologie der Algen des nackten Gesteins in den Alpen im Jura und im schweizerischen Mittelland, Beitr. Kryptog. Flora Schweiz 9, 1-560.

Kann E (1963) Ökologische Untersuchungen des eulitoralen Algenaufwuchses im Lago Maggiore, Lago di Mergozzo und Lago d' Orta. Mem. Ist. Ital. Idrobiol., 16, 153-187.

Kann E (1978) Systematik und Ökologie der Algen Österreichischer Bergbäche, Arch. Hydrobiol./Suppl. 53,4, 405-643.

Kolbe RW (1932) Grundlinien einer allgemeinen Ökologie der Diatomeen, Ergebnisse der Biologie 8, 221-348, Berlin.

Komárek J (1978) Oddelenie Cyanophyta - Sinice. In Hindák F Slovenske Sladkovodne Riasy p. 238-283,Pedagogicke Nakladatel'stvo, Bratislava.

Kondratjeva NV (1968) Viznachnik Prisnovodnikh Vodorostej Ukrayinskoyi RSR I. Sinjo-zeleni Vodorosti - Cyanophyta, 2. Hormogoniophyceae, 523 pp., Akad. Nauk. Ukrayinskoyi RSR, "Naukova Dumka" Kiev.

Kondratjeva NV (1975) Morfogenez i osnovonije puti jevoliutsii gormogonijevikh vodoroslej (otdel Cyanophyta, klass Hormogoniophyceae) (Morphogenesis and main ways of hormogon algae evolution) 302 pp, Izdat. "Naukova Dumka", Kiev.

Lange-Bertalot H and Bonik K (1978) Zur systematisch - taxonomischen Revision des ökologisch interessanten Formenkreises um *Navicula*

mutica Kützing, Bot. Mar. 21, 31-37.

Lemmermann E (1910) Algen. I. In Kryptogamenflora der Mark Brandenburg, Bd. 3, 712 pp., Schizophyceae, 1-256, Leipzig.

Mac Entee FJ, Schreckenberg SG and Bold HC (1972) Some observations on the distribution of edaphic algae, Soil Sci. 114, 171-179.

Mann DG (1977) The diatom genus *Hantzschia* Grunow - An appraisal, Nova Hedwigia, Beih. 54, 323-354.

Novichkova - Ivanova LN (1980) Potsvennije vodorosli. Fitotsenozov Saharo - Gobinskoi pustinnoi oblasti. (Soil algae of Sahara - Gobi desert region) Akad. Nauk SSSR, (Bot. Inst. "Nauka") 256 pp., Leningrad.

Oestrup E (1918) Fresh-water Diatoms from Iceland. In Rosenvinge and Warming Botany of Iceland 2,1, nr. 5, 1-96, 5 pls. (Arb. Bot. Have København, nr. 86).

Paasche E, Johansson S and Evensen DL (1975) An effect of osmotic pressure on the valve morphology of the diatom *Skeletonema subsalsum* (A. Cleve) Bethge, Phycologia 14, 205-211.

Patrick RM and Reimer CW (1966) The diatoms of the United States, exclusive of Alaska und Hawaii. Vol. I. Ac. Nat. Sci. Philad. Monogr. 13, 1-688 + 64 pls.

Petersen JB (1935) Studies on the biology and taxonomy of soil algae, Dansk. Bot. Ark. 8,1-183.

Potts M and Friedmann EI (1981) Effects of water stress on Cryptoendolithic Cyanobacteria from hot desert rocks, Arch. Microbiol. 130, 267-271.

Sampaio J (1959) Cianofitas da flora Portuguesa, Publ. Inst. Bot. Dr. Gonçalo Sampaio Fac. ciênc. Univ. Porto, sér. 2, N. 44, 5-16 + 8 pls.

Schmid AM (1976) Morphologische und physiologische Untersuchungen an Diatomeen des Neusiedler Sees: II. Licht - und rasterelektronenmikroskopische Schalenanalyse der umweltabhängigen Zyklomorphose von *Anomoeoneis sphaerophora* (Kr.) Pfitzer, Nova Hedwigia 28, 309-351.

Schmid AM (1979) Influence of environmental factors on the development of the valve of diatoms, Protoplasma 99, 99-115.

Schultz M (1971) Salinity - related polymorphism in the brackish - water diatom *Cyclotella cryptica*, Canad. J. Bot. 49, 1285-1289.

Shtina EA (1961) Zonality in the distribution of soil algae communities, International Congr. Soil 7 (Madison) 2, 630-634.

Singh RN and Tiwari DN (1970) Frequent heterocyst germination in the blue-green alga *Gloeotrichia ghosei* Singh, J. Phycol. 6, 172-176.

Skuja H (1964) Grundzüge der Algenflora und Algenvegetation der Fjeldgegenden und Abisko in Schwedish - Lappland, Nova Acta Reg. Soc. Sci. Upsal. ser. 4. 18,3, 1-465. Uppsala.

Stanier YR, Kunisawa R, Mandel M and Cohn-Bazire G (1971) Purification and properties of unicellular blue-green algae (Order Chroococcales), Bacteriol. Rev. 35, 171-205.

Starks TL, Shubert LE and Trainor FR (1981) Phycological Reviews 6. Ecology of soil algae: A review, Phycologia 20, 65-80.

Starmach K (1962) Glony zyiace na sciezkach w

110

nadrzecznych wierzbinach (Algae found growing
in foot paths in willow thickets near the
Mszanka and Raba rivers, Polish Western Carpathians), Fragm. Florist. et Geobot. 8, 81-88.

Starmach K (1966) Cyanophyta - Glaucophyta
(sinice - Glaucofity). In Flora Slodkowodna
Polski 2, 1-807. Polska Akad. Nauk. Warszawa.

Starmach K (1975) Glony w Wawozie Szopczań-
skim w Pieninach. (Algae in the Szopczański
Gorge in the Pieniny Mts.),Fragm. Florist. et
Geobot. 21, 537-549.

Starmach K (1981) Algen, besonders *Cyanophyceae* im Tal Dolina Bialego (Tatragebirge),
Fragm. Florist. et Geobot. 27, 275-317.

Staub R (1961) Ernährungsphysiologisch - aut-
ökologische Untersuchungen an der plankti-
schen Blaualge *Oscillatoria rubescens* DC.
Schweiz. Z. Hydrol. 23, 82-199.

Stewart WDP (1973) Nitrogen fixation. In
Carr NG and Whitton BA eds. The Biology of
Blue-green algae, Bot. Monogr. Vol. 9, 260-278,
Blackwell Sci. Publ., Oxford, London, Edinburgh,
Melbourne.

Venkataraman GS (1975) The role of blue-
green algae in tropical rice cultivation. In
Stewart WDP ed. Nitrogen fixation by free-
living micro-organisms p. 207-218. Cambridge
University Press. Cambridge, London, New York,
Melbourne.

Explanation of plates

Figs 1-2. *Oscillatoria limosa* Ag. Fig. 1:
Trichomes; note the absence of sheaths. Fig.2:
status *phormidioides*; sheaths coated with particles of soil.

Figs 3-5. *Gloeocapsa montana* Kütz. em. Hollerb. Fig. 3: status *polydermaticus*. Fig. 4:
Colonies with barely visible sheath stratification. Fig. 5: The same after treatment with
Alcian blue.

Figs 6-8. *Oscillatoria margaritifera* Kütz.
ex Gom. Fig. 6: Isolated trichome. Fig. 7:
Formation of hormogones at the start of desiccation. Fig. 8: Fragmentation of trichomes
after prolonged desiccation.

Scale: Figs 1-5=20 µm; Figs 6-8=30 µm.

Figs 9-13. *Scytonema crustaceum* Ag. Fig. 9:
Thalli. Fig. 10: Filaments with intercalary and
basal heterocysts. Fig. 11: Filament with an intercalary heterocyst (arrow). Fig. 12: status
oscillatorioides; note the long trichomes with
cylindrical cells in the form of hormogones reminiscent of *Oscillatoria*. Fig. 13a,c: Short
trichomes with basal heterocysts and bag-shaped
sheaths corresponding to stage *Diplocoleopsis*
and also to *Tolypothrix elenkinii* f. *saccoideo -
fruticulosa* Hollerb. Fig. 13b: Isolated trichome.

Scale: Fig. 9=2cm; Fig. 10=20 µm; Figs 11-13=
10 µm

Fig. 14. *Cylindrospermum licheniforme* (Bory)
Kütz. Fig. 14a: Isolated trichome. Fig. 14b: The
same after desiccation.

Fig. 15. *Cyanothece cedrorum* (Sauv.) Komarek.

Figs 16-18. *Nostoc commune* Vauch. Fig. 16:
Thalli (arrows). Fig. 17: Colony; at arrows dege-
nerating trichomes due to incipient desiccation.
Fig. 18: Colony preserving only heterocysts
(arrows): summer collection.

Scale: Figs 14-15=10 µm; Fig. 16=5cm; Figs 17-
18=20 µm

Figs 19-22. *Hantzschia amphioxys* (Ehr.) Grun.
Fig. 19: valve with normal formation of fibulae.
Fig. 22: Irregular arrangement of fibulae in the
form of elongated punctae at arrow.

Fig. 20. *Navicula nivalis* Ehr.

Fig. 21. *Navicula nivaloides* Bock.

Figs 23-24. *Achnanthes brevipes* Ag. pseudoraphe
valve. Fig. 23: Teratological form with irregularly arranged striae. Fig. 24: normal individual.

Fig. 25. *Navicula mutica* Kütz.

Figs 26-29. *Synedra crystallina* (Ag.) Kütz.,
valve view. Figs 26,27,29: with irregularly arranged striae. Figs 27-29: SEM.

Scale: 10 µm.

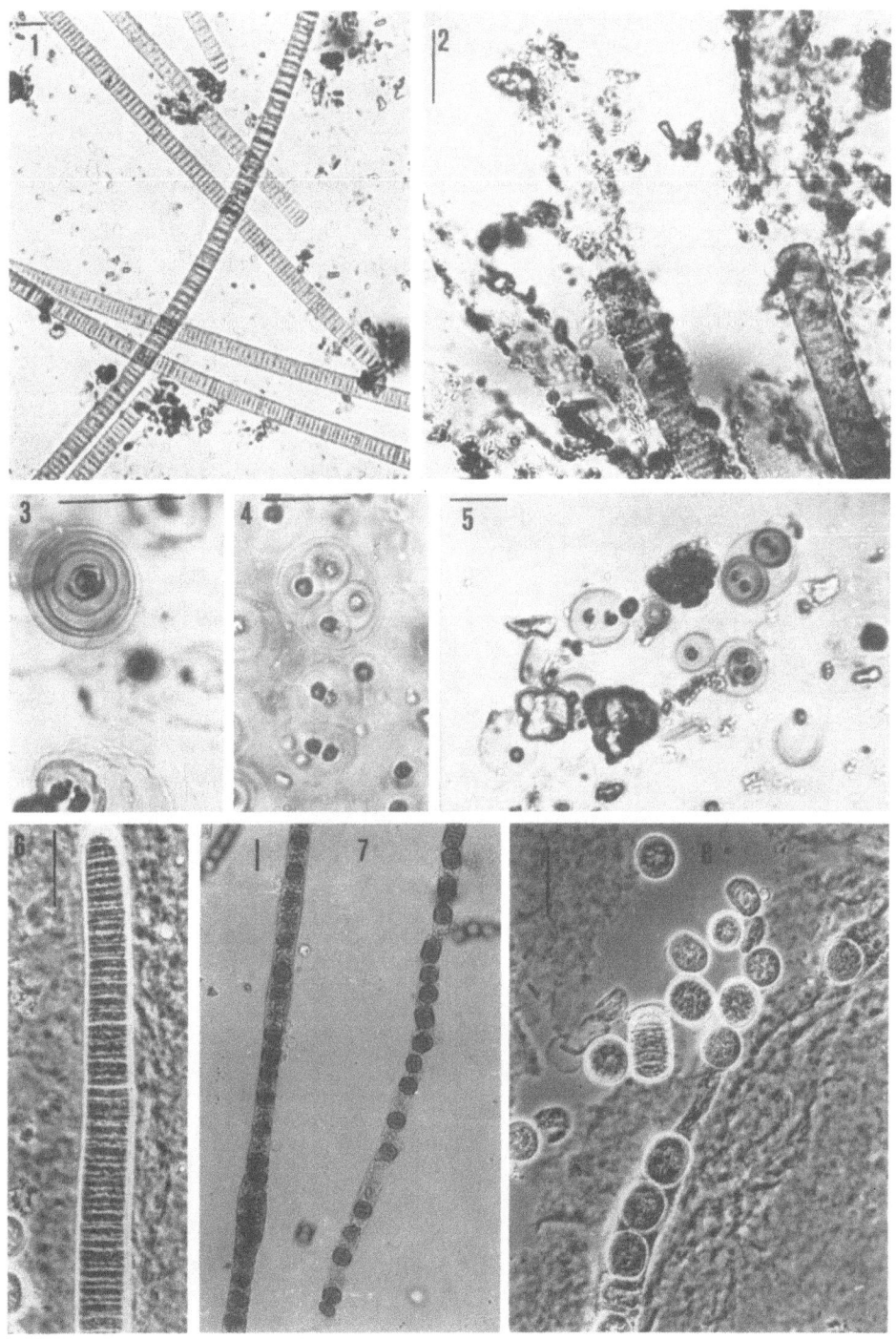

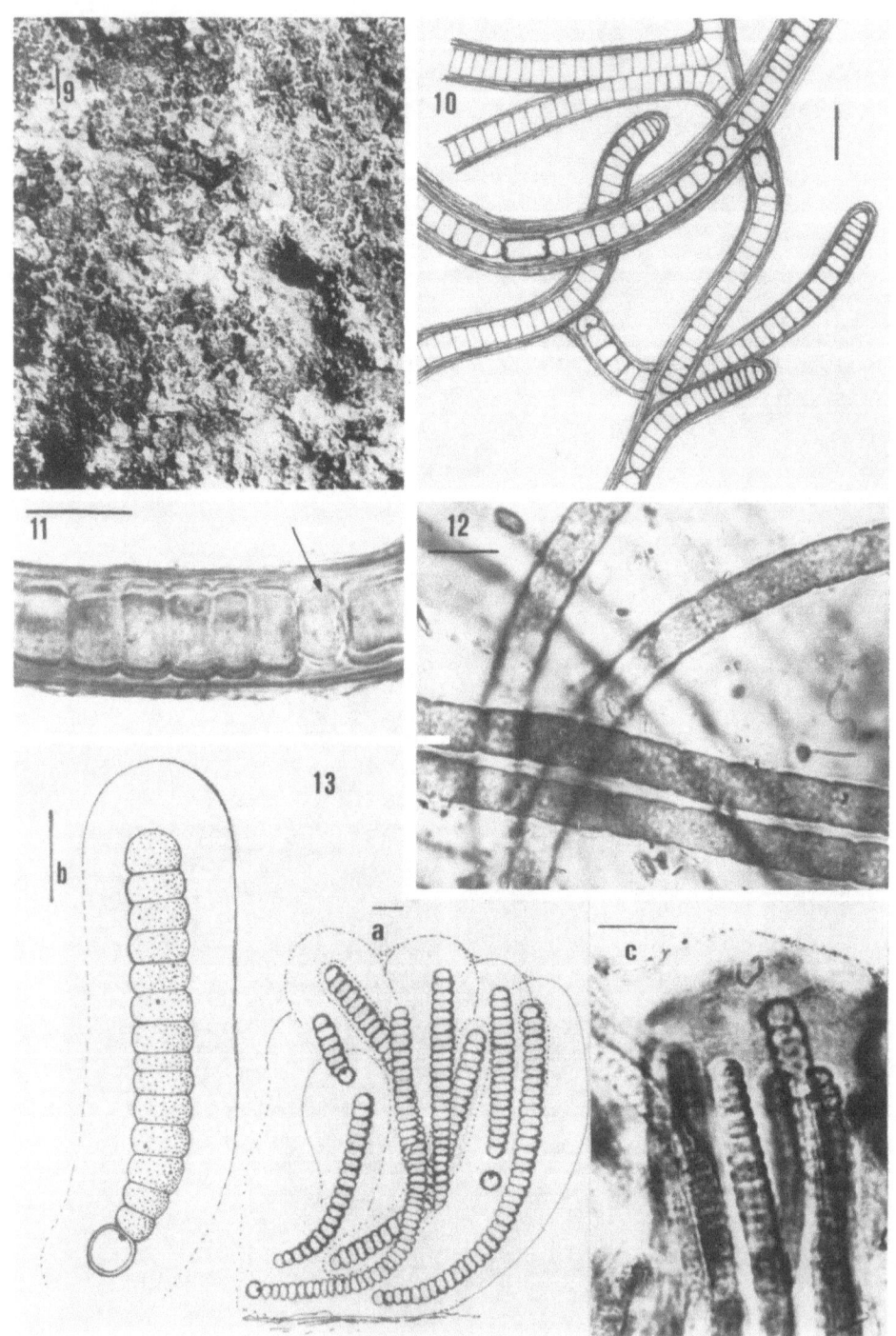

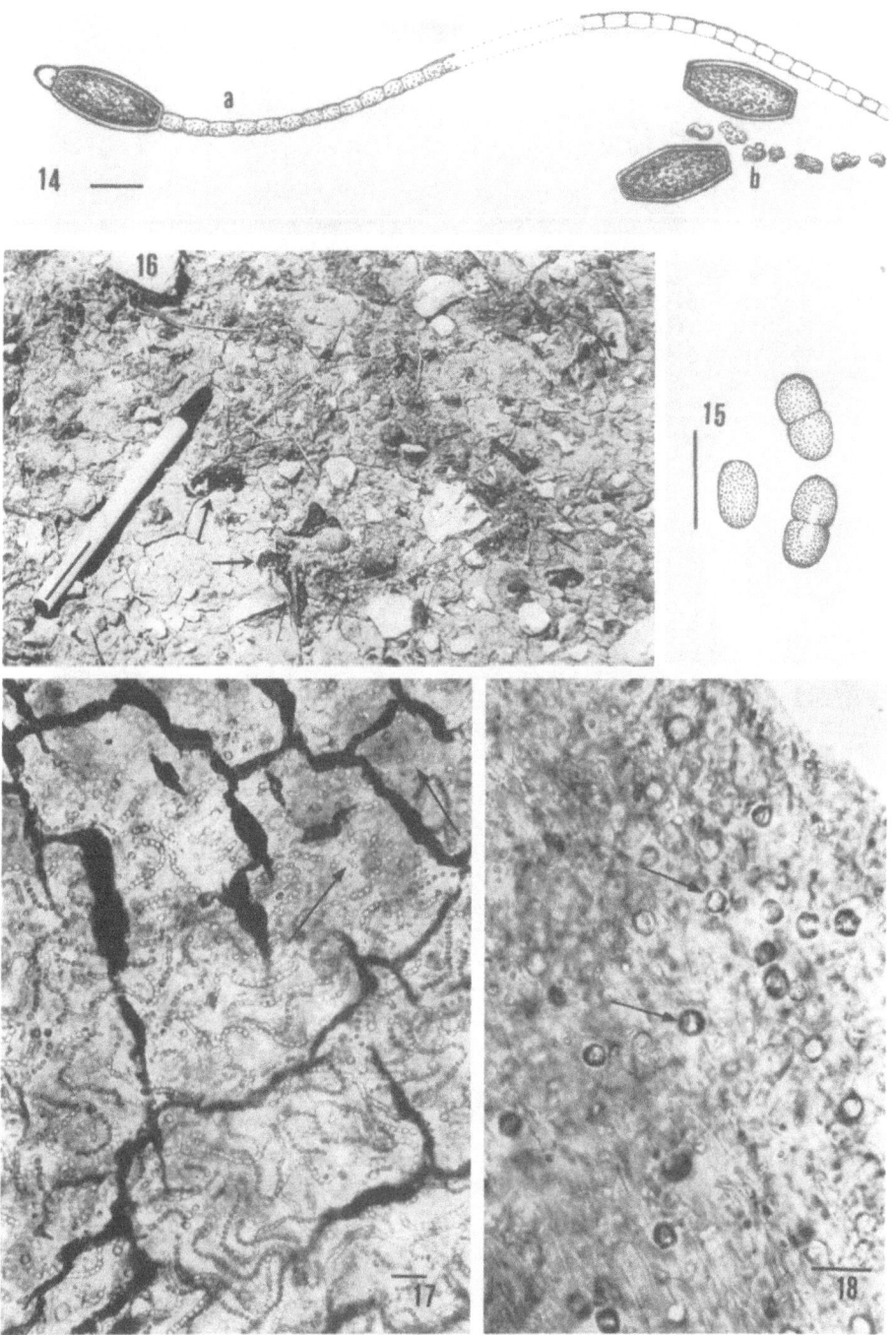

114

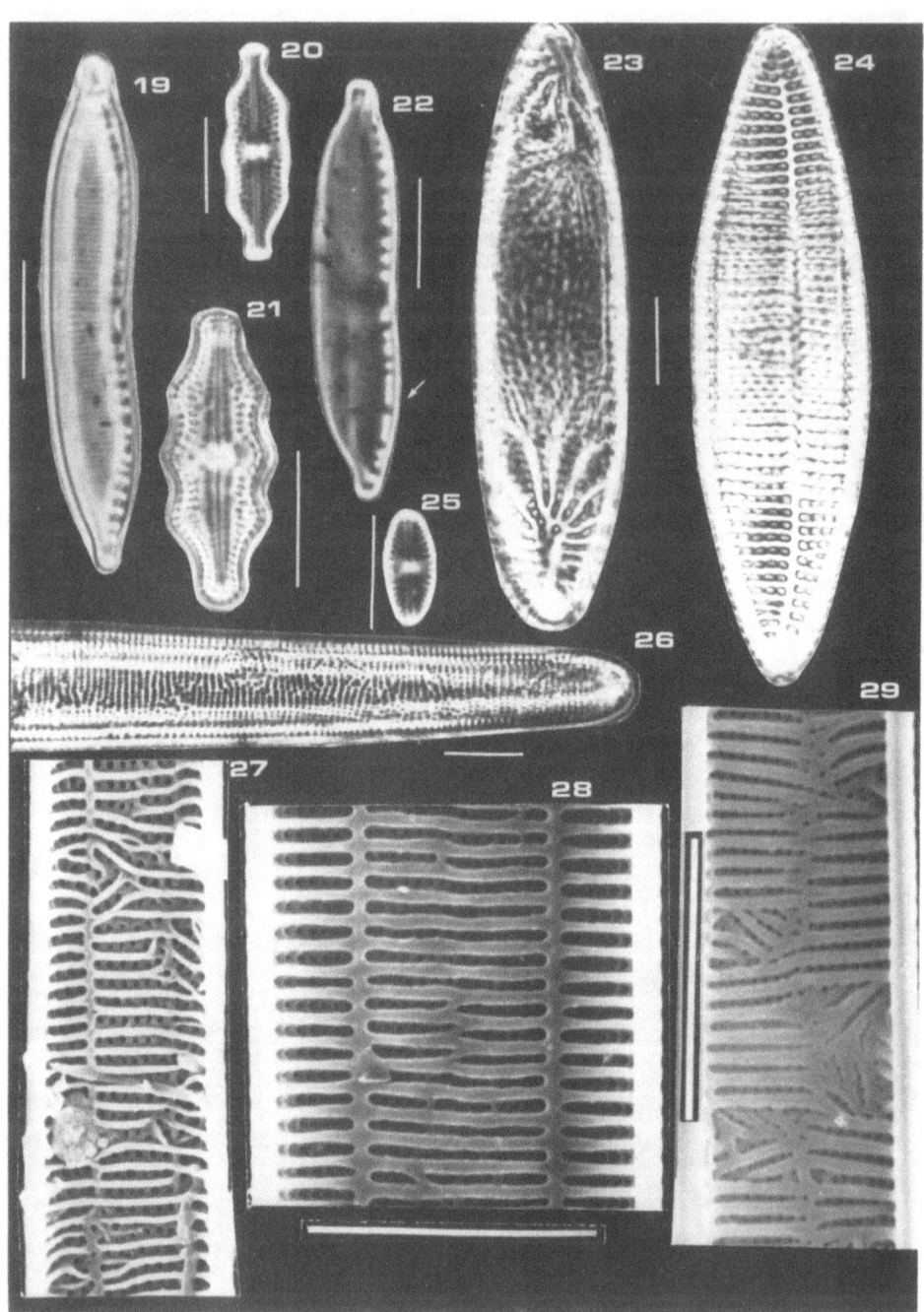

PART III

ADAPTATIONS TO TEMPERATURE

SPECIES-SPECIFIC RESPONSES TO TEMPERATURE IN ACID METABOLISM AND GAS EXCHANGE PERFORMANCE OF MACARONESIAN SEMPERVIVOIDEAE

RAINER LÖSCH

Botanisches Institut, Olshausenstr.40-60, D 2300 Kiel, FRG

1. INTRODUCTION

In the Macaronesian archipelagoes speciation in the course of an adaptive radiation has occurred since Tertiary times within the genus *Sempervivum* s.l.(Lems, 1960). As a result, nowadays about 70 endemic species can be found in the Canary Islands, Madeira, and the Cape Verde Islands, classified into the genera *Aeonium, Aichryson, Greenovia, Monanthes* (Praeger, 1932/67). These can be found in various habitats ranging from Mediterranean-type environments of the lowland zones through crevices of rocks within laurel forests up to rocky places of the subalpine *Pinus canariensis* forests and reaching an altitudinal distribution border at about 2400 m. Many species are restricted to only one island.

Floristics and chorology of the group are well documented (e.g.Bramwell,Bramwell, 1974; Voggenreiter, 1972), and the sociology of the *Aeonium* and *Greenovia* species growing in Tenerife has been studied in detail (Voggenreiter, 1974). Descriptions of their habitat situations are given by Burchard (1929).

In order to quantify the physiological peculiarities leading to the ecological differentiation of the species in the course of their evolution, field and laboratory studies of their production and stress biology were undertaken. A first survey of their resistance to heat and cold stress is given elsewhere (Lösch, Kappen, 1981; 1983). Here for selected members of the alliance the temperature-dependent and species-specific ability to perform either crassulacean acid metabolism (CAM) or C_3 photosynthesis shall be documented. Data about the CO_2 gas exchange are focused on the *Aichryson* branch of the phylogenetic tree of the alliance (Lems, 1960; Voggenreiter, 1974). A comprehensive treatment of the gas exchange behavior of all species will be published elsewhere (Lösch, in preparation).

2. METHODS, PLANTS, AND LOCATIONS

Daily courses of habitat microclimate as well as leaf water content and titratable acidity of various Semperviva were measured in various places at Tenerife in spring. These locations with their investigated species are as followed:

(1) A sun-exposed Mediterranean-type succulent shrub area at the Montana de Taco near Los Silos/nw-Tenerife, 50 m above sea level: *Aeonium haworthii, Ae. urbicum.*

(2) West-facing walls densely covered by *Aeonium holochrysum* and *Ae. urbicum* within the urban area of Puerto de la Cruz,50 m a.s.l.

(3) A nw-facing ravine near Icod de los Vinos, 200 m a.s.l.:*Aeonium tabulaeforme, Ae. holochrysum.*

(4) A south-facing mountain slope near El Bailadero,Anaga mountains, 800 m a.s.l.: *Aeonium ciliatum, Ae. canariense, Ae. cuneatum, Monanthes laxiflora.*

(5) A north-exposed barranco (=a broad ra- - vine) close to Agua Mansa (central ridge), 1200 m a.s.l.: *Greenovia aurea, Aichryson laxum, Ai.punctatum, Aeonium spathulatum, Monanthes pachycaulon.*

Leaf sap acidity was measured according to Lüttge, Ball (1974). An aliquot of the sap was added to a phosphate buffer solution of pH 7.8 and titrated back with 0.01 n KOH (portable pH-meter WTW pH-digi 88; field balance of 0.05 g exactness (Kern); field burette (Bürkle)). Independently of the titration the pH of the undiluted sap was measured immediately after grinding the leaves. Parallel to the acidity determinations habitat microclimate was assessed in proper intervals from sunrise to sunset. A digital thermometre (Tastotherm D 700) with a NiCr-NiAl-thermocouple served for the leaf temperature measurements. Investigations during the night were done where this was technically possible, otherways diurnal courses of deacidification were assessed.

The diurnal courses of temperature and - approximately - humidity and light intensity which were measured at the natural habitats were simulated later in the laboratory in Kiel/Germany in a climatized chamber (BBC York), and the acidity measurements were repeated under these controlled conditions. Species growing naturally under the tested conditions and species not occurring under the respective climate were used. All plants were kept within the climatized chamber under the respective temperature regime one week prior to the measurement day.

Another set of experiments dealt with the expression of CAM or C_3 behavior in the gas exchange performance under various temperature conditions: CO_2 and H_2O gas exchange were measured in well watered plants preconditioned in a climatized chamber with a 12:12 hour light rhythm and 18°C during the light phase and 12°C during the dark period. For measurements

a twig of them was enclosed into a fully conditioned gas exchange chamber (Walz) installed in a walk-in climatized chamber (BBC) which was conditioned to the measuring temperatures. The CO_2 turnover was measured in the open air stream by an IRGA (BINOS 2,Leybold-Heraeus), the transpiration by humidity control with dewpoint mirrors and trapping the transpired water with a water vapor trap Light (500-550 $\mu E \ m^{-2} s^{-1}$) was provided by two HQIL lamps at a 12:12 hour rhythm, and the photosynthetically active radiation (PAR) received within the chamber was measured with a LiCor quantum sensor. Temperature during the dark phase was constantly kept at 10°C. Light phase temperatures were 10, 15, 20, 25, and 30 °C, respectively, lasting constantly during the 12 hours of each light phase. Dew point temperature was kept at 7°C

3. RESULTS

Under natural conditions in spring *Aeonium holochrysum*, *Ae.haworthii*, and *Ae. urbicum* show daily fluctuations in acidity of more than 200 $\mu eq \ H^+ g^{-1}fw$ (equivalent to 3 $meq \ H^+ g^{-1}dw$ in *Ae. holochrysum* and about 5 $meq \ H^+ g^{-1}dw$ in *Ae. haworthii* and *Ae.urbicum*). Fig.1 shows a 36 h run of such acidity fluctuations in *Ae. holochrysum* that can be taken as representative for the species investigated at the lowland sites. In Fig.2 the results of measurements at the two other lowland sites are depicted. The daily deacidification becomes most intensive when leaf temperatures raise drastically after exposure to direct sunlight. It ceases already in the late afternoon hours giving way to new acid accumulation, when the period of direct insolation exceeds about four hours. In ravine habitats where plants are exposed to the sun for shorter times only (like at the place near Icod, Fig.2, left hand side) a drop in pH of the leaf sap and an increase of the titratable

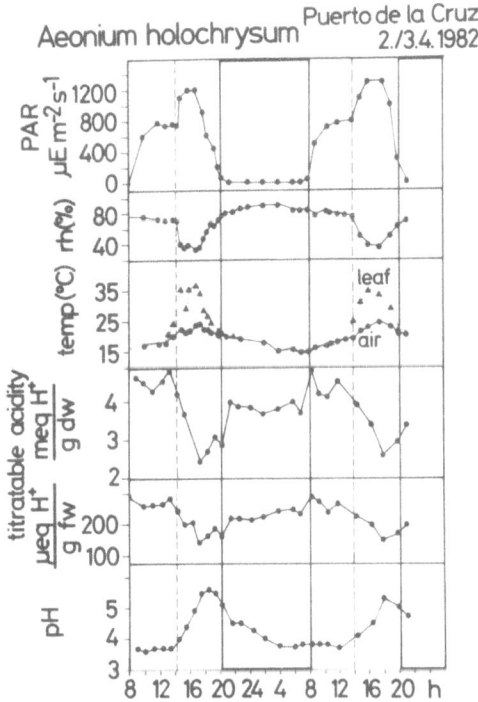

FIGURE 1. Diurnal courses of pH, titratable acidity, and micrometeorological parameters in *Aeonium holochrysum*, location 2. The beginning of direct insolation of the rosettes is indicated by a dotted line, the night period by a black bar.

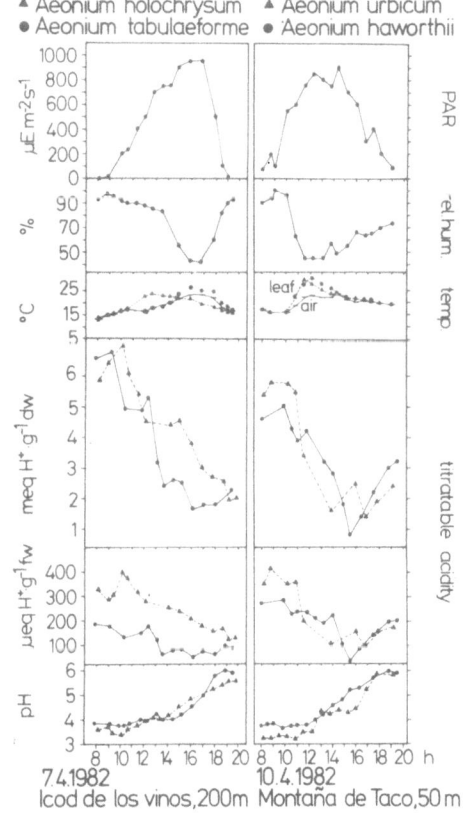

FIGURE 2. As Fig.1, but for the indicated species growing at locations 1 resp. 3.

acidity occurs only at sunset. The acidity fluctuations of *Aeonium tabulaeforme*, which colonizes characteristically shady slopes of rocks, amount to 150 µeq H$^+$ g^{-1}fw. This lower value compared with the ones of the other species is a consequence of the enormous water content of *Ae. tabulaeforme* (about 2700 % of the dry weight compared with 1500 - 1700 % in *Ae. holochrysum*, *Ae. haworthii* and *Ae.urbicum*). No difference in acidity fluctuations can be found between the lowland species, if the titratable acidity is related to dry weight.

The described pattern indicative for CAM was measured also in *Ae. ciliatum* at the laurel forest investigation site in the Anaga mountains (Fig.3). *Monanthes laxiflora*, *Ae. canariense* and *Ae. cuneatum* showed acidity fluctuations of less than 100 µeq H$^+$ g^{-1}fw

(or 1 - 2 meq H$^+$ g^{-1}dw). Probably, CAM mechanism can work in these species, but it does not seem to be very essential for them at this habitat in spring.

Nearly no fluctuations of titratable acidity can be found in species (*Aichryson laxum*, *Ai. punctatum*, *Greenovia aurea*, *Aeonium spathulatum*, *Monanthes pachycaulon*) growing at the Agua Mansa site **chosen** as to be representative for ravines within the montane to subalpine pine forest belt (Fig.4).

The microclimatic situations of the habitats at lower elevations with plants performing CAM and of the barranco site at 1200m with species showing no acidity fluctuations differ especially in the temperature conditions. Near sea level in spring they do not fall short of 14°C in the succulent *Aeonium*

120

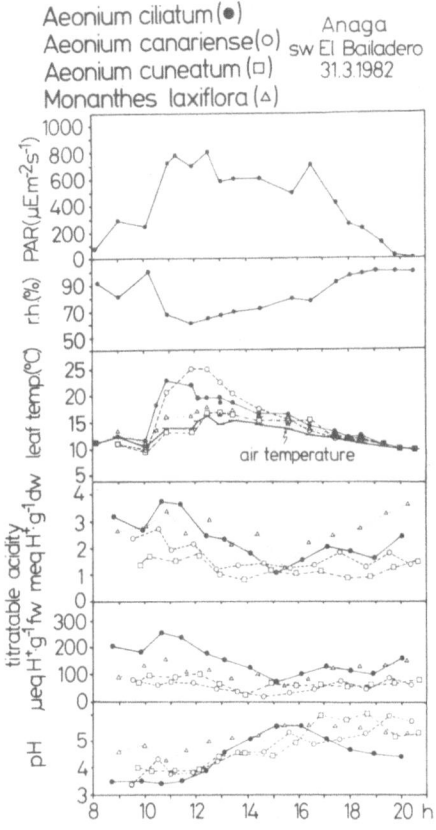

Aeonium ciliatum (●)
Aeonium canariense (○)   Anaga
Aeonium cuneatum (□)   sw El Bailadero
Monanthes laxiflora (△)   31.3.1982

FIGURE 3. As Fig.1, but for the indicated species growing at location 4.

leaves being subjected to daily temperature fluctuations of up to 16 C. At the n-exposed mountain site big *Greenovia*-leaves heated by sun reached diurnal temperature spans of 11 C. The nocturnal minimum temperature at this site was 5°C, however. The environmental conditions of the plants at 800 m a.s.l. range between these extremes. At this site under identical climatic conditions a species-specific assortment of acidity metabolism can be found.

Whether or not the Crassulacean pathway of photosynthesis is realized may depend therefore either on the environmental conditions or on species-specific peculiarities.

Acidity fluctuations of greenhouse-cultivated Canarian Semperviva followed the same patterns as in the respective natural habitats, when subjected in a climatized chamber to an appropriate daily course of temperature light, and humidity conditions. The proton concentrations are not totally identical with those measured in the field under similar circumstances. This is probably due to differences in the water relations the well watered potted plants having higher water contents in general. It also was impossible to simulate the high radiation yields of the Mediterranean-type natural stands with the artificial light available (up to 300 $\mu E\ m^{-2}s^{-1}$). Nevertheless the principal patterns of the diurnal courses of titratable acidity could be compared (Fig.5). *Aeonium ciliatum* underwent significant acidity changes under the simulated climate representative for the southern slope of the Anaga mountains. With the cold conditions prevailing on the north slopes of the pine forests its acidity re-

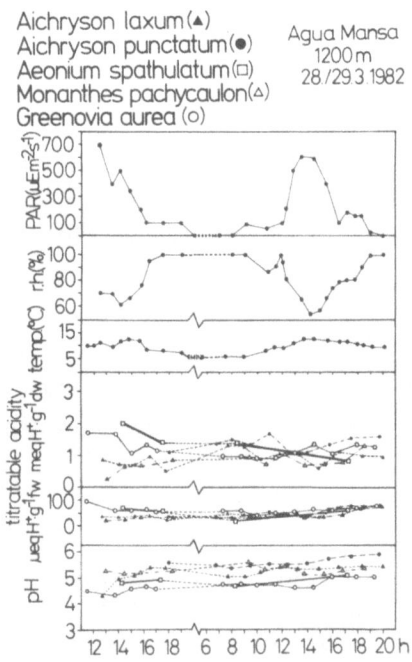

Aichryson laxum (▲)
Aichryson punctatum (●)   Agua Mansa
Aeonium spathulatum (□)   1200 m
Monanthes pachycaulon (△)   28./29.3.1982
Greenovia aurea (○)

FIGURE 4. As Fig.1, but for the indicated species growing at location 5.

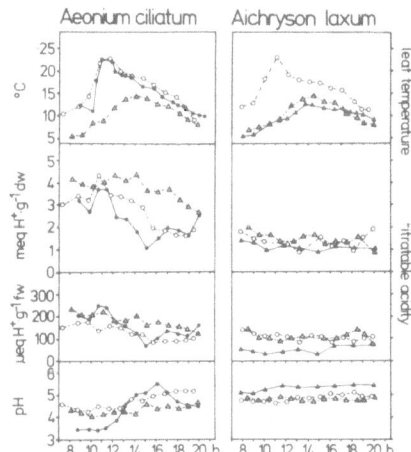

FIGURE 5. Diurnal courses of pH, titratable acidity, and leaf temperatures of *Aeonium ciliatum* and *Aichryson laxum* as measured at their natural habitats (● ; ▲) and under simulated conditions within a climatized chamber. ○ : Simulation of the temperature conditions of location 4 - Anaga Mountains, the growing place of *Ae. ciliatum* - and the corresponding plant parameters. △ : Simulation of the temperature conditions of location 5 - Agua Mansa, the growing place of *Ai. laxum* - and the corresponding plant parameters.

mained constantly high for half the light period, but levelled down finally to rather low proton contents. Well watered *Aichryson laxum*, contrarily, did not change to a typical CAM rhythm even with the warmer and more fluctuating temperature regime prevailing in the habitat of *Ae. ciliatum*. Under the various spring temperature regimes of Tinerfenian mountain habitats well watered *Aichryson laxum* apparently does not make use of the acid metabolism (According to unpublished investigations it can be shifted, however, to do so by drought conditions). *Aeonium ciliatum*, on the other hand, always behaves as an obligate CAM plant.

The degree of species-specific tendency to CAM was tested by gas exchange measurements on the species of the genus *Aichryson* (Fig.6). This genus is considered as one branch within the phylogenetic tree of the southern Semperviva, next allied to the *Aeonium* section Goochia. Both originate probably from ancestors whose type nowadays is represented by *Aeonium* species of the section Holochrysa. In the two relatively arid eastern Canary islands, Fuerteventura and Lanzarote, two shrubby species occur, *Ai. tortuosum* and *Ai. bethencourtianum*. The other ten members of the genus are herbaceous annual to triennial plants and grow in more or less humid habitats between 300 and 1300 m at the western Canary islands and in Madeira. Most probably *Ai. laxum* and *Ai. punctatum* occurring on all five of the western Canaries are the basic types from which the other forms can be derived.

When subjected to a program of different temperature regimes with constantly 10 °C at night, and with day-time temperatures of 10, 15,20,25, and 30 °C, respectively, the gas exchange behavior of these species varies considerably. Common to all species is the tendency to increase nocturnal CO_2 uptake or at least to decrease nocturnal CO_2 loss when temperatures are higher during day-time. But, while dark CO_2 fixation occurs in *Aeonium* species of the basal section Holochrysa (example in Fig.5: *Ae. holochrysum*) even at 10 °C, this is not the case in any of the *Aichryson* .

Members of two evolutionary lines within the genus *Aichryson* utilize the CAM mechanism effectively at higher temperatures. The one group are the shrubby *Aichryson* species restricted to the arid eastern islands. Generally being of low productivity a reasonable amount of their total carbon gain occurs by nocturnal CO_2 uptake. At higher temperatures the gas exchange pattern of *Ai. tortuosum* describes that of a full CAM plant (Kluge,Ting, 1978). The *Ai. punctatum* group realizes also CAM at higher temperatures, but with a much higher overall productivity. Along this developmental axis a pronounced tendency has

122

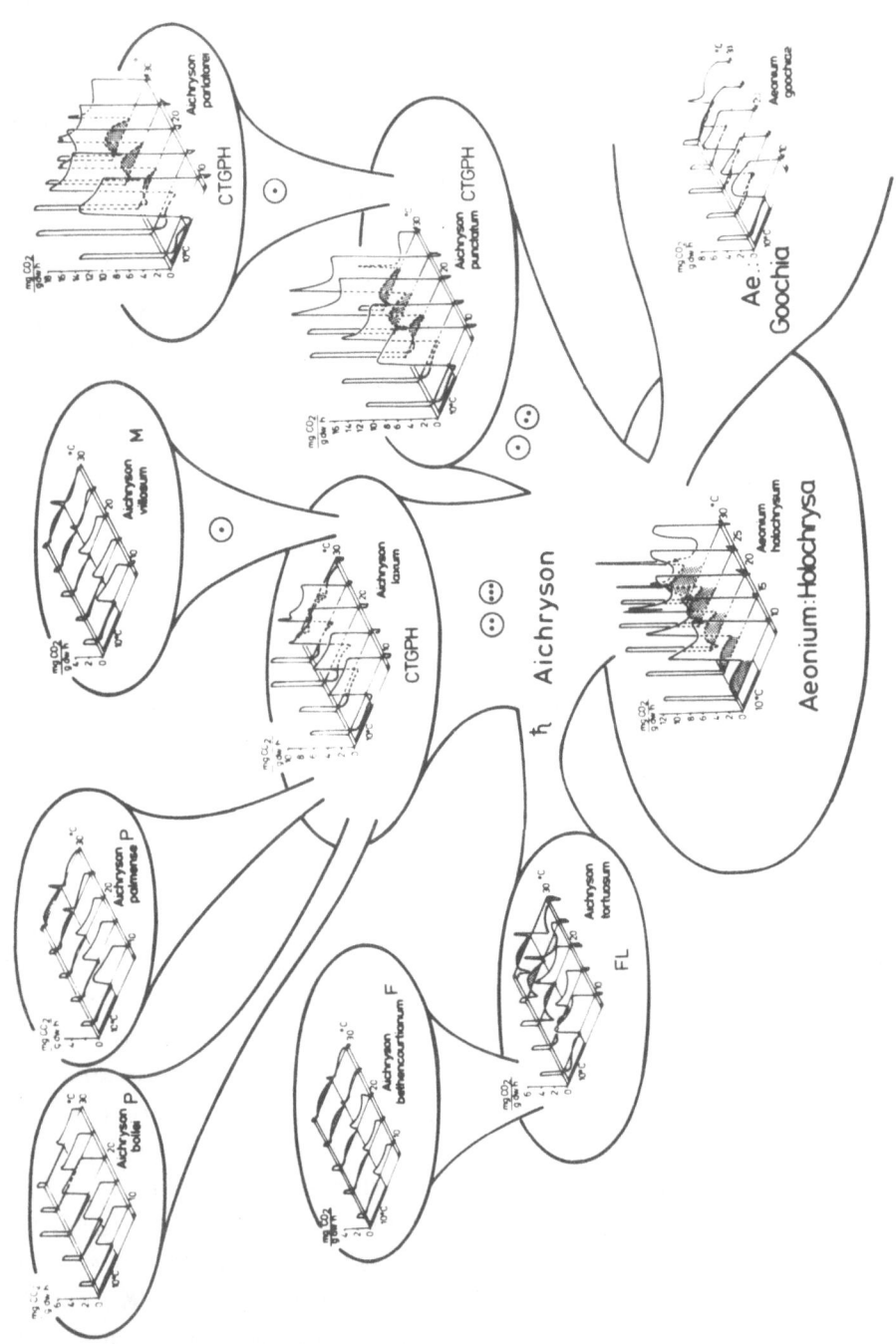

FIGURE 6. Phylogenetic tree of the genus *Aichryson* (investigated species) with the diurnal courses of net CO_2 gas exchange under different light phase temperatures inserted.

Initials indicate species distribution: C : Gran Canaria, T : Tenerife, G : Gomera, P : La Palma, H : Hierro, L : Lanzarote, F : Fuerteventura, M : Madeira. Cross-hatched areas: dark CO_2 uptake; vertically hatched areas: CO_2 loss .

developed to shorten the ontogenetic life cycles, so that finally *Ai. parlatorei* is a small, short-lived, but fast growing plant with the abundance of individuals of weedy annuals.

The CAM pathway has lost much of its importance in the evolution of *Ai. laxum* and the species that can be related to it for morphological reasons. In well watered *Ai. laxum* and *Ai. bollei* plants it is not possible to give rise to dark CO_2 uptake even with temperature regimes most favorable for CAM behavior in other species. That the mechanism has not been lost totally in this developmental line can be seen when these plants are subjected to drought stress. Even with well watered individuals the ability to perform CAM is realized by *Ai. palmense* and *Ai. villosum*. Morphologically being closely related to *Ai. laxum* they utilize nocturnal CO_2 fixation at higher temperatures in order to meet part of their carbon uptake. In general, the three investigated *Aichryson* species that are supposed to be derived from *Ai. laxum* show lower productivities than the more widespread species. The latter again ranges intermediately between the low productive shrubby species and the high productive *Ai. punctatum* branch.

The differences within the genus *Aichryson* in daily CO_2 uptake - expressed on a leaf dry weight basis - and the temperature optima of their productivity are compared in Fig.7. The carbon gain due to nocturnal CO_2 fixation occurring - if realized - at 20 °C and higher temperatures is relatively uniform among the *Aichryson* species. But in all of them and also in *Aeonium holochrysum* (that may represent the ancestor type) the overwhelming amount of carbon gain is due to C_3 photosynthesis.

Despite of *Ai. parlatorei* none of the

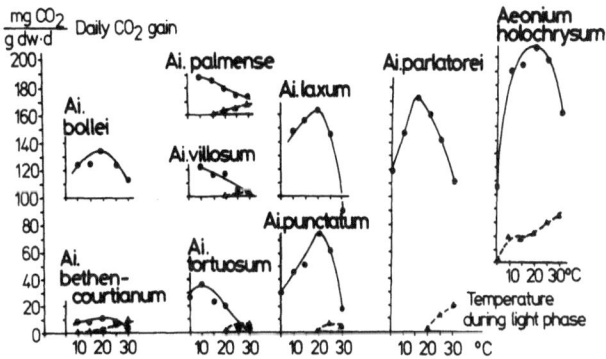

FIGURE 7. Total CO_2-uptake or -loss during 24 hours in *Aichryson* species depending on various daytime temperatures (dark phase temperature: 10 °C). Shaded areas: CO_2-uptake under dark conditions.

Aichryson species approaches the high daily CO_2 uptake rates shown by *Aeonium holochrysum*. Among the others the temperature-dependent rates of the widespread *Ai. laxum* and *Ai. punctatum* are very similar; the CO_2 fixation rates of both exceed those of the remaining species.

Aichryson punctatum, laxum, bollei, bethencourtianum, and *parlatorei* work most efficiently between 15 and 20 °C. In the other species maximal rates of photosynthesis are already reached around 10 °C. For *Ai. palmense* and *Ai. villosum* this might be in accordance with the prevailing habitat conditions. In the distribution area of *Ai. tortuosum* adequate microhabitat conditions only in shaddy crevices are feasible. Though temperatures of 30 °C during daytime being a disadvantage for the productivity of all *Aichryson* species, *Ai. bollei* copes with such a high temperature relatively best. In *Ai. laxum* 30 °C during the light phase result in a daily net CO_2 loss. In the shrubby species (*Ai. tortuosum* and *Ai. bethencourtianum*) the nocturnal CO_2 fixation yields most of the daily carbon gain under such circumstances.

4. DISCUSSION

Among the Macaronesian Sempervivoideae apparently a high flexibility in the photosynthetic pathways has been developed . This is already indicated by the wide range of ^{13}C values that can be found in Tinerfenian Semperviva from different habitats (Tenhunen et al., 1983). The results by this method agree with the data presented here in giving evidence for an obligate CAM performance of *Aeonium urbicum*, *Ae. haworthii*, *Ae.ciliatum*, and for C_3 metabolism normally to be found in *Aichryson laxum* (and also mostly in *Greenovia aurea*). While widely differing ^{13}C values for *Ae. holochrysum* indicate a high flexibility in photosynthetic performance of this species, all field measurements and laboratory treatments applied here give evidence to the fact, that this species is a CAM-plant. Nevertheless, as in the genus *Aichryson* studied here in detail, many Canarian Semperviva - and among them also *Ae.holochrysum* - cover much of the net carbon gain by daytime CO_2 uptake.

Possibly, seasonal periods in preponderance of the light or the dark phase CO_2 uptake play an important role. Seasonal patterns of acid metabolism and gas exchange are known e.g. in *Opuntia basilaris*/Cactaceae (Szarek, Ting, 1974) and in *Tillandsia usneoides*/Bromeliaceae (Martin et al., 1981). While annual courses of productivity rates and diurnal acidity fluctuations in the cactus respond sensibly to the water relations of this desert plant, this is not as pronounced in Spanish moss growing in a more humid climate. In this epiphyte CO_2 uptake becomes restricted mainly by low temperatures. The Crassulaceae studied here attain peak productivity rates at lower temperatures than the subtropical cacti and bromeliads. Their daily performance of gas exchange under controlled conditions realizes the CAM pattern in some species only weakly. But if realized the general shapes of the CAM-determined diurnal courses of gas exchange under either low or high daytime temperatures are comparable with those of members of the northern "hardy Semperviva" (Wagner, Larcher, 1981; Schuber, Kluge, 1979).

Under low to moderate daytime temperatures the C_3 mechanism of CO_2 fixation apparently is favored. Thus, at the natural habitats of most *Aichryson* species a reasonable degree of growth may be based thoroughly on C_3 photosynthesis during the relatively cold and wet winter months. CO_2 dark fixation will be subsidiary to this carbon gain. During the hot, dry season existence and metabolism in particular of the perennial, shrubby *Aichryson* may be sustained chiefly by the acid metabolism mode. The perennial herbaceous species possibly derive more profit from this photosynthetic pathway too, when the habitat situation becomes more arid. The ability for dark CO_2 uptake of the annual *Aichryson* species is remarkable,for in most other cases CAM can be found only as a strategy of existence of long lived plants.

Studies about the realisation of CAM with the scope on evolutionary tendencies have been done to date especially with the Bromeliad family (Medina, 1974; Ortlieb, Winkler, 1977) CAM seems to have been evolved within this family concomitantly with the colonization of more arid habitats by advanced members of all three subfamilies of the Bromeliads. With in the Crassulaceae the ability to perform CAM seems to be a basic feature available already to the Tertiary ancestors of the Macaronesian Semperviva. It works in members of the Central Europaean hardy Semperviva (Osmond et al., 1975; Schuber, Kluge, 1979;

Wagner, Larcher, 1981) as well as in many species of the southern, tender group (Praeger, 1932/67) of the subfamily (Neales, 1968; Tenhunen et al., 1983; this study). Within several branches of the last-named alliance it does not seem to be of essential ecological importance for each species. This can be seen from the behavior of several *Aichryson* species as well as within the *Aeonium* section Goochia. Also within this branch of the phylogenetic tree a great variability in the degree of CAM utilisation can be found with the morphologically least advanced extant species, *Aeonium goochiae*, photosynthesizing almost exclusively by the C_3 mode (Fig.6).

Within the Macaronesian Sempervivoideae niche colonization processes have occurred whereby relatively short-lived herbaceous forms of the genus *Aichryson* covering their carbon requirements by C_3 photosynthesis can compete more successfully for less extreme habitats than the aridity-adapted, ancestral CAM-species. Other branches of the phylogenetic tree originating from these (Holochrysa-like) Tertiary precursors have found adequate environments in the more arid lowland parts of the Atlantic islands just by extending this water-saving metabolism (nowadays found in particular among members of the sections Holochrysa and Urbica). Others again (e.g. some species of the *Aeonium*- section Canariensia) preserved a high flexibility in their modes of carbon fixation utilizing the one or the other possibility depending on the peculiarities of the diverse environments on the various islands.

Acknowledgement. I am indebted very much for help and discussions to Prof.L.Kappen and for support during the field work to Prof.W.Wildpret de la Torre/La Laguna. Ms.C.Lasalle skillfully photographed the artwork; Ms.Perschmann and Ms.Pissarek helped in the calculations of the gas exchange data. The care taken by Mr.Hesselbarth and Mr.Schuhmacher for the greenhouse collection of the Canarian Semperviva is also gratefully acknowledged.

SUMMARY

Acid metabolism under natural and controlled conditions and gas exchange under defined temperature regimes were investigated in Macaronesian Sempervivoideae. Diurnal changes in titratable acidity were great in all *Aeonium* species investigated at lowland sites. No fluctuations in acidity could be detected in *Greenovia*, *Aeonium*, *Aichryson*, and *Monanthes* species growing in cold and wet places of the high mountain area of Tenerife. Species colonizing warmer mountain slopes with greater temperature fluctuations showed a species-specific assortment in the intensity of diurnal acid metabolism. Also the investigations under controlled conditions gave evidence for a combined action of species-specific peculiarities and environmental conditions affecting the acid metabolism of the plants. Diurnal courses of CO_2 gas exchange under different temperatures during the light phase and 10 °C at night showed an increase of CAM pattern with increasing temperatures in the daytime. In well watered *Aichryson* species the decisive share of the carbon uptake occurred by C_3 photosynthesis. The phylogenetic tree of the genus *Aichryson* can be divided into three branches. Two shrubby species colonizing the arid eastern Canary Islands show low productivity. At higher temperatures they behave as full CAM plants. All other species are annual to triennial herbs growing in more humid habitats. Among them, nocturnal CO_2 uptake in *Aichryson laxum* and its allies is minimal with medium overall productivity rates. In the evolutionary line represented by *Aichryson punctatum* additionally to a high light phase photosynthesis also nocturnal CO_2 uptake occurs. In these species a tendency is obvious to shorten the individual life-cycles, accompanied by high productivity rates.

126

REFERENCES

Bramwell D and Bramwell Z (1974) Wild flowers of the Canary Islands. London and Burford, Stanley Thornes.

Burchard O (1929) Beiträge zur Ökologie und Biologie der Kanarenpflanzen. Bibliotheca Botanica 98: 262 pp. + 78 plates. Stuttgart, Schweitzerbart.

Kluge M and Ting IP (1978) Crassulacean acid metabolism: Analysis of an ecological adaptation. Ecol.Stud.30. Berlin-Heidelberg-New York, Springer.

Lems K (1960) Botanical notes on the Canary Islands. II. The evolution of plant forms in the islands: *Aeonium*. Ecology 41, 1-17.

Lösch R and Kappen L (1981) The cold resistance of Macaronesian Sempervivoideae. Oecologia 50, 98-102.

Lösch R and Kappen L (1983) Die Temperaturresistenz makaronesischer Sempervivoideae. Verh.Ges.Ökol., Jahretagung 1981 Mainz; Göttingen - in press.

Lüttge U and Ball E (1974) Proton and malate fluxes in cells of *Bryophyllum daigremontianum* leaf slices in relation to potential osmotic pressure of the medium. Z.Pflanzenphysiol. 73, 326-338.

Martin CE, Christensen NL and Strain BR (1981) Seasonal patterns of growth, tissue acid fluctuations, and $^{14}CO_2$ uptake in the Crassulacean acid metabolism epiphyte *Tillandsia usneoides* L.(Spanish Moss). Oecologia 49, 322-328.

Medina E (1974) Dark CO_2 fixation, habitat preference and evolution within the Bromeliaceae. Evolution 28, 677-686.

Neales TF, Patterson AA and Hartney VJ (1968) Physiological adaptation to drought in the carbon assimilation and water loss of xerophytes. Nature 219, 469-472.

Ortlieb U and Winkler S (1977) Ökologische Differenzierungsmuster in der Evolution der Bromeliaceen. Bot.Jahrb.Syst. 97, 586-602.

Osmond CB, Ziegler H, Stichler W, Trimborn P (1975) Carbon isotope discrimination in alpine succulent plants supposed to be capable of Crassulacean acid metabolism (CAM). Oecologia 18, 209-217.

Praeger LR (1932/1967) An account of the *Sempervivum* group. London, The Royal Horticultural Society - Reprint: Lehre, Cramer.

Schuber M and Kluge M (1979) Crassulaceen-Säurestoffwechsel (CAM) bei mitteleuropäischen Sukkulenten: Ökologische Untersuchungen an *Sempervivum*-Arten. Flora 168, 205-216.

Szarek SR and Ting IP (1974) Seasonal patterns of acid metabolism and gas exchange in *Opuntia basilaris*. Plant Phys. 54, 76-81.

Tenhunen JD, Tenhunen LC, Ziegler H, Stichler W and Lange OL (1983) Variation in carbon isotope ratios of Sempervivoideae species from different habitats of Teneriffe in the spring. Oecologia - in press.

Voggenreiter V (1972) Pflanzenverbreitungstypen auf Tenerife. Geobotanisch-arealkundliche Untersuchungen. I. *Aeonium* Webb & Berth. Cuad.Bot.Canar. 16, 1-8.

Voggenreiter V (1974) Geobotanische Untersuchungen an der natürlichen Vegetation der Kanareninsel Tenerife. Dissertat.Bot. 26. Lehre, Cramer.

Wagner J and Larcher W (1981) Dependence of CO_2 gas exchange and acid metabolism of the alpine CAM plant *Sempervivum montanum* on temperature and light. Oecologia 50,88-93.

SOIL TEMPERATURE EFFECTS ON CARBON EXCHANGE IN TAIGA TREES

WILLIAM T. LAWRENCE AND WALTER C. OECHEL (Systems Ecology Research Group HA-564, San Diego State University, San Diego, California, USA 92182)

SUMMARY

The tree communities of the taiga of interior Alaska form a mosaic-like pattern of vegetation types, their dominant species depending largely on time since last disturbance. One marked change, however, is a nearly exceptionless replacement of hardwoods by evergreen species with time. It is hypothesized that this replacement is due to unfavorable changes in the carbon balance of the hardwoods brought about by a gradual decline in soil temperatures at a site as forest floor organic materials build up, insulating the soil during the growing season. To test for a soil temperature effect on the carbon exchange of taiga hardwoods, both field (maximum photosynthesis) and laboratory (photosynthesis, root respiration) experiments were done under varied soil temperature regimes. Seedlings of paper birch, aspen, balsam, poplar, and alder were selected as test species since they are the hardwood pairs replaced by evergreens in the uplands and on the floodplain respectively. It was found that in the laboratory total, growth, and maintenance root respiration as well as photosynthesis in some species was affected by soil temperature. A survey of maximum field rates of photosynthesis also showed reductions in gas exchange on sites with cold soils.

INTRODUCTION

An immediate first impression of the taiga of interior Alaska is one of a large scale mosaic of various deciduous and evergreen tree communities. These patterns, from the juxtaposition of distinct vegetation types, are most easily discerned in early or late season when the hardwoods stand out in contrast to evergreen canopies. On the river floodplain, the vegetation pattern is most affected by disturbance in the form of erosion by the meandering rivers that destroys existing plant communities while the slower moving waters leave bare sandy areas for colonization elsewhere. In the uplands, and to a lesser degree on the floodplain, fire is the factor of disturbance which allows the introduction of pioneering plant communities on sites where both canopy and forest floor have been removed. Disturbance can be either large or very small scaled, as witnessed by the narrow stringers of hardwoods within evergreen forests where only small areas have undergone species replacement.

On both floodplain and upland sites, the post-disturbance changes in vegetation composition feature first canopy dominance by hardwoods with later replacement by white or black spruce. This change of species through time, coupled with frequent and patchy disturbance, leads to the mosaic-like pattern observed in the taiga. The research reported here was stimulated by an interest in the biotic and physical factors governing the eventual replacement of hardwoods by evergreens in virtually all taiga sites. It was of special interest to see if a single factor could account for the dramatic changes observed.

Due to the relative uniformity of site environment during the hardwood replacement, it is not immediately apparent what parameters are best correlated with the change in canopy dominants. Given the above ground uniformity among sites, our attention focused on below ground environments where soil temperature is perhaps the most dominant factor. Viereck (1970), has worked with floodplain communities where he found that growing season soil temperatures were very

128

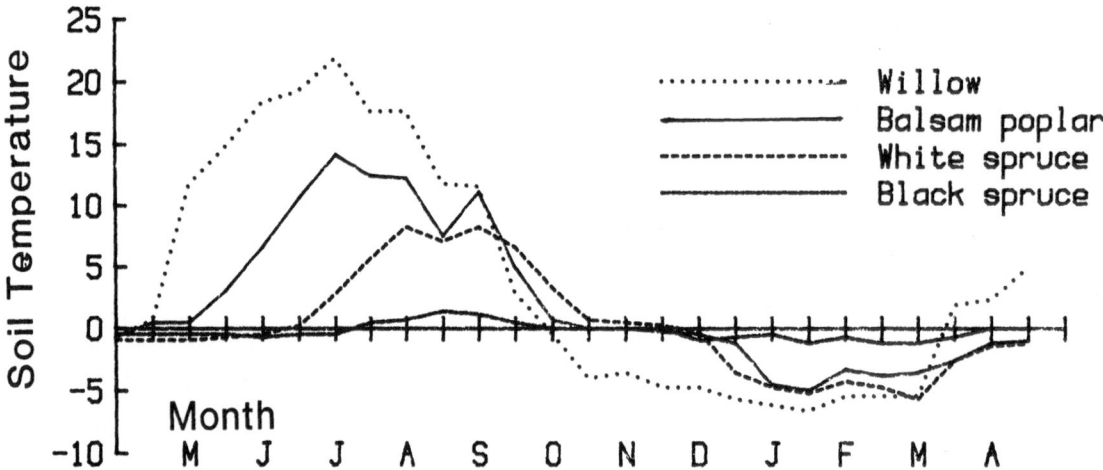

Figure 1. Seasonal course of soil temperature (°C) at 10 cm depth in willow, balsam poplar, white spruce, and black spruce stands in interior Alaska adjacent to the Chena River. Months are may through April (After Viereck 1970).

specific for certain taiga communities (fig. 1). The hardwoods as a group have the highest growing season soil temperatures in the summer, but soil heating begins earlier in the season, just as winter soil temperatures are those of the hardwood sites, where the soils are cooler longer. Of the four sites depicted in fig. 1, the willow stand had soil temperatures above 0° C. at 10 cm depth for nearly 160 days, the balsam poplar for about 150 days, the white spruce for less than 90 days. Such trends hold well for all similar taiga communities.

The basic cause for the differentiation of species specific soil temperature regimes is the site energy budget and changes in canopy and forest floor cover. After any kind of disturbance, the cover of canopy and forest floor are altered to some degree. Following erosion on the floodplain, the newly deposited and colonized areas are completely free of both litter and canopy. In the initial stages, canopy development plays the most important role in soil temperature mediation,

as it is the only interceptor of solar radiation. In later floodplain, and post-fire upland sites, the depth and accumulation rate of forest floor litter and moss most influences the site's soil temperature throughout the year. Blackened surfaces of recently burned areas warm rapidly, thawing underlying soils. Likewise, a destroyed, or reduced forest floor litter layer also increases soil warming (Van Cleve and Viereck, 1981; Bliss 1957). Canopy closure has little effect once a moss understory is present. Van Cleve, Zasada (1976) found that a thinning 60% of the original basal area in a white spruce forest had no effect on soil temperature. As a average figure, the development of a single centimeter of forest floor decreases the growing season heat sum (cumulative degrees above 0° C. at 10 cm depth from 20/V to 10/IX) by 37° (Van Cleve, Viereck 1981).

In their review of nutrient and successional patterns in the Alaskan taiga, Van Cleve, Viereck (1981) described the basic patterns of species replacement in floodplain, cold, and warm upland sites. On the floodplain, the pioneering hardwoods are willow and alder, eventually overtopped by balsam poplar. The poplars are themselves replaced by a white spruce forest in 175 to 200 years barring further catastrophic disturbance. White spruce may be replaced by black spruce in time, but due to frequent fires, such old sites are not readily encountered. In the cold upland site, black spruce is almost exclusively self-replacing, since fire is seldom intense enough to denude and warm the permafrost dominated soils. On warmer upland sites, the pre-fire white spruce or hardwood forest is replaced by aspen and birch groves after fire, with white spruce dominating the canopy and replacing the hardwood component by about 200 years since last disturbance.

Given this scenario of hardwood replacement by evergreens in both floodplain and upland sites with an observed growing season soil temperature decline, we decided to look for a possible physiological explanation for that replacement. Soil temperature seemed to be the key, since it changes so dramatically between hardwood and evergreen stands. Van Cleve (unpublished data) found that in an analysis of 25 intensively samples sites, the growing season soil heat sums averaged 1258 degree days in all hardwood stands, but only 725 degree days in black and white spruce stands.

Broad ranges of studies with native northern species and agronomic plants have shown soil temperature to affect plant water relations (Anderson, McNaughton 1973; Kaufmann 1977; Kramer 1940; Running, Reid 1980), nutrient uptake (Chapin 1977; Raper et al. 1978), root and shoot growth rates (Billings et al. 1977; Chapin 1977; Ellis and Kummerow 1982), root and shoot

gas exchange (Billings et al. 1977; Higgins, Spomer 1976; Lawrence, Oechel 1982a,b; Lediq et al. 1976, Raper et al. 1978), and germination (McDonough 1979; Zasada, Viereck 1975). These citations are by no means all inclusive, but will give an idea of potential effects of cold soil. We felt that an analysis of the integrated effect of soil temperature on individual taiga plants would best reveal any interspecific difference. In our opinion, the most integrated measure of a plant's status is its carbon balance. Whether the net carbon acquisition is negative or positive, and to what degree, is tantamount to gauging whether or not a plant will survive or perish. Without a positive carbon balance, plants can neither grow nor reproduce, as carbon skeletons are obviously required for all syntheses of new tissue (Penning de Vries et al. 1974).

For this research, our hypothesis was that declining soil temperatures on-site have detrimental effects on the carbon balance of the hardwoods present, leading eventually to their replacement by evergreen species that are better adapted to carbon accumulation under the cooler soil temperatures. In the uplands post-disturbance forests of birch (Betula papyrifer) and aspen (Populus tremuloides) are replaced by white spruce (Pica glauca). On the floodplain, forest of balsam poplar (Populus balsamifera) intermixed with alder (Alnus tenuifolia) are also replaced by white spruce. Given these upland and floodplain instances of hardwood replacement by a single evergreen species, these four plants were chosen as typical of interior Alaska for the experiments to test the effect of soil temperature on seedling carbon exchange.

To best measure the carbon status of plants experimentally, we chose to monitor both above and below ground CO_2 exchange of seedlings across a range of soil temperatures like those commonly observed in the field. Below ground root respiration, including both total and maintenance (Ledig et al. 1976; McCree 1970) components were measured. Above ground the light and temperature response of photosynthesis and dark respiration of leaves wer determined.

PROCEDURE

Laboratory Work - Plant Culture

Seedlings of the four species were grown in sand in ca. 15 cm pots in a growth room under 22 h daylength at 250 uE $m^{-2} d^{-1}$ and day:night air temperatures of 25:20 °C. Full strength Hoagland's solution was administered weekly. Seedlings were thinned to 3/pot and were ready for experimental use at 3-4 months of age. (For complete methodology and results see also Lawrence and Oechel 1982 a,b).

Root Temperature Control

A coil of copper tubing was fit to tightly grip the pots, and was installed in an insulated plastic foam ice chest. This assemblage, with a sealed lid through which the potted plant tops protruded, allowed close temperature control of the potted soil during the gas exchange measurements. Control was exerted by flowing a coolant from a water bath through the coiled tubing around the pot while in the insulated root chamber. Thermocouples were used to monitor the soil temperature, and manual adjustments were made to keep the soils at the experimental temperatures of 5, 15, and 25°C. During any experiment, six root chambers were in use, yielding six replicates from a single species. Roots were chilled below ambient (25°C) only during

the one week experimental runs at each of the 3 control temperatures.

Gas Exchange of Roots

Rather than using excises roots or forcing air through a column of gravel with living root material, both techniques with some problems (Billings et al. 1977), we chose to measure the efflux of CO_2 from an undisturbed soil mass, that of a pot containing the roots of the species under study. We used infrared gas analysis for carbon dioxide measuration by adapting a system ordinarily used for photosynthetic work (Oechel, Lawrence 1979) by connecting the sealed root temperature control chambers to the existing gas ports. Using the root chamber, the potted soil was effectively sealed in its own volume of air, but the tops were left intact and under the normal growth room regime. Split corks allowed the plant tops to pass through the top of the root chamber. This technique had little effect on the potted plant, but allowed the air to be flowed around the potted soil mass so that the CO_2 flux of the roots could be measured.

Photosynthesis

Our standard techniques (Oechel, Lawrence 1979) were followed in deriving the temperature and light curves of photosynthesis and dark leaf respiration for the plants under study. At each of the 3 soil temperature treatments (5, 15, 25°C) photosynthesis and dark respiration were determined and air temperatures of 5, 10, 15, 20, 25, and 30°C. Light curves were run using incandescent light from 25 to 1200 uE $m^{-2} s^{-1}$.

Field Work - Maximum Photosynthesis

A small non-temperature controlled, hand

held cuvette was used in the field to label leaves with $^{14}CO_2$ in air. Plant tissue was killed in a methanol - phenethylamine mix, and later combusted in an Oxamat biological oxidizer. Site soil and air temperatures and photon flux density (uE m^{-2} s^{-1}) were measured prior to each maximum photosynthesis (Pmax) labelling. Photosynthesis was calculated from the disintegration per minute of each sample determined with liquid scintillation spectrophotometry. Samples were made of as many species as possible across naturally occurring gradients of soil temperature.

RESULTS AND DISCUSSION

Laboratory Work - Plant Culture

At the fixed soil temperature at which the plants were grown (25°C), the greatest average root mass in a pot of three individuals was the 15.3 grams of birch at the end of the 4 month culture period. The other three species, aspen, balsam poplar, and alder were ranked 13.0, 10.0, and 7.6 grams dry weight per pot respectively. Total leaf areas (dm^2) were similarly ordered, 12.2 for alder, 11.7 for birch, 9.4 in aspen, and 7.0 for balsam poplar.

Root Respiration

Total and maintenance root respiration (Hansen 1978; Ledig et al. 1976; McCree 1979; Penning de Vries 1975) were measured at all three of the soil temperature treatments for alder, aspen, and balsam poplar (Table 1). Only total root respiration was measured in birch. As expected for most biological processes, the total root respiration increased over the 20°C range with average Q_{10} values of 2.0,

much like those found by Billings et al. (1977) for arctic plants. Highest root respiration was that of alder at all soil temperatures.

Maintenance respiration is that necessary to maintain protein turnover, active nutrient uptake from the soil solution, and tissue function and integrity (Ledig et al. 1976; Penning de Vries 1975). Without protein turnover, there is little possibility of adaptation to new environmental conditions (on a mambrane basis), and without nutrient uptake, all processes, including photosynthesis (Hansen 1978; Raper et al. 1978) are eventually reduced. We measured maintenance respiration by darkening the tops of the plants for 6 - 12 h, which reduces carbohydrate concentration in the plant and eliminates growth, or constructive respiration, which is that supplying intermediates for creation of new biomass (Ledig et al. 1976).

By subtraction of total - maintenance root respiration, growth respiration is calculated. A significant (P>0.95) statistical difference between total and maintenance root respiration among the 3 soil temperature treatments was found in aspen at all treatment levels, but only at 5 °C in alder, and only at 15, and 25°C soil temperatures in balsam poplar. These results allow the inference of root growth at all temperatures in aspen, but only below 15°C in alder and only above 5°C in balsam poplar. Such results are unusual, showing species differences in adaptation to cold soils, but have been observed as differential growth rates of roots in other arctic plants (Billings et al. 1977; Chapin 1977; Ellis, Kummerow 1982). The interpretation must be cautious however, since the root tissue used in the experiment was essentially that of the culture period, 25°C. The reduced soil tempe-

Table 1. Summary of total (Rt), maintenance (Rm), and growth (Rg) root respiration in the taiga hardwoods under the three soil temperature treatments. Mean ($\bar{x}$) and standard errors of the mean ($s\bar{x}$) are calculated for the replicate measurements of Rt and Rm. Growth root respiration is calculated as $\bar{x}$ Rt $-$ $\bar{x}$ Rm. The Q_{10}'s (temperature coefficient) of the $5 - 15^{o}$ and $15 - 25^{o}$ C. temperature intervals, as well as the average Q_{10} $5 - 25^{o}$ are under the 5, 15, and 25° C. treatment columns respectively. No maintenance respiration was measured for paper birch. Significant differences between Rt and Rm (growth respiration) within a soil temperature treatment are indicated by ** (P>0.95).

		TR 5			TR 15			TR 25		
		Rt	Rm	Rg	Rt	Rm	Rg	Rt	Rm	Rg
Alder	$\bar{x}$	0.24 **	0.20	0.04	0.56	0.53	0.03	1.11	0.97	0.14
	$s\bar{x}$	0.019	0.026		0.040	0.064		0.118	0.101	
	Q10	2.33	2.65		1.98	1.83		2.16	2.24	
Aspen	$\bar{x}$	0.24 **	0.15	0.09	0.35 **	0.27	0.08	0.70 **	0.43	0.27
	$s\bar{x}$	0.009	0.008		0.011	0.013		0.019	0.041	
	Q10	1.46	1.80		2.00	1.59		1.73	1.70	
Balsam Poplar	$\bar{x}$	0.15	0.15	0.02	0.46 **	0.38	0.08	0.75 **	0.62	0.14
	$s\bar{x}$	0.002	0.009		0.021	0.044		0.026	0.034	
	Q10	3.07	2.53		1.63	1.63		2.35	2.08	
Birch	$\bar{x}$	0.19			0.49			0.75		
	$s\bar{x}$	0.005			0.017			0.030		
	Q10	2.58			1.53			2.06		

rature treatments were short term, on the order of 1 week each. The effect of temperature preconditioning can be substantial in acclimating the root system to new sets of environmental conditions (Markhart et al. 1979).

Photosynthesis

Photosynthesis versus air temperature curves for the three soil treatments (fig. 2) show consistently significant (P>0.95) soil temperature effects only in alder and birch. Photosynthesis at 5°C soil treatment in alder is greater than at other soil temperatures at air temperatures from 5 to 20°C. In birch 5°C soil temperature treatment gave significantly lower photosynthetic rates than for the other treatments at 20 to 35°C air temperature.

These differences can be explained in terms of two limiting factors. In birch, measurements showed significant reductions in conductance above 15°C air temperature as soil temperatures decreased.

Evidently the cooler soils reduced water uptake by increasing root resistance, thereby promoting partial stomatal closure and reduced carbon fixation. Such effects have been previously noted (Anderson, McNaughton 1973; Kaufmann 1975; Markhart et al. 1979; Running, Reid 1980). In the case of alder, photosynthesis is greatest at low soil temperatures where root resistance would be greatest. Here the carbon exchange of the roots may be the key, as we also found root growth respiration only at the same 5°C as well. If carbon were in any way limited at other soil temperatures, there could be a reduction in available energy for root processes such as extension and active ion uptake. Raper et al. (1978) has found such effects due to a feed back loop where lack of carbohydrate in the roots reduces active nitrogen uptake and reduction, eventually reducing photosynthesis. Such may be the case with alder. Changes in source - sink relationships (Hansen 1978) may also increase the apparent photosynthetic rate as root respiration is affected by changing

133

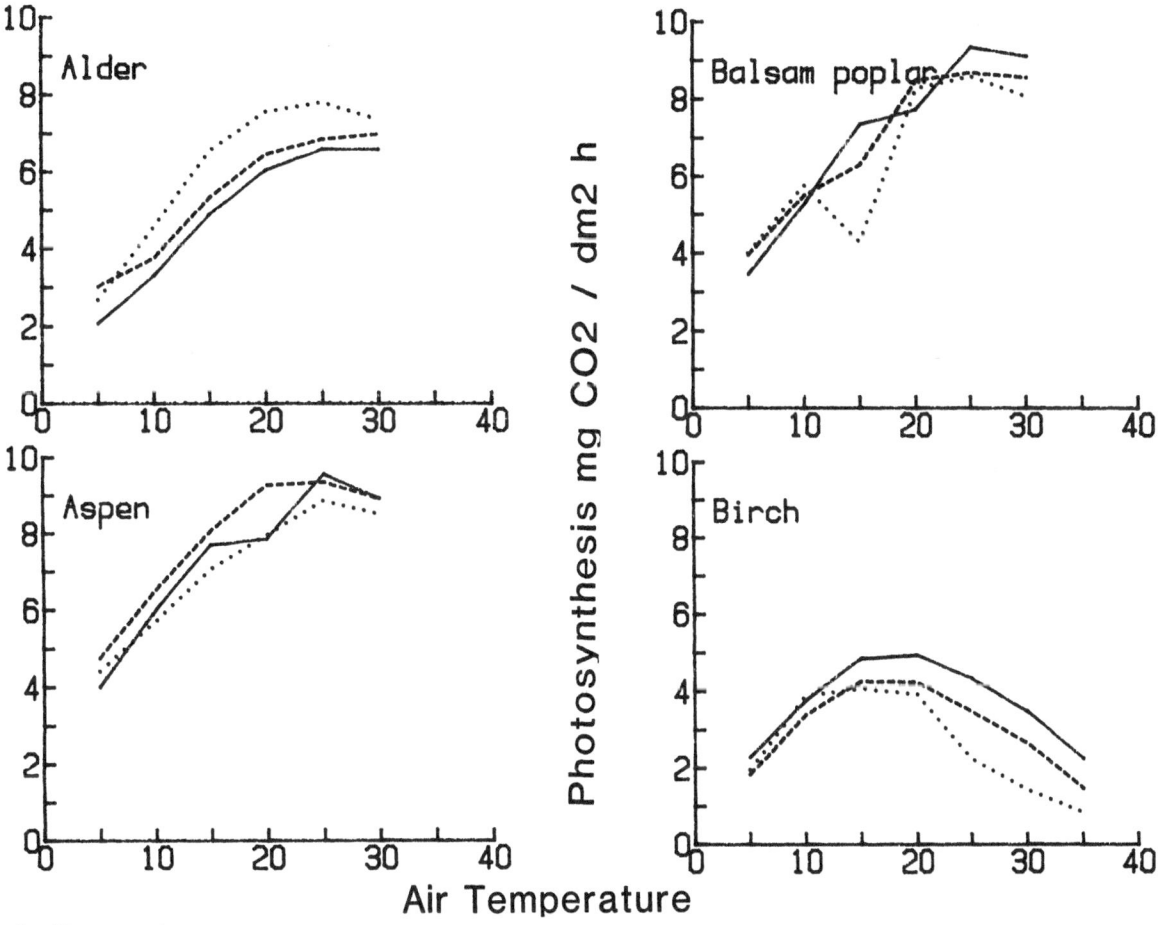

Figure 2. Photosynthetic temperature curves for the hardwood seedlings under the three soil temperature regimes at a photon flux density of 600 uE m^{-2} s^{-1}. Gas exchange was measured at soil temperature treatments of 5°, 15°, and 25° (indicated by dotted, dashed, and solid lines). Data points are means of replicate measurements from six individuals of each species.

soil temperature.

Finally, leaf dark respiration was significantly and consistently affected by soil temperature only in birch where it was reduced by both the 5 and the 25°C treatment at all air temperatures.

Field Maximum Photosynthesis

Data from the field maximum photosynthesis (Pmax) $^{14}CO_2$ sampling among the dominant tree species across naturally occurring soil temperature gradients is still provisional, but some trends are clear. An effort was made to measure both seedlings and mature trees on both cold and warm

Table 2. Maximum observed photosynthetic rates of selected hardwood species in the field along naturally occurring soil temperature gradients. Photosynthetic units are mg CO_2 dm^{-2} h^{-1} (+ 1 SE). Indicated soil temperature is that at 10 cm depth in both warm and cool sites.

Species	Tsoil	Seedlings	Tsoil	Trees
Aspen	22.4	15.7 (0.96)	23.1	10.1 (0.97)
	15.6	6.2 (0.83)	15.6	5.6 (0.78)
Poplar	22.3	15.2 (0.66)	23.1	11.9 (1.09)
	--	-- --	14.1	9.3 (0.10)
Birch	22.3	25.0 (0.94)	16.6	10.3 (1.08)
	15.6	3.4 (0.30)	11.3	4.8 (0.42)
Alder	22.4	16.9 (0.71)	15.4	14.4 (1.07)
	9.1	9.6 (1.79)	9.2	10.9 (0.36)
Willow	22.5	23.5 (1.62)	--	-- --

soils. Of course, not all species occur on both sites (which supports the hypothesis to some extent). In aspen, the cooler soil reduced seedlings Pmax (Table 2) by 60%, and that of trees by 45%. No balsam poplar seedlings were found on cool soils, but adult Pmax was reduced by 39% at cool sites. Birch seedlings were most affected by cooler soils, as was the case in our laboratory work. Their Pmax was reduced 86%, but only 32% in trees. In alder similar results were found, with Pmax reductions of 43 and 40% respectively in seedlings and trees. However, no increase in Pmax was found on cold soils as was the case in lab studies.

The field experiments were completely uncontrolled as opposed to those of the laboratory. The sites were chosen more or less arbitrarily so that both seedlings and trees of as many species as possible could be sampled at once. No response surfaces of photosynthesis were made as neither light nor air temperature were controlled. The root - shoot ratios, nutrient relations, past environmental history, water relations, and total biomass of all the sampled trees were also unknown, so direct comparisons with the seedling results from the laboratory are made difficult at best.

CONCLUSIONS

Root Respiration

Gas exchange increases exponentially with temperature increase in the four species studied. Q_{10}'s average 2.0 from 5 to 25°C soil temperature. Growth respiration is detectable at all soil temperature treatment levels in aspen, but only at 15°C and above in balsam poplar, and below 15°C in alder. The lack of growth of pop-

lar at low temperatures and alder at high temperatures may represent specific adaptive capabilities and root growth optima for those species.

Photosynthesis

Soil temperature had no effect on photosynthesis in aspen and balsam poplar. In alder, photosynthesis is highest at 5°C soil temperature, probably due to effects of reduced ion uptake and/or changing sourse - sink relationships. Birch shows reductions in photosynthesis at high air and low soil temperaturew where high root resistance to water uptake elicits classical water stress symptoms of reduced conductance through partial stomatal closure and concomitant reduction in gas exchange.

Field Photosynthesis

In hardwoods, trees and seedlings on cooler soils routinely show reduced Pmax values compared to similar individuals on warmer sites. Both the minimum and maximum observed reductions in Pmax from all species occurred in birch, 86% and 32% respectively in seedlings and trees. Reductions in other species were on the order of 40%. These values are relative only since the field Pmax is that under uncontrolled conditions and represents a reaction to an entire matrix of environmental variables.

LITERATURE CITED

Anderson, J.E. and S.J. McNaughton. 1973 Effects of low soil temperatures on transpiration, photosynthesis, leaf relative water content, and growth among elevationally diverse plant populations. Ecology 54:1220-1233.

Billings, W.D., K.M. Peterson, G.R. Shaver, and A.W. Trent. 1977. Root growth, respiration, and carbon dioxide evolution in arctic trundra soil. Arctic and Alpine Res 9:129-137.

Bliss, L.C. 1957. Succession on river alluvium in northern Alaska. Amer Midl Nat 58:452-469.

Chapin, F.S. III. 1977. Temperature compensation in phosphate absorption occurring over diverse time scales. Arctic and Alpine Res. 9:139-148.

Elis, B.A. and J. Kummerow. 1982. Temperature effect on growth rates of Eriophorum vaginatum roots. Oecologia 54:136-137.

Hansen, G.K. 1978. Utilization of photosynthates for growth, respiration, and storage in tops and roots of Lolium multiflorum. Physiol Plant 42:5-13.

Higgins, P.D. and G.G. Spomer. 1976. Soil temperature effects on root respiration and the ecology of alpine and subalpine plants. Bot Gaz 137:110-120.

Kaufmann, M.R. 1977. Soil temperature and drying cycle effects on water relations of Pinus radiata. Can J Bot 55:2413-2418.

Kramer, P.J. 1940. Root resistance as a cause of decreased water absorption at low temperatures. Plant Physiol 15:63-79.

Lawrence, W.T. and W.C. Oechel. 1982a. Effects of soil temperature on the carbon exchange of taiga seedlings. I. Root respiration. Can J For Res (in review).

Ibid. 1982b. Effects of soil temperature on the carbon exchange of taiga seedlings. II. Photosynthesis. Can J For Res (in review).

Lawrence, W.T. 1982. Photosynthesis of taiga trees along naturally occurring soil temperature gradients. Can J For Res (in preparation).

Ledig, F.T., A.P. Drew, and J.G. Clark. 1976. Maintenance and constructive respiration, photosynthesis, and net assimi-

lation rate in seedlings of pitch pine (Pinus regida Mill.). Ann Bot 40:289-300.

Markhart, A.H., E.L. Fiscus, A.W. Naylor, and P.J. Kramer. 1979. Effect of temperature on ion transport in soybean and broccoli systems. Plant Physiol 64:83-87.

McCree, K.J. 1970. An equation for the rate of respiration of white clover plants grown under controlled conditions. p. 221-229 In I. Sctlik (ed) Prediction and measurement of photosynthetic productivity. PUDOC, Wageningen, Netherlands.

McDonough, W.T. 1979. Quaking aspen - seed germination and early seedling growth. USFS Res Paper INT-234. Ogden, Utah.

Oechel, W.C. and W.T. Lawrence. 1979. Energy utulization and carbon metabolism in Mediterranean scrub vegetation of Chile and California: I. Methods: A transportable cuvette field photosynthesis and data aquisition system and representative results for Ceanothus greggii. Oecologia 39:321-335.

Penning de Vries, F.W.T. 1975. The cost of maintenance processes in plant cells. Ann Bot 39:77-92.

Penning de Vries, F.W.T., A.H.M. Brusting, and H.H. van Laar. 1974. Products, requirements, and efficiency of biosynthesis: A quantitative approach. J theor Biol 45:339-377.

Raper, C.J., Jr., D.L. Osmond, M. Wann, and W.W. Weeks. 1978. Interdependence of root and shoot activities in determining nitrogen uptake rate of roots. Bot Gaz 139:289-294.

Running, S.W. and C.P. Reid. 1980. Soil temperature influences on root resistence of Pinus contorta seedlings. Plant Physiol 65:635-640.

Van Cleve, K. and L. Viereck. 1981. Forest succession in relation to nutrient cycling in the boreal forest of Alaska. p. . 185 - 211 In D.C. West. H.H. Shugart,

136

D.B. Botkin (eds) Forest succession. Concepts and applications. Springer-Verlag,
New York.

Van Cleve, K. and J.C. Zasada. 1976.
Response of 70-year-old white spruce to
thinning and fertilization. Can J For Res
6:145-152.

Viereck, L.A. 1970. Forest succession
and soil development adjacent to the
Chena River in interior Alaska. Arctic
and Alpine Res 2:1-26.

Zasada, J.C. and L.A. Viereck. 1975.
The effect of temperature and stratification on germination in selected members
of the Salicaceae in interior Alaska.
Can J For Res 5:333-337.

INFLUENCE OF THE ENVIRONMENT ON COLD HARDENING AND WINTER SURVIVAL OF FORAGE AND CEREAL SPECIES WITH CONSIDERATION OF PROLINE AS A METABOLIC MARKER OF HARDENING

R. PAQUIN (Research Station, Agriculture Canada, 2560, Hochelaga Blvd., Ste. Foy, Quebec, Canada G1V 2J3)

1. INTRODUCTION

Due to stresses such as cold, drought, heat, salinity, deseases and insects, only 7.6% of the earth surface can be used to grow crops. Moreover, each year these stresses reduce or destroy completely several crops on cultivated lands in several parts of the world (Weiser, 1970). For example, in the United States of America from 1963 to 1968, annual losses in crops due to cold stress reached 341 million dollars on 1.42 million hectares.

In Quebec, the proportion of arable soils is only 2.5% of the Province area and, almost every year, heavy losses due to adverse winter conditions are observed in many crops including forage and cereal species.

A few years ago, the Department of Agriculture of Quebec sponsored a contest for farmers to increase alfalfa production in the Province. Using remote sensing techniques in 1974, we inspected the fields of selected participants in the contest and showed that in areas best suited for alfalfa cultivation, several stands were completely destroyed by winterkill (Paquin et al. 1977). Most investigations on cold hardening and frost resistance of plants have been and still are conducted in growth chambers where light, temperature, moisture and aeration are controlled. Although the pursuit of experiments in an artificial medium offers several advantages, it is quite different from a natural medium where environmental influence, especially climate and soil, is much more variable. Interactions between climate and soil and variability in climatic conditions from year to year make interpretation of results difficult. In addition, digging out plant samples from frozen soils under a considerable snow cover renders the field work particularly hazardous and difficult during autumn and winter. It is not therefore surprising that little progress has been made on the physiological and biochemical adaptation of plants growing in northern countries.

It seems essential that research be conducted not only in the laboratory, but also under field conditions if one wishes to understand and solve problems that arise from the effects of hostile climate on the physiology of plants.

2. DEFINITIONS AND METHODS

Very often, cold resistance is confused with winter survival. Cold resistance can be defined as the capacity for a given plant to resist low temperature stresses and is related to the genetic background of the species. Winter survival is a function of interactions of several factors which result in a plant resisting or succombing to climatic stresses. Cold resistance is one of those factors affected by climate, plant and soil conditions. Therefore, speaking of winter survival refers to a succession or sequence of stresses to which plants must resist to stay alive during the cold season (Levitt, 1980; Steponkus, 1978, 1980). The capacity for a plant to resist the sequence of stresses constitutes its adaptation and adaptation varies from one region to another.

The percentage of plants that resist stresses borne during the cold season can be used as a measure of winter survival. The frost resistance of a plant, though in part a genotypic character, also varies with a plant growth, age and the freezing conditions. This complicates considerably the determination of cold resistance. But, within limits that are arbitrarily fixed such as the duration and temperature of the freezing test, the hardiness or frost resistance of different cultivars can be estimated and compared. This

definition of cold or frost resistance is thus
synonymous with frost survival and is expressed
in the results as the lethal temperature for 50%
of a plant population (LT_{50}). This terminology is
similar to the LD_{50} (lethal dose) used in toxicity
studies and antibiotic actions.

Natural features of the experimental sites,
cultivation and plant management techniques,
measurements of snow depth, soil moisture, air
and soil temperature as well as the freezing test
and frost survival (LT_{50}) used in this study have
already been described elsewhere (Paquin and
Pelletier, 1980).

Two methods have been developed to dig out plant
samples from frozen soils. The modified chain-
saw (Paquin and Pelletier, 1975) is easy to handle
and more rapid but wears out more quickly than
the tubular drill, due to stones present in soils
(Bolduc, 1976).

A modified method of proline determination in
plant extracts has previously been described
(Paquin and Lechasseur, 1979).

2.1. Cold acclimation of alfalfa

Figure 1 shows the results of six years of experiments from three Quebec locations on the influence
of the environment on the cold acclimation of
alfalfa (*Medicago media* Pers.). Cold hardening
of alfalfa begins when mean air temperature
reaches $10^{\circ}C$ but accelerates when air temperature
is close to 5° or $6^{\circ}C$. Hardening coincides generally with the first night frosts. This increase
varies from year to year accordingly. In 1975,
the first night frost in La Pocatière occurred
October 8, while in 1978, it occurred on September
17.

Cold hardening stops and loss of hardening or dehardening of alfalfa is observed if air temperature rises above $10^{\circ}C$ for a few days after the
first frost. During this warm period called
"Indian Summer" in North America, shoots and buds
of the crown begin to grow again, reducing food
reserves. A good example occurred in 1977 when
the air temperature reached a maximum of $20^{\circ}C$

between October 20 to November 11 (Fig. 1). All
cultivars showed a decrease in frost resistance,
but the non-hardy Caliverde responded to that
period more rapidly than the semi-hardy Saranac
and the hardy Rambler.

Once initiated, cold hardening seems to depend
mainly on air temperature and not on changing or
short photoperiod. This is made evident by the
fact that cold hardening continues to increase
under the snow, in December and January, in the
absence of photosynthesis and at the moment where
soil temperature at the crown level stays below
the freezing point.

The maximum frost resistance (LT_{50}) is generally
reached between mid-January and mid-March for
alfalfa, though there is variation from season
to season and region to region. The maximum
depends also on the cold hardiness of the cultivars; it varied from -17 to $-25^{\circ}C$ for the hardy
Rambler at La Pocatière and from -10 to $-14^{\circ}C$ for
the non-hardy Caliverde. At the St-Hyacinthe site,
which enjoys a milder climate, the cultivars did
not harden as much as in La Pocatière because
frosts occurred later in the season and there is
a slower decrease in the mean air temperature.

The maximum frost resistance reached by the hardy
Rambler under field conditions coincided with
the maximum attained in a growth cabinet after
4 weeks hardening at $1^{\circ}C$, about -21 to $-22^{\circ}C$
(Paquin, 1977). This adds to the credibility of
the experiments performed under the laboratory
conditions.

Dehardening or loss of cold resistance begins
with the rewarming of soil temperature above the
freezing point, after snow melt. Dehardening of
the non-hardy Caliverde occurred earlier than
that of the semi-hardy Saranac and the hardy
Rambler (Fig. 1).

Contrary to our expectations, cold dehardening
lasted 4 to 6 weeks under field conditions. This
helped alfalfa to resist late spring frosts. In
laboratory conditions, 2 to 3 days at 15 to $20^{\circ}C$
are more than sufficient to cause a complete cold

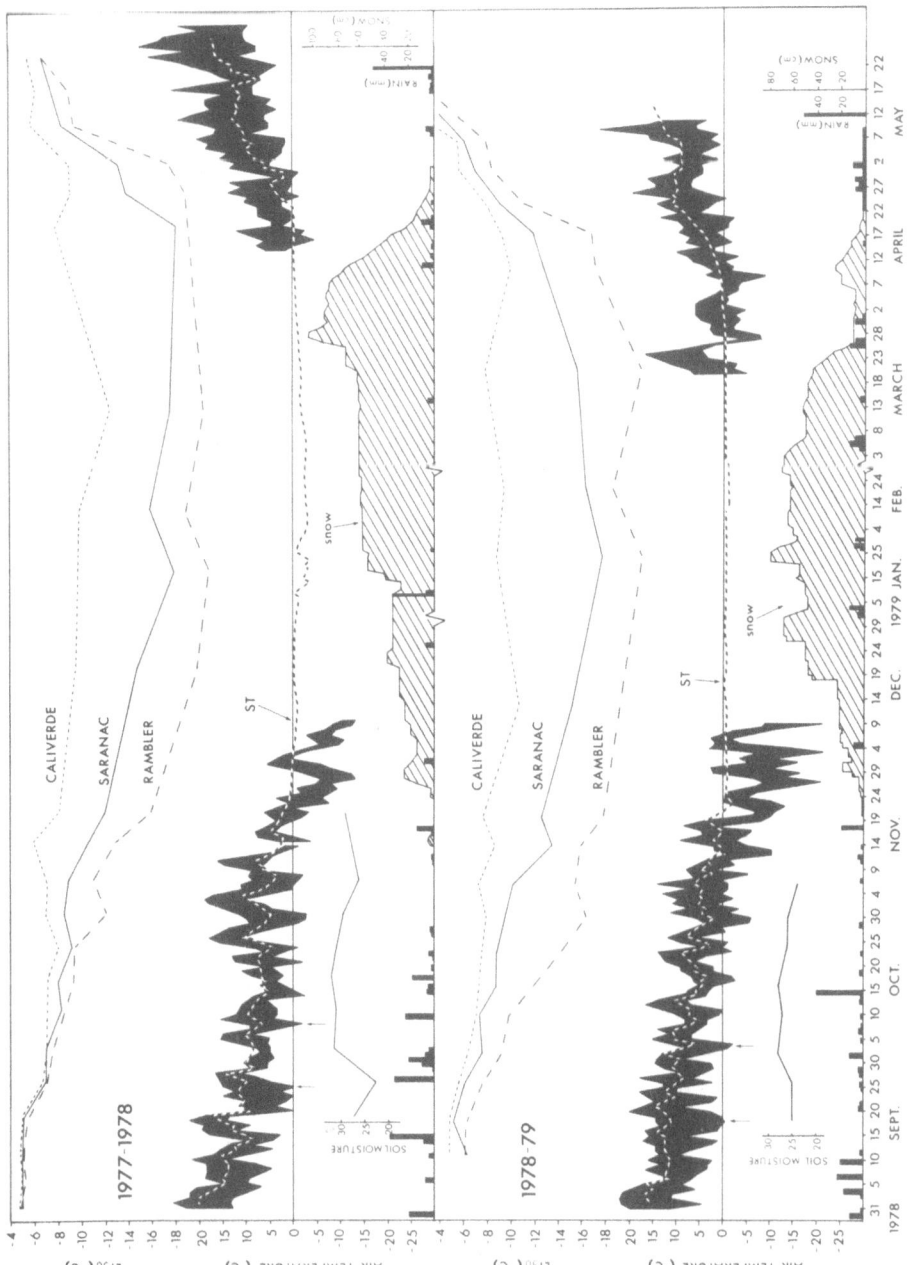

FIGURE 1. Hardening and dehardening (LT$_{50}$) of alfalfa cultivars grown at La Pocatière, air and soil temperature (ST) at the plant level, soil moisture content and rainfall and snowfall accumulations in 1977-78 and 1978-79. Arrows on temperature curve indicate the first night forsts.

dehardening of the cultivars. Dehardening under field conditions is dependent on the air and soil temperatures. This explains why dehardening occurs earlier in St-Hyacinthe than in La Pocatière, where the rewarming is slower. However, damages to crops by late frosts in the spring are more frequent in milder regions because of the earlier start of the growing period.

The first question that arises when one has to face the problem of alfalfa winter survival in a location where snow piles up is this: does the soil temperature around the crowns at 2 to 3 cm depth decrease enough to kill the alfalfa? Observations have shown that at the crown level, the soil temperature does not drop below -5 to -6°C. This temperature is reached generally in late November or early December in the absence of snow or when snow cover is too thin to provide an insulating layer. Once the snow cover has thickened enough, the thermal capacity of the soil compensates for the heat loss in the A horizon creating a soil temperature equilibrium at -1 to -3°C at the crown level. Because of the snow cover, soils remain frozen at -1 to -3°C to a depth of 8 to 15 cm throughout the winter. From these observations, it can be concluded that under a good snow cover, the soil temperature does not fall enough to kill alfalfa stands.

2.2. Cold acclimation of other forage plants and cereals

Experiments on the influence of the environment upon the winter survival of other forage plants and cereals began in 1980 and were performed with the most hardy cultivars we could find for the purpose of comparison.

The winter 1980-81 in Quebec was rather exceptional for most of the snow cover disappeared in February. Such winter conditions had not been observed for a century. Fortunately, a continuous lowering of the mean air temperature during the fall, without a rewarming period ("Indian Summer"), caused a deep cold hardening. Timothy (*Phleum pratense* cv. Engmo), rye (*Secale cereale* cv.

Cougar), triticale (*Triticale cereale* cv. Wintri) and winter wheat (*Triticum aestivum* cv. Kharkov) reached a LT$_{50}$ maximum of less than -30°C, that of brome (*Bromus inermis* cv. Saratoga) was -29°C, while for alfalfa (*Medicago media* Pers. cv. Rambler) used as a standard for comparison, it was -22°C.

Because of the thin layer of snow, the soil temperature was lower than normal and reached -10°C in February 1981. In spite of the abnormal conditions of that winter, dehardening did not occur earlier than in the preceding years.

During the fall of 1981, the climate was almost identical to that of the preceding year, with a continuous and steady decline in the air temperature and no rewarming period (Fig. 2). Consequently, the maximum LT$_{50}$s reached by these species were close to those of the preceding year; being -30°C and less for timothy and rye, -28°C for triticale and wheat, -29°C for brome and -24° to -26°C for alfalfa. Of the species under investigation, alfalfa was therefore the least cold hardy and timothy the most hardy.

In general, dehardening of all species appeared much less continuous than cold hardening. The variations observed in dehardening might result from incomplete observations due to difficulties in sampling at spring time or from other mortality factors independent of direct cold stress including diseases, root rotting, frost upheavals and competition.

Brome lost its frost resistance much more rapidly than all other species investigated while timothy remained frost resistant latest in the spring. Alfalfa and the cereals were intermediate in losing their frost resistance. We also observed that hardy cultivars such as Rambler lost their frost resistance later than the non-hardy Caliverde (Fig. 1). This argues for an efficient response mechanism, perhaps mediated through phytohormones. The mechanism deserves further attention.

The maximum of frost resistance does not occur at the same time in different regions (Fig. 3). At St-Hyacinthe, where the climate is milder it was

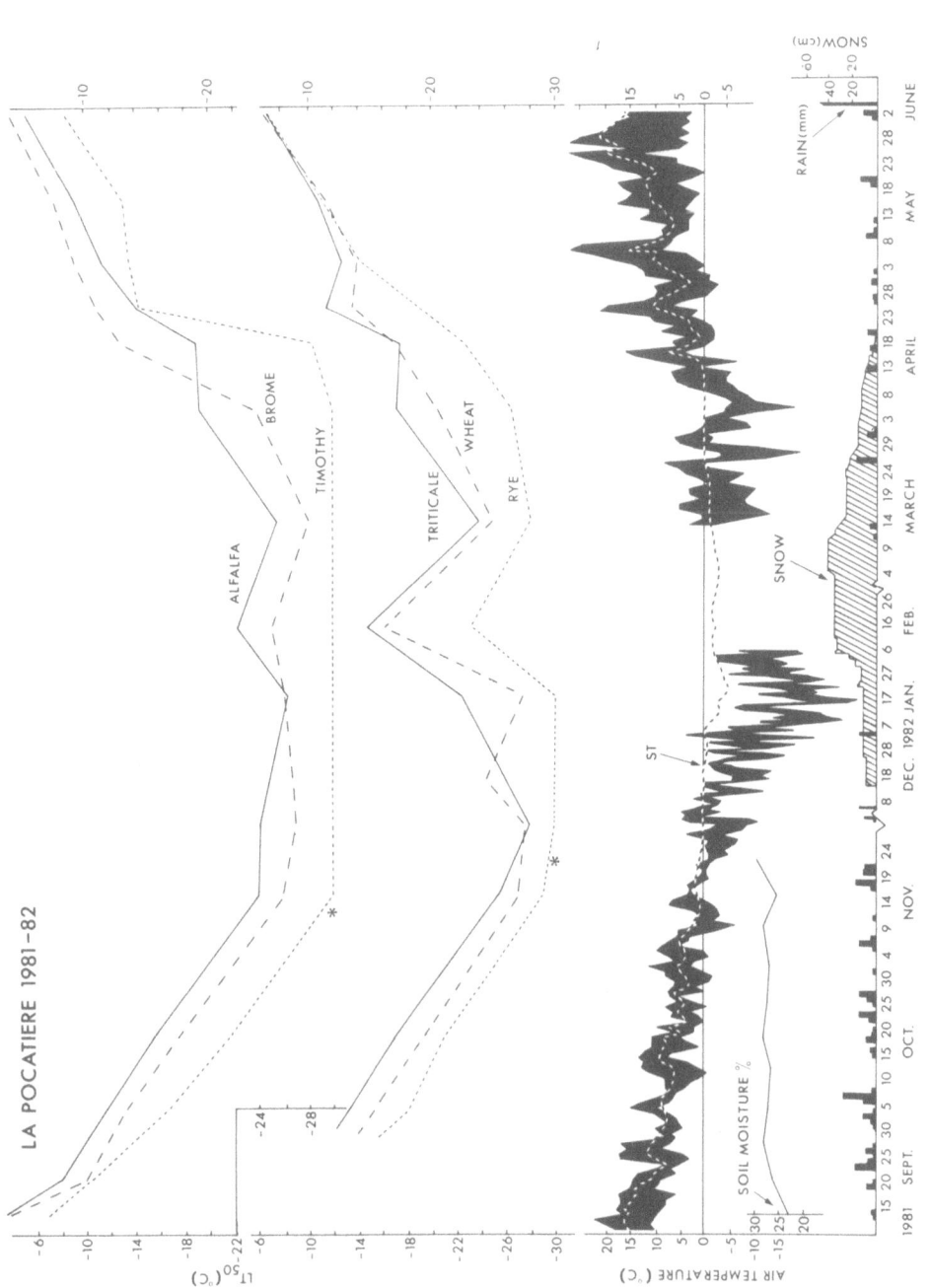

FIGURE 2. Hardening and dehardening (LT_{50}) of forage and cereal species grown at La Pocatière, air and soil temperature (ST) at plant level, soil moisture content and rainfall and snowfall accumulations in 1981-82. The symbol ＊ indicates the lowest temperature used in the laboratory experiments.

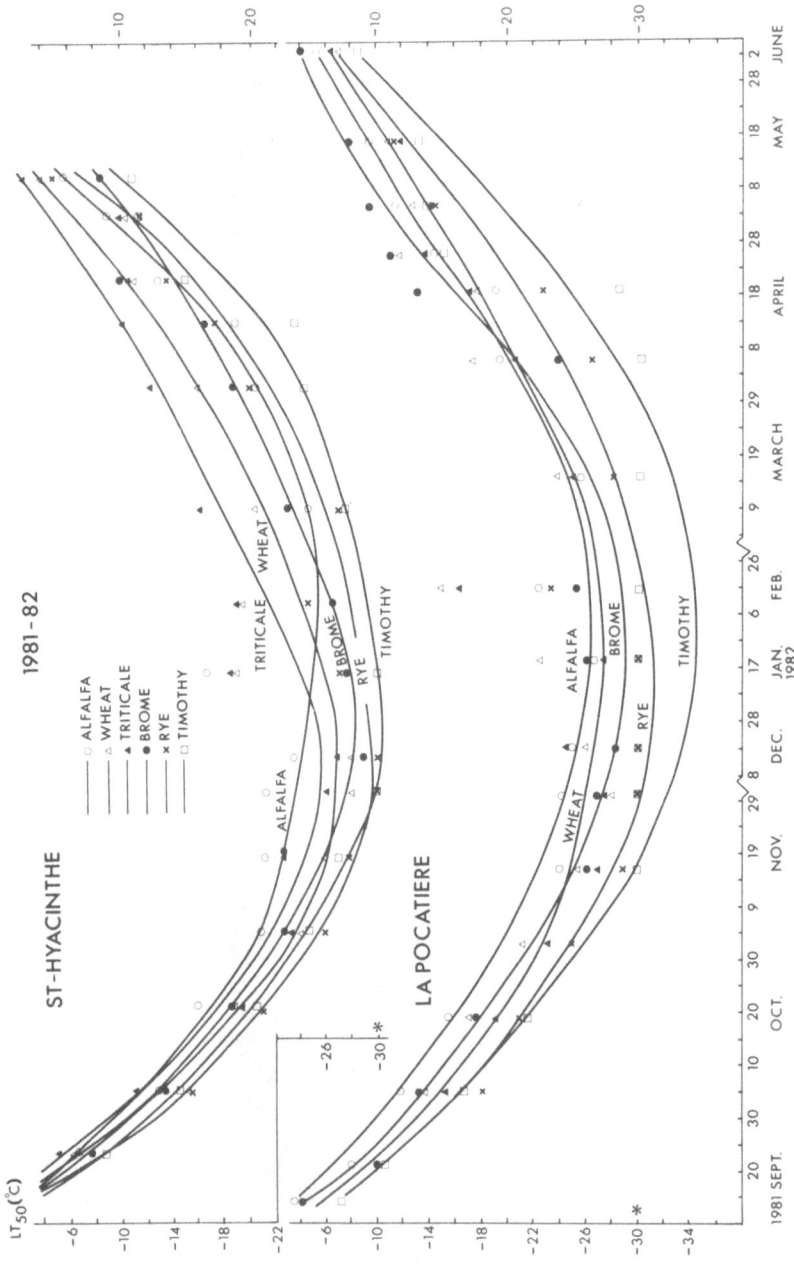

FIGURE 3. Hand drawn curves showing the evolution of frost resistance (LT_{50}) of forage and cereal species grown at La Pocatière and St-Hyacinthe in 1981-82. The symbol ✱ indicates the lowest temperature used in the laboratory experiments.

reached by the end of December for all species except alfalfa. At La Pocatière, where the climate is a little cooler, the maximum occurred one month later except for alfalfa. In both regions, alfalfa had its maximum later in the season than the other plants, though as for the other plants, the maximum at La Pocatière occurred after the maximum at St-Hyacinthe. In addition, plants from La Pocatière were capable of withstanding lower temperatures (measured as LT_{50}) than the same species from S -Hyacinthe. The relationship between the timing of maximum frost resistance and the lowest temperature a plant can withstand appears complex and further research will be needed to clarify it.

2.3. Relations between cold resistance and air and soil temperatures

The scatterograms of cold resistance (LT_{50}) versus air and soil temperatures show that the relationships are linear and upon calculation give a first order equation (Paquin and Pelletier, 1980). The correlation coefficients show also that the relations between the cold hardening or frost resistance acquisition and the air and soil temperatures at plant level are highly significant and close to unity (Table 1). Consequently, it is possible to predict the degree of frost resistance reached by plants from data of the mean air or soil temperature for one week preceding the date of prediction. These means are used in a previously established equation describing the regression of cold hardening to air and soil temperatures. The period used to calculate the equation extended from the beginning of cold hardening that occurs normally at the end of September in Quebec, to the first days of December when soils begin to freeze. Table 2 gives an example where frost resistance (LT_{50}) was estimated in 1981 for La Pocatière, using the data obtained in 1980. On the dates selected for the prediction, November 2 and 30, the mean air temperatures for the preceding weeks were $3.6^{\circ}C$ and $-2.6^{\circ}C$ respectively. The deviations observed between the LT_{50} estimated from the data

TABLE 1. Correlation coefficients between air temperature (AT), soil temperature (ST) and frost resistance (LT_{50}) of forage and cereal species at La Pocatière (PO) and St-Hyacinthe (HY) from 1980 to 1982.

SPECIES	LOCATION	YEAR	FALL HARDENING		SPRING DEHARDENING	
			AT	ST	AT	ST
WHEAT	PO	80-81	.99[A]	.99	.92	.82*
"	"	81-82	.96*	.96*	.97*	.82NS
"	HY	81-82	.92*	.95*	.91NS	.80NS
	TOTAL		.91	.93	.68	.52*
BROME	PO	80-81	.96	.97	.97	.90*
"	"	81-82	.98	.96	1.00	.73NS
"	HY	81-82	.96	.97	.98*	.94NS
	TOTAL		.86	.90	.88	.71
TIMOTHY	PO	80-81	.97	.98	.88*	.94
"	"	81-82	.99	.98	.71NS	.38NS
"	HY	81-82	.97	.97	.98*	.98*
	TOTAL		.90	.93	.78	.72
ALFALFA	PO	80-81	.99	.99	.89*	.89*
"	"	81-82	.99	.98	.87NS	.87NS
"	HY	81-82	.97	.98	.99*	.94NS
	TOTAL		.92	.94	.89	.87
RYE	PO	80-81	.99	1.00	.96	.94
"	"	81-82	.98	.97	.87NS	.61NS
"	HY	81-82	.98	.98	.99*	.86NS
	TOTAL		.90	.93	.82	.78
TRITICALE	PO	80-81	.99	1.00	.88*	.84*
"	"	81-82	.97*	.96*	.84NS	.84NS
"	HY	81-82	.89*	.91*	.74NS	.54NS
	TOTAL		.90	.93	.51NS	.40NS

[A] NS = NOT SIGNIFICANT; * = SIGNIFICANT.
ALL OTHER FIGURES ARE VERY SIGNIFICANT.

of 1980 and those found in 1981 after freezing tests are in general inferior to $2^{\circ}C$. The closest agreement between predicted and experimentally determined values occurred on November 30, near the end of autumn as cold hardening in the plants achieved its maximum. If soil temperature is used instead of air temperature, the difference is again in general inferior to $2^{\circ}C$ though slight differences exist between the values generated by soil temperature and those obtained using air temperature. It also appears possible to predict the LT_{50} of a species grown in St-Hyacinthe from the equations derived from the data obtained for the same species at La Pocatière in 1980. Confirmation of this statement by field experiments in the near future should allow the prediction of frost

TABLE 2. Frost hardening (LT_{50}) of forage and cereal species at La Pocatière in fall 1981. Calculations based on correlation with air temperature obtained in 1980 at the same location.

Species	AT °C	LT_{50} °C		
		Calculated	Found	Deviation
Alfalfa	3.6	-16.7	-20.0	3.3
	-2.6	-22.1	-24.0	2.9
Brome	3.6	-20.2	-21.0	0.8
	-2.6	-27.1	-27.0	0.1
Triticale	3.6	-22.6	-21.2	1.5
	-2.6	-30.6	-28.0	2.6
Wheat	3.6	-20.7	-23.3	2.6
	-2.6	-27.4	-27.7	0.3
Rye	3.6	-22.5	-25.0	2.5
	-2.6	-29.6	-30.0	0.4
Timothy	3.6	-23.7	-25.0	1.3
	-2.6	-31.4	-30.0	1.4

resistance (LT_{50}) for one cultivar grown in several regions of the Province from an equation obtained at one place.

However, the equation must be calculated for each cultivar of a species because of their difference in cold hardiness. Moreover, correlation coefficients decrease with decreasing cultivar hardiness (Paquin and Pelletier, 1980). In such cases, greater deviations estimated and observed LT_{50} values should be expected.

Correlation coefficients of dehardening to air or soil temperatures determined during the spring were less significant. Technical difficulties in sampling and mortalities caused by factors other than freezing appear to be the most plausible explanation.

2.4. Effects of soil moisture on frost resistance of alfalfa

Laboratory studies showed decreases of soil moisture from 100% to 30% of the water holding capacity increased the frost resistance (LT_{50}) of non-hardened alfalfa by 4 to 7°C (Paquin and Mehuys, 1980). Identical results were obtained

with alfalfa hardened at 1°C for 2 to 4 weeks at similar soil moisture treatments (Fig. 4). Drought stress appears to have additive effects on cold stress in alfalfa. However, the latter has a greater effect than the former, causing an increase in the LT_{50} of 10 to 15°C, and sometimes 20°C.

Freezing of plants in insulated pots to simulate field conditions did not influence the frost resistance of alfalfa in spite of the delay in the freezing of high soil moisture treatments (Paquin and Mehuys, 1980).

In the course of the field investigations on plant acclimation, the water content of the Kamouraska clay on which experiments were carried out varied from 20 to 35% (gravimetric basis) in spite of occasional heavy rains during fall. These figures re-

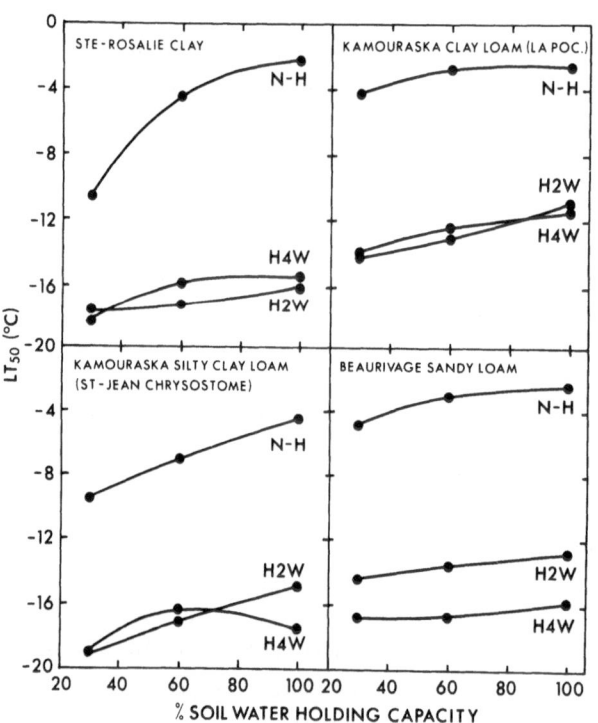

FIGURE 4. Cold tolerance (LT_{50}) of alfalfa grown in field soils without a cold-hardening period (N-H) or with a 2wk (H2W) or a 4-wk (H4W) hardening period, at different soil moisture levels and frozen as such.

present 35 to 65% of the maximum water holding capacity for this soil. Contrary to the expectations generated by the previous experiments (Fig. 4), the effects of soil moisture on the LT_{50} under the field condition were only 1 to $2^{o}C$, a slight response when compared to the effects of cold hardening. The results were confirmed by further experiments with irrigated plants grown on the same type of soil (Figs. 5 and 6). Soil compaction and slow water percolation in the Kamouraska clay would explain the 15% variation observed in the water content during the fall. Drainage had little effect on the soil moisture which remained around 20%. Consequently, drainage had little influence on plant cold hardening.

2.5. Effects of ice and snow on winter survival of alfalfa

Ice formation on frozen soils before final snow coverage (end of November or first days of December) was catastrophic for alfalfa (Fig. 5). Frost resistance decreased rapidly after ice formation and plants were completely destroyed within 4 to 8 weeks.

Under ice, the soil temperature at crown and bud level decreased to a freezing point critical for plant frost resistance. This point depends on soil moisture content, the soil temperature and the duration of the plant exposure to a given temperature. At $-8^{o}C$ with a soil moisture of 17%, the period necessary to kill 50% of plant population is approximately 4 weeks. This period is reduced to 2 weeks at $-10^{o}C$, at the same moisture content (Table 3). At a soil moisture of 35%, the period is reduced to 7 to 8 days at $-8^{o}C$ and to 24 to 36 hours at $-10^{o}C$. In a saturated soil, the period is 1 to 5 days at $-8^{o}C$ and 8 to 14 hours at $-10^{o}C$. Pomeroy et al. (1975) have already shown that the freezing period becomes more important when stress temperature is near the lethal temperature. For example, winter wheat could resist $-6^{o}C$ for 120 hours while at $-12^{o}C$ only 40% of population survived the same exposure period. Only 10% were still alive after 48 hours at $-16^{o}C$.

TABLE 3. Effects of freezing at -8 and $-10^{o}C$, and of soil moisture on frost survival of the alfalfa cultivar Rambler expressed as the time necessary to kill 50% of the population.

SOIL MOISTURE	$-8^{o}C$		$-10^{o}C$	
%	I	II	I	II
25 (16.7) [A]	33.3 D [B]	26.9 D	11.7 D	13.6 D*
50 (33.4)	7.6 D	8.6 D	13.6 H	40.0 H
100 (66.7)	1.1 D	4.9 D*	8.5 H	13.9 H
FLOODING	1.0 D	4.1 D*	3.3 H	2.1 H

[A] LEFT: % OF WATER HOLDING CAPACITY;
IN PARENTHESIS: WATER EXPRESSED AS % OF SOIL DRY WEIGHT.

[B] TIME IN DAYS (D) OR IN HOURS (H) FOR 50% KILLING.

* EXTRAPOLATED.

Because it is a good cold conductor, ice can rapidly cause a decrease in soil temperature to a lethal point. For instance, soil temperature at 5 cm depth (crown level) in alfalfa stands covered with ice in December 1980 reached $-15.5^{o}C$ January 13, 1981, and remained below $-11^{o}C$ from January 8 to January 22 (Fig. 5). In 1977, soil temperature at crown level reached $-17^{o}C$ on December 13, in icy stands of alfalfa and $-13^{o}C$ on January 13. In stands not covered by ice, soil temperature remained above -5 to $-6^{o}C$ on the same dates. In 1979, soil temperature decreased to $-11^{o}C$ on Devember 20, to $-12.5^{o}C$ on January 25 and to $-14^{o}C$ on March 3 in icy stands of alfalfa. In 1981, soil temperature dropped to $-9.5^{o}C$ on January 22 and remained below $-8.0^{o}C$ for 10 days, from January 18 to 28 (Fig. 6). All plants were killed in these two years by the end of December or in January, 4 to 8 weeks after coverage by ice.

Comparison of the field studies with results obtained in artificial conditions (Table 3) leads to the conclusion that ice caused a decrease in soil temperature which when coupled to the dura-

146

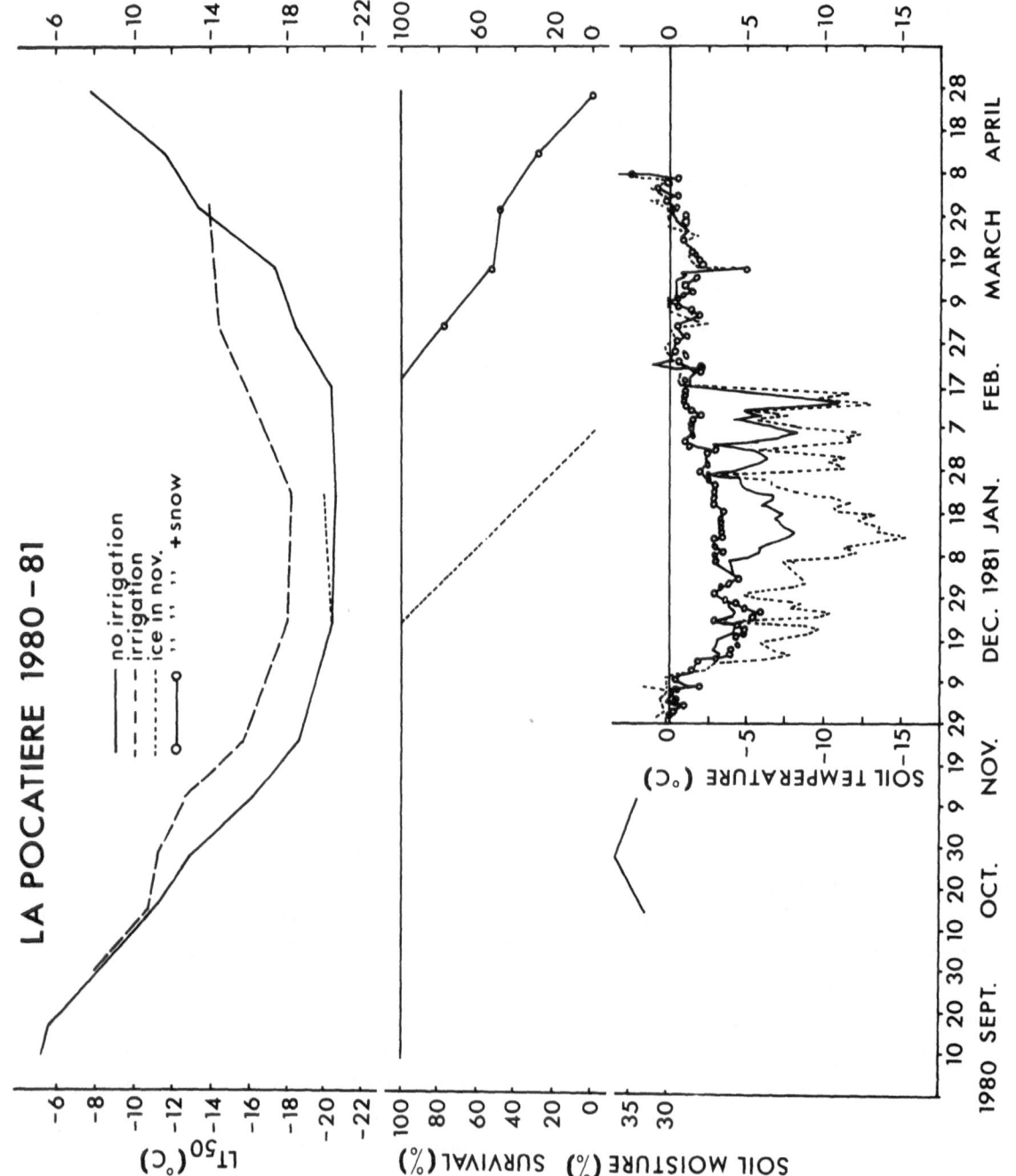

FIGURE 5. Effects of irrigation, ice and snow on soil temperature and on winter survival (%) and frost resistance (LT$_{50}$) of the alfalfa Vernal grown at La Pocatière in 1980-81. Soil moisture variation in %.

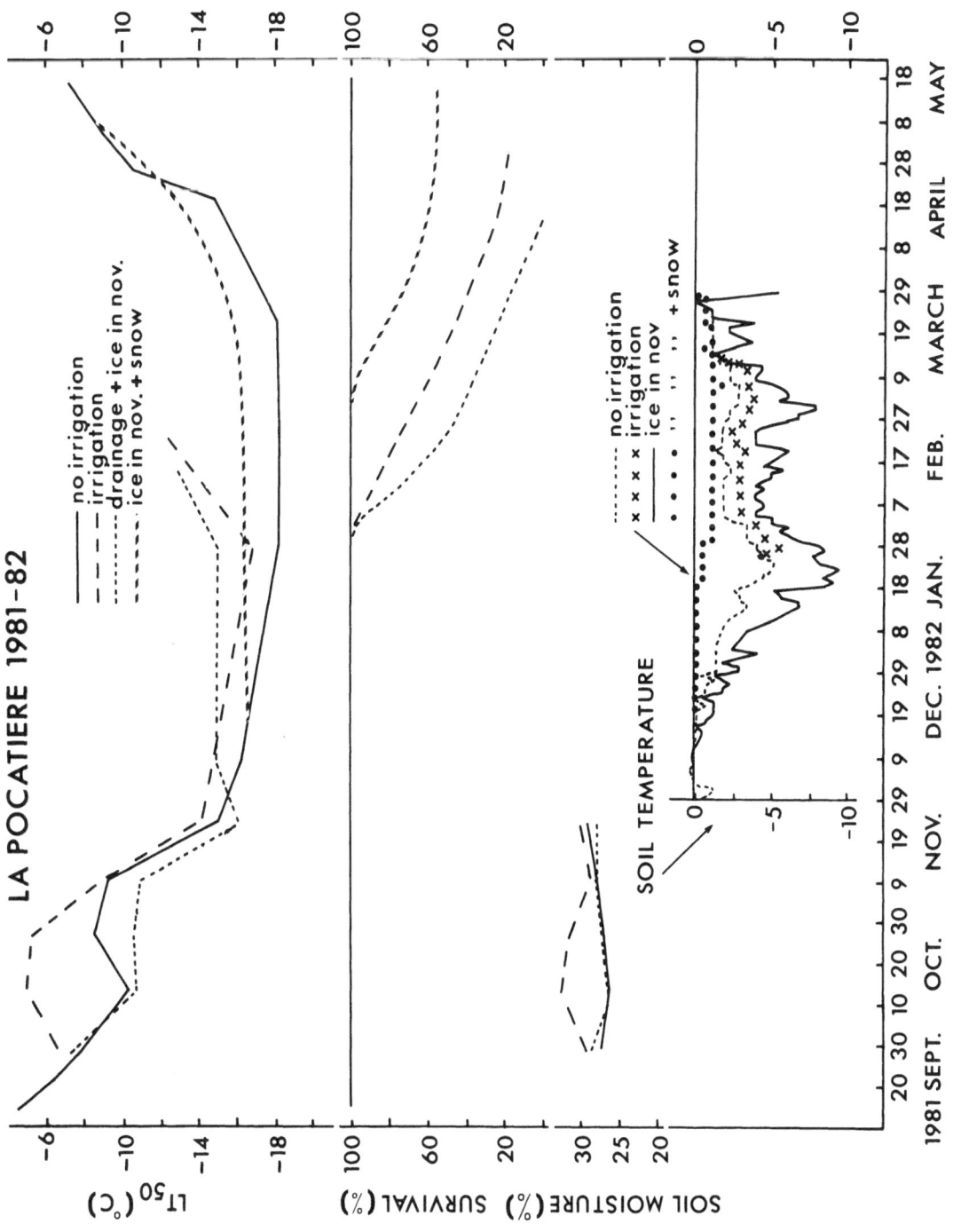

FIGURE 6. Effects of irrigation, drainage, ice and snow on soil temperature and on winter survival (%) and frost resistance (LT_{50}) of the alfalfa Vernal grown at La Pocatière in 1981-82. Soil moisture variation in %.

tion of exposure was well below the lethal point for alfalfa. Even when the frost resistance of alfalfa (LT_{50}) reaches a maximum as low as -22 to $-25^{o}C$, it cannot withstand them for more than a few hours.

Whenever snow covers ice, plants are better able to survive the damaging effects of ice (Fig. 6). In such cases, soil temperature is near the freezing point, similar to that of the check stand, but damages caused by the presence of ice are reduced and occur late in the season. A warming period is observed almost every year during winter, between December and February. This period may be accompanied by heavy rains lasting one to four days (Figs 2 and 3). In such cases, rains increase the density of snow and may run through it to form an ice cover on frozen soil. The soil in general remains frozen. Contrary to expectation, ice cover formed in January or February by this means or simulated artificially caused little winterkill of alfalfa (Figs 5 and 6). Soil temperature is thus little influenced if snow cover is thick enough to prevent ice from conducting cold temperatures to the soil.

2.6. Effect of cold hardening on the free proline content of forage and cereal species

During the growth stage, the free proline content of plant tissues is normally less than one micromole per gram of fresh weight ($\mu mole.g^{-1}$ fr. wt). The turnover of the amino acid is probably very rapid.

Any stress that slows down or stops growth such as drought, cold, salinity, heat, diseases or insects, causes an accumulation of free proline. Several plant species have been found to respond to general stresses in this way (Aspinall and Paleg, 1981). It seems therefore quite logical to try and relate the augmentation of proline to the resistance of plants undergoing a stress and to use this as a measure of that resistance. Several authors have shown that the augmentation of the proline content could be used as a measure of drought or cold resistance and could serve in

a selection program (Le Saint, 1966; Singh et al., 1972; Draper, 1975; Pinter et al., 1979). However, other investigators argue that the augmentation varies too much and occurs only in severe stresses, thus rendering the technique unsuitable as a measure of plant resistance (McMichael and Elmore, 1977; Chu et al., 1978; Yelenosky, 1979; Hanson et al., 1979).

During controlled laboratory hardening at 1.5 and $5.0^{o}C$, free proline accumulated in leaves and crowns of alfalfa (*Medicago media* Pers. cv. Rambler). The accumulation was significantly correlated with the frost hardening (LT_{50}) of the cultivar (Fig. 7) (Paquin, 1977). The correlation was less evident with the proline content of the roots or when hardening took place at $10^{o}C$. Proline accumulated also in leaves, crowns and roots of three cultivars of timothy (*Phleum pratense* L.) hardened for 4 weeks at $1.5^{o}C$ (Paquin and St-Pierre, 1980).

Proline accumulation occurred later than frost resistance acquisition when plants underwent two laboratory cycles of cold hardening and dehardening, indicating that the accumulation could be a consequence rather than a cause of the hardening. At $1.5^{o}C$, proline accumulated rapidly in leaves of alfalfa, but slowly in crowns and roots. In contrast, hardening at 5 and $10^{o}C$ caused a rapid accumulation of proline in crowns and roots (Paquin, 1977). Several authors have suggested proline is synthesized in leaves and transported to roots (e.g. Stewart et al., 1966; Tully et al., 1979). The hardening temperature could therefore have an effect not only on proline accumulation, but also on its transport. In fact, the cooling to $1^{o}C$ of a plant section at the base of a stem modified the proline distribution, while cooling to $5^{o}C$ produced no modification (Vézina and Paquin, 1982). It appears that there is a critical point between $1^{o}C$ and $5^{o}C$ below which the translocation of proline is hindered. (Table 4). On the other hand, proline did not accumulate in isolated alfalfa roots held at $1^{o}C$ nor did these roots develop frost resistance (Vézina and Paquin,

TABLE 4. Effect of cooling a stem section on free proline distribution in alfalfa.

PLANT PART	PROLINE (μMOL.G^{-1} DRY WT)			RATIO (%)	
	21°C [1]	5°C	1°C	5/21°C	1/21°C
LEAVES	45.9 [2]	45.2	21.0	98	46
STEM					
- ABOVE C.S. [3]	75.9	75.1	51.1	99	67
- UPPER S.C.	44.4	48.2	29.7	109	67
- LOWER S.C.	52.9	50.6	8.4	96	16
- BELOW C.S.	69.4	69.5	35.0	100	51
CROWNS	133.0	120.8	99.8	91	75

[1] TEMPERATURE INSIDE THE COOLING-JACKET.

[2] MEAN OF TWO EXPERIMENTS. DIFFERENCES IN THE PROLINE CONTENT BETWEEN PLANT PARTS AT 1.0°C AND BETWEEN TREATMENTS AT 1.0, 5.0 AND 21.0°C WERE VERY SIGNIFICANT, AT LESS THAN 0.01 LEVEL.

[3] C.S. = COOLED SECTION; S.C. = SECTION COOLED.

obtained with the winter wheat Kharkov (*Triticum aestivum*). Leaves therefore appear to be necessary for proline synthesis, with proline metabolism and transport also being influenced by low temperatures.

In a field study on cold acclimation of alfalfa grown in three regions of Quebec, analysis showed that the hardy cultivars accumulated more proline than the non-hardy ones (Fig. 8) (Paquin and Pelletier, 1981). Given the high correlation coefficients obtained, it appeared that proline concentration in crowns, site of frost resistance, could serve as a measure of cold hardening in the fall and dehardening or loss of resistance in the spring. However, variations from year to year and region to region make the measure unreliable.

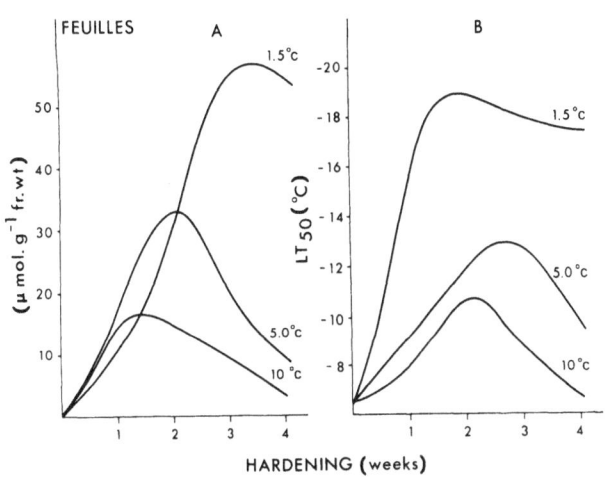

FIGURE 7. Effects of hardening temperatures on (A) the free proline content of leaves and (B) the frost resistance (LT$_{50}$) of alfalfa.

1982). Moreover, proline did not accumulate in crowns or roots of alfalfa kept at 21°C when the aerial part of the plant was maintained at 1°C (Vézina and Paquin, 1982). Identical results were

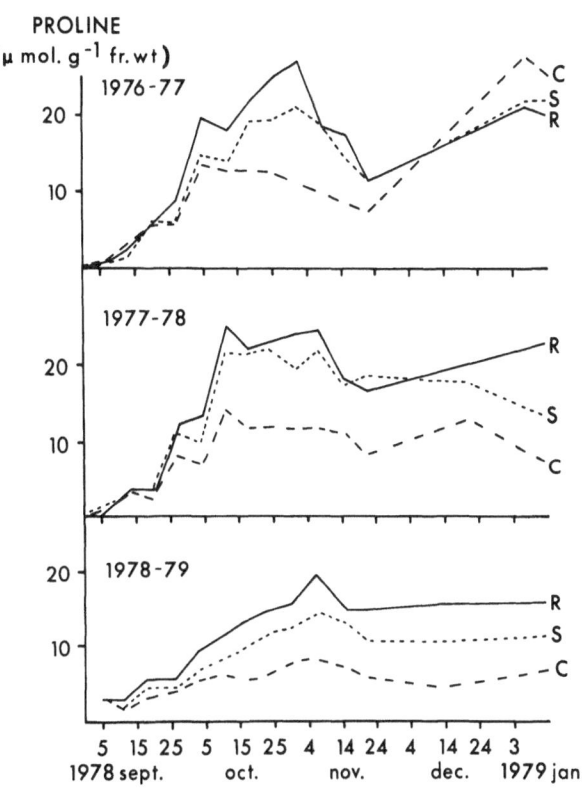

FIGURE 8. Free proline content of the hardy cultivar Rambler (R), the semi-hardy Saranac (S) and the non-hardy Caliverde (C) of alfalfa grown at La Pocatière from 1976 to 1979.

Moreover, because the proline accumulation ceases after leaves fall in November, its determination cannot be used to measure the maximum of frost resistance which occurs between mid-December and end of January (Paquin and Pelletier, 1980). The unreliability of proline as a measure of freezing resistance was confirmed in field experiments with 6 forage and cereal species. Timothy, which is more hardy than brome and alfalfa, accumulated less proline that the other two species (Fig. 9). However, rye, a more hardy species than wheat and triticale, accumulated more proline than the latter two. Once more, variations were observed in the proline content: in 1981, timothy accumulated as much proline as brome and alfalfa at La Pocatière while, in the same location, rye had a greater proline content than wheat and triticale. At St-Hyacinthe, the cereal species showed the same accumulation. The high standard deviation observed in the proline content between samples is an indice of a high variability between individuals of the same species and therefore makes the proline accumulation less reliable as a measure of frost resistance. The high standard deviation might be explained by the small number of plants dug out from frozen soils and by the poor health of some plants.

In the course of these investigations, a surge in the proline level in the crowns was observed towards the end of the winter, just before dehardening and regrowth. This observation apparently contradicts most of the studies that concluded to the requirement of light and photosynthesis for proline synthesis (Le Saint, 1966; Stefl et al., 1978; Yelenosky, 1979). The surge in proline level and regrowth occurred earlier for brome, generally in February, compared to the other species investigated.

3. DISCUSSION AND CONCLUSIONS

Once initiated, cold hardening of plants depends more on temperature than on photoperiod, though this does not exclude a role for light and photosynthesis in the process. Cold hardening and frost resistance (LT_{50}) of crowns and roots increased after leaves were destroyed by frost, normally in November, and continued to increase under the snow to a maximum level in January. The maximum level of frost resistance observed under field conditions was similar to that obtained in growth chambers.

Correlation coefficients between frost resistance (LT_{50}) and soil and air temperatures were near unity. The linear type relationships can be used to predict the degree of frost resistance attained by a plant from the mean air temperature. The prediction is valid until snow covers the soil, generally in December. The first attempts at prediction were successful but this type of investigation can be profitably pursued to develop more precision in modeling equations.

Will it be possible to predict from air or soil temperature the maximum of frost resistance for all the plant species considered? This will depend on the accuracy of further experimental assays. When soil is covered with snow, soil temperature becomes stable at -1 to $-3^{o}C$ and the increase in frost resistance no longer correlates with air and soil temperature. Our formerly linear equation of the first degree for the LT_{50} data versus time should then change to a parabolic (second order) equation (Fig. 3). Our results confirm those of Baadshaug (1973) who showed that, under snow, soil temperature remains near the freezing point, independent of the soil type. Moreover, according to Aase and Siddoway (1979), 7 cm of snow, in the absence of ice, is enough to prevent winterkill of wheat at temperatures of -25 to $-40^{o}C$. Soil temperature cannot then be used after snow fall to express the relationship with frost resistance. However, it is quite possible that the level of cold hardening reached by plants when snow begins to fall helps to determine the maximum cold resistance which occurs later. The second order equation could then be coupled with other factors to form a model useful to predict the winter survival of a species and to help to spread its cultivation in suitable regions.

151

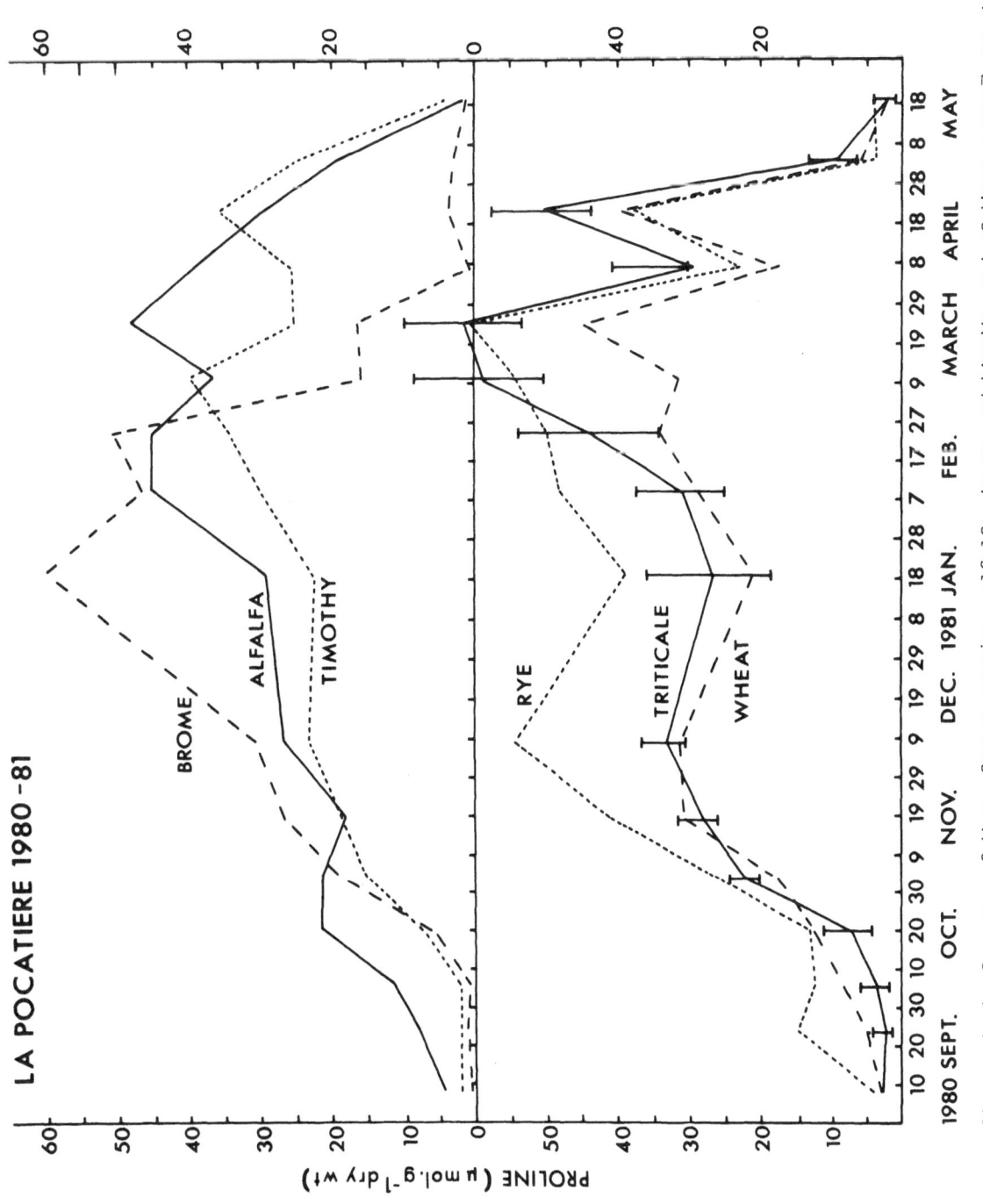

FIGURE 9. Free proline content of crowns of three forage species, alfalfa, brome and timothy and of three cereal species, rye, triticale and wheat, grown at La Pocatière in 1980-81. Vertical lines indicate standard deviations.

Ice formation on soil before snow coverage causes mortality of plants within 4 to 8 weeks. Ice, which is cold conductive, causes a decrease in soil temperature often below that of the critical point of frost resistance. The critical point depends on temperature, duration of exposure to that temperature and the cold hardening ability of a given cultivar. Even though hardened plants can withstand temperatures as low as -30 and -35°C for a few hours (2 to 6 hours), their exposure to moderate temperatures of -8 to -10°C for a longer period, for example several days, can prove fatal. Pomeroy et al. (1975) came to the same conclusion. The hypothesis that soil freezing is the main mortality factor during winter is supported by the observation that when snow covers the ice sheet on soil early enough to prevent a lowering of soil temperature to the critical point, the winterkill is reduced and occurs later in the season. Moreover, ice formation in January or February under snow causes little winterkill. These results can hardly be reconciled with the root asphyxiation hypothesis as the main cause of winterkill (Andrews and Pomeroy, 1981). However, soil freezing joined to root asphyxiation occurring under ice cover could hasten winterkill. In such cases, the type of ice, flaky, porous or hard, could be important. Soil type and compaction could also influence the winterkill (Baadshaug, 1973). On heavy and compacted soils, winterkill is always greater than on light soils.

Non-hardy cultivars such as Caliverde responded more rapidly to a rise in air temperature during cold hardening than hardy one. Moreover, some species such as brome have been observed to start growth earlier in the spring than others. It is likely that dormancy and/or phytohormones play a role in the frost resistance acquisition and in winter survival. This subject remains to be explored.

Free proline content increases in tissues of most species undergoing cold, drought, salinity, heat and disease stresses (Wilding et al., 1960; Barnett and Naylor, 1966; Draper, 1972; Perdrizet, 1974; Liu and Hellebust, 1976). In actively growing plants, the turnover of the amino acid is rapid and its concentration stays low, less than one micromole per gram of fresh weight. If in some cases proline determination can be used as a measure of cold hardening and other stresses (Singh et al., 1972; Paquin, 1977; Pinter et al., 1979; Paquin and Pelletier, 1981), it still leads to controversy when applied as a criterion in a selection program for plant resistance (Chu et al., 1978; Yelenosky, 1979; Hanson et al., 1979). Under field conditions, free proline content of plants is not necessarily related to cold hardiness (Fig. 8). The timothy Engmo that was more hardy than the winter wheat Kharkov, the brome Saratoga and the rye Cougar, accumulated less proline during cold hardening.

However, proline accumulation in crowns and roots of alfalfa appears to be necessary to the frost resistance acquisition. The translocation of the amino acid from leaves to roots is influenced by the hardening temperature. Most of the investigation until now led to the conclusion that proline is synthesized in leaves and requires light (Le Saint, 1966; Stefl et al., 1978; Yelenosky, 1979). However, this process cannot explain how increases in the free proline content of alfalfa crowns occur just before the plant starts to grow at the end of the winter. Two hypotheses could be advanced: first, proline could be translocated from roots to the crown where meristematic tissues and buds become active and, second, the crown or root tissues synthesize the amino acid from a precursor such as glutamic acid. Most of the investigations on proline metabolism to date have been performed with leaf tissues. It would be interesting to investigate whether the roots are capable of synthesizing this amino acid in the absence of leaves.

It is well known that, at least under our climatic conditions, the winter wheat is more hardy than alfalfa but does not survive as well during winter. Morphological differences such as the alfalfa root

system which is more developed than that of winter
wheat could be postulated as an explanation. But
this cannot be applied to timothy which has a
root system similar to that of winter wheat and
shows better winter survival than alfalfa. The
most logical way to explain the differences
observed in the winter survival of these species
appears to be their food reserves (starch, sugars
fructosans, amino acids, etc.) in relation to
dormancy and respiratory metabolism at low temperature. Jung and Smith (1961) have already observed
that the increase in the total sugars (non-struc-
tural carbohydrates) in alfalfa roots during fall
and winter depends on starch conversion to soluble
sugars. According to these authors, the frost
resistance of alfalfa begins to decrease when the
total sugars become less than 15%. We have obtained
results which contradict these findings (Paquin
and Lechasseur, 1982) and are currently conducting
experiments to clarify the question.

ACKNOWLEDGMENTS

We wish to thank G. Pelletier and P. Lechasseur
for technical assistance, the staff of the Ste.
Foy Research Station and the Experimental Farm
of La Pocatière for their help in the field work,
Dr H.J. Hope and Mr F. Whoriskey for manuscript
revision. We are also indebted to Mrs. M. Bernier-
Cardou who performed statistical analysis of the
results, Mr J. St-Cyr for the drawings and figures
and Mrs. C. Brochu for typing the manuscript.

REFERENCES

Aase JK and Siddoway FH (1979) Crown-depth soil
temperatures and winter protection for winter
wheat survival, Soil. Sci. Soc. Am. J. 43, 1229-
1233.
Andrews CJ and Pomeroy MK (1981) The effect of
flooding on cold hardiness and survival of winter
cereals in ice encasement, Can. J. Plant Sci. 61,
507-513.
Aspinall D and Paleg LG (1981) Proline accumulation: physiological aspects. In Paleg and Aspinall, eds. The physiology and biochemistry of
drought resistance in plants, pp. 205-241, Academic Press, N.Y.
Baadshaug OH (1973) Effects of soil type and
soil compaction on the wintering of three grass
species under different wintering conditions,
Acta Agr. Scand. 23, 77-86.
Barnett NM and Naylor AW (1966) Amino acid and
protein metabolism in Bermuda grass during water
stress, Plant Physiol. 41, 1222-1230.
Bolduc R (1976) Technique pour échantillonner
les racines de plantes dans le sol gelé et enneigé,
Can. J. Plant Sci. 56, 633-638.
Chu TM, Jusaitis M, Aspinall D and Paleg LG
(1978) Accumulation of free proline at low temperatures, Physiol. Plant 43, 254-260.
Draper SR (1972) Amino acid changes associated
with low temperature treatment of Lolium perenne,
Phytochem. 11, 639-641.
Draper SR (1975) Amino acid changes associated
with the development of cold hardiness in perennial
ryegrass, J. Sc. Fd. Agric. 26, 1171-1176.
Hanson AD, Nelson CE, Pedersen AR and Everson EH
(1979) Capacity for proline accumulation during
water stress in barley and its implications for
breeding for drought resistance, Crop Science 19,
489-493.
Levitt J (1980) Responses of plants to environmental stresses.I. Chilling, freezing and high
temperature stresses. Academic Press Inc. N.Y.
Le Saint AM (1966) Observations physiologiques
sur le gel et l'endurcissement au gel chez le
Chou de Milan. Thèse Doct. Sc. (Etat), Paris,
série A, n° 4669.
Liu MS and Hellebust JA (1976) Effects of
salinity changes on growth and metabolism of the
marine centric diatom Cyclotella cryptica, Can.
J. Bot. 54, 930-938.
McMichael BL and Elmore CD (1977) Proline
accumulation in water stressed cotton leaves,
Crop Science 17, 905-908.
Paquin R and Pelletier H (1975) Technique de
prélèvement de plantes dans un sol gelé, Can.
J. Plant Sci. 55, 327-330.
Paquin R (1977) Effet des basses températures
sur la résistance au gel de la luzerne (Medicago
media Pers.) et son contenu en proline libre,
Physiol. Vég. 15, 657-665.
Paquin R, Ladouceur G, Desrosiers R and Mack A
(1977) Etude sur la survie de la luzerne au
Québec au moyen de photos couleurs et infrarouges
à des échelles de 1:6000 à 1:40000, Proc. IVe Symp.
Can. sur la Télédétection 2, 506-515.
Paquin R and Lechasseur P (1979) Observations
sur une méthode de dosage de la proline libre dans
les extraits de plantes, Can. J. Bot. 57, 1851-1854.
Paquin R and Méhuys GR (1980) Influence of soil
moisture on cold tolerance of alfalfa, Can. J.
Plant Sci. 60, 139-147.
Paquin R and Pelletier H (1980) Influence de l'
environnement sur l'acclimatation au froid de la
Luzerne (Medicago media Pers.) et sa résistance
au gel, Can. J. Plant Sci. 60, 1351-1366.
Paquin R and St-Pierre JC (1980) Endurcissement,
résistance au gel et contenu en proline libre de
la fléole des prés (Phleum pratense L.), Can. J.
Plant Sci. 60, 525-532.
Paquin R and Pelletier G (1981) Acclimatation
de la Luzerne (Medicago media Pers.) au froid.
1. Variations de la teneur en proline libre des

feuilles et des collets, Physiol. Vég. 19, 103-117.

Paquin R and Lechasseur P (1982) Acclimatation naturelle de la luzerne (*Medicago media* Pers.) au froid.II. Variations de la teneur en sucres totaux des feuilles et des collets, Acta Oecologica Oecol. Plant. 3, 27-38.

Perdrizet E (1974) Effect of chlorophyll deficiency on proline metabolism in higher plants, Fiziologiya Rastenii 21, 47-53.

Pinter L, Kalman L and Palfi G (1979) Determination of drought resistance in maize (*Zea mays* L.) by proline test, Maydica 24, 155-159.

Pomeroy MK, Andrews CJ and Fedak G (1975) Cold hardening and dehardening responses in winter wheat and winter barley, Can. J. Plant Sci. 55, 529-535.

Singh TN, Aspinall D and Paleg LG (1972) Proline accumulation and varietal adaptability to drought in barley: a potential metabolic measure of drought resistance, Nature New Biol. 236, 188-190.

Stefl M, Trcka I and Vratny P (1978) Proline biosynthesis in winter plants due to exposure to low temperatures, Biol. Plant. (Praha) 20, 119-128.

Steponkus PL (1978) Cold hardiness and freezing injury of agronomic crops, Adv. in Agronomy 30, 50-98.

Steponkus PL (1980) A unified concept of stress in plants in Basic Life Sciences. Genetic Engineering of Osmoregulation 14, 235-255, Plenum Press, N.Y.

Vézina L and Paquin R (1982) Effet des basses températures sur la distribution de la proline libre dans les plantes de Luzerne (*Medicago media* Pers.), Physiol. Vég. 20, 101-109.

Weiser CJ (1970) Cold resistance and injury in woody plants, Science 169, 1269-1278.

Yelenosky G (1979) Accumulation of free proline in citrus leaves during cold hardening of young trees in controlled temperature regimes, Plant Physiol. 64, 425-427.

Wilding MD, Stahmann MA and Smith D (1960) Free amino acid in alfalfa as related to cold hardiness, Plant. Physiol. 35, 726-732.

PART IV

ADAPTATIONS TO LIGHT

ADAPTATIONS OF UNDERSTOREY SPECIES TO EXIST IN TEMPERATE DECIDUOUS FORESTS

P. ELIÁŠ

Institute of Experimental Biology and Ecology, Slovak Academy of Sciences,
Bratislava, Czechoslovakia

1. INTRODUCTION

Temperate-zone forests are complex ecosystems with high species diversity and
complex vertical and horizontal structure, including some biogeohorizons and layers. Three basic forest layers are usually distinguished: tree, shrub and herb
or ground layer. Each of the layers exhibits particular environmental conditions
and has many plants, animals and microorganisms belonging to several life forms.

Leaf canopy of tall trees is the structural unit of forests which form the particular and characteristic environment of
the forest stand interior, beneath which
the understorey species develop and grow.
Existence of these plants in the frequently unfavourable specific conditions is
not possible without genetically fixed
and/or environmentally induced adaptations to limited or stress factors.
In this review I attempt to evaluate several adaptive features of understorey
plants and introduced some mechanisms of
adaptation to certain stress factors.

2. PARTICULAR ENVIRONMENT OF FOREST STAND
   INTERIOR

Light, soil drought and mineral-nutrient
deficiency we will consider to be basic
stress factors limiting the presence or
absence and distribution of plant species
in forest stands. The environment under
the tree canopy is primarily characterized by light-limited conditions which are
clearly manifested in the growing period.

Leaf canopies act as selective absorption
filters. Plants growing beneath them are
therefore subjected to a light environment
which differs from open habitats in two characteristic respects; namely, low total photon fluence rate, and high proportion of
far-red light (for details, see Smith,1982).
In deciduous forests the transmissivity of
tree canopy exhibits strong seasonal dynamics (Fig. 1). Therefore, we may distingui-
she two different periods or phases of
light environment: 1) shade phase in summer
period and 2) light phase in fall autumn,
winter and in spring, under the leafless
tree canopy.
Radiation regime in a forest stand is also
characterized by very large horizontal variability, namely uneven distribution of
light flux density on a soil surface (see
e.g. Evans, 1956; Anderson et al., 1969;
Alekseiev, 1975; Hutchison, Matt, 1977).
The variability results from tree canopy
heterogeneity and it is manifested in the
occurrence of sunflecks. The penetration of
direct solar radiation through canopy open-
ings is strongly controlled by the interaction of solar elevation and variation of
canopy density with angular elevation (Alekseiev, 1975; Hutchison, Matt, 1977). The
sunflecks differ by their irradiance intensity, duration, size and frequency.
Light conditions in the forest stand are
influenced also by age of trees which form
them. Light conditions are, in general, better in younger forest stands, less favourable in middle-aged stands, and again more

158

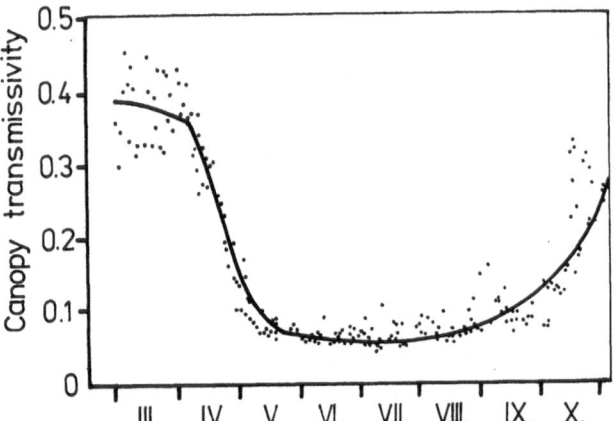

FIGURE 1 Seasonal variations in the canopy transmissivity of an oak-hornbeam forest stand in Báb, SW. Slovakia (by Žuffa, 1975).

favourable in old, matured stands (see e.g. Draskovits, 1975). In a successional series from clearings to closed forest stands, the development of the canopy results in reduction of light intensity, temperature fluctuation, vapour pressure deficit, and windspeed.

Temperature and air humidity in forest interior are closely connected with the complex radiation regime in a forest stand. It is known that in forest stands air temperature is relatively lower and air humidity higher, in comparison with free areas. However, in old, matured stands air humidity decrease to 40% of relative humidity (RH) on hot sunny days but in rainy or cloudy days it is close to 90% RH (Eliáš, unpublished). Windspeed in the forest interior is reduced but in specific topographic situations (hills, slopes etc.) it may be relatively high.

Soil drought we consider to be of secondary importance as a stress factor in forest stands. In warm regions of the temperate zone, the summer drought period also occurs frequently in deciduous forests, mainly when the soil water table is unavailable to the root systems of trees and

understory species. In the Slovak lowlands momental soil-moisture storage decreases in this period under the values of physiological availability by ground-layer species. The high soil-water deficits which occur in this period are not reduced to the balance level even when there is summer rain (Tužinský, 1976; Eliáš, 1978).

Soil-moisture content in forest stands also exhibits typical seasonal dynamics. In the spring period, the soil-water storage is close to maximum capillary capacity but in the summer period it decreases to the wilting point in some forest communities (Tužinský, 1976).

Horizontal variations in soil moisture are caused by soil properties, relief, snow-cover and its persistance in spring period, canopy structure, tree species composition etc.

The third basic stress factor in forests is the mineral nutrient-content in forest soils. Soil properties are determined first by mineral substrate but the forests strongly affect them by the quality and quantity of the litter fall which form humus horizons.

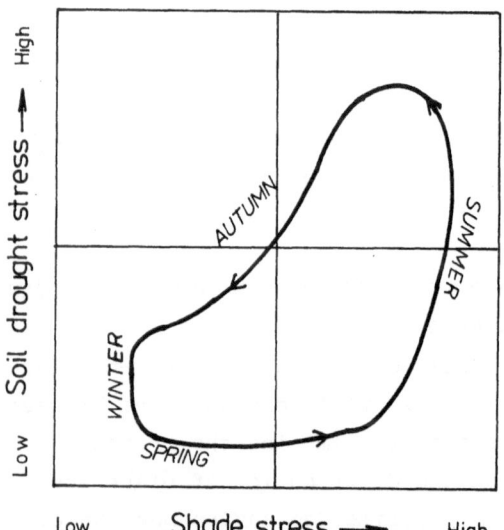

FIGURE 2 Seasonal changes in the occurrence and severity of two stresses in deciduous forests of the temperate zone.

TABLE 1. Occurrence of the stress factors in various ecological types
of temperate-zone deciduous forests

Forest type	Mineral-nutrient stress	Soil moisture stress	Light stress
Flood-plain forests	low	low	moderate
Mesophytic oak and beech forests	low	moderate	high
Xerothermic oak forests	low	high	low
Acidophilous forests	high	high	low
Scree-woods	low	low	high

Occurrence and severity of the stress factors differ in various forest stands or forest communities (Tab. 1). Compensations of unfavourable conditions in one direction by favourable conditions in other directions are evident. For example, soil drought in xerothermic oak forests is compensated by better light conditions and favourable mineral-nutrient content. Similarly, in acidophilous oak and beech forests the nutrient deficiency is compensated by better light conditions.

Variations in the basic stress factors of forest stands are large and may be factored into following types:

i) vertical distributions of stress factors in stand
ii) seasonal distributions of stress factors in stand
iii) horizontal distributions of stress factors in stand
iv) successional distribution of stress factors in stand, i.e. changes in the first three sources of variation during the successional development of the stand.

In this paper, we will consider mainly stress factor occurrence in ground layer and in shrub layer. These two low layers are occupied by understory species.

With regard to seasonal distribution of stress factors in forest stands we distinguish two main seasonal periods:

a) summer, stress-present period with limited light and soil-moisture content
b) spring, stress-absent period with unlimited light and soil-moisture content but occasional low temperature stress, and third
c) winter, low-temperature period (see Fig. 2).

3. UNDERSTOREY SPECIES IN DECIDUOUS FORESTS

Life forms are the morphological expression of the adaptation of organisms to their environment (Boarkman, 1980). Understory species may be designated in most cases as perennials: herbs with underground-storage organs and woody plants, shrubs and young individuals of trees (seedlings, saplings etc.). Mosses, lichens and algae occur also in ground layer but are omitted in this paper.

In deciduous forests of temperate zones, herbs are predominantly caulophytes, i.e. the species with perennial organs of stem origin and with adventitious roots as perennial organs. Rhizophytes and rhizocaulophytes represent only an insignificant portion, about 10% of forest flora (Plašilová, 1970). The herbs are typical forest plants, and except for some species of xerothermic forests,they do not occur outside of forest stands. Both forbs and grasses are present in the forest understory with grasses being most abundant in xerothermic stands. Underground organs gans with storage reserves play an important role in stress avoidance and stress tolerance of the species (Eliáš, 1978).

Shrub species occurring in forest stands are not typical forest plants: they grow better in forest margins forming "Mantel" and frequently form particular scrubs in open agricultural landscapes. In closed forest stands the species often do not flower or do so only with reduced intensity.

Tree seedlings and young individuals of trees (saplings etc.) occur in ground and shrub layers where they grow in competition with herbs and tall trees. Under a closed tree canopy they grow very slowly and never grow up to the height of tall trees unless light conditions improve. They only survive and grow up at most to shrub-size individuals. They do not prosper and frequently die away and have been classified by foresters as "suppressed trees" or trees of "0" -class.

Species composition of forest stands is a result of the occurrence and severity of stress factors. In mesophytic beech forests characterized by deep shade, mesophytes and sciophytes, large-leaf herbs predominate and only few grasses occur. On the other hand, in xerothermic oak forests characterized by more favorable light conditions but simultaneously high severity of soil drought, grasses and herbs with small leaves predomi-

nate.

Species composition of a forest stand may be viewed as a stage of successional development from clearings to the mature forest stand. Shrub species may be considered from this perspective as relics of previous successional stages of the forest. They prosper in clearings. After development of the tree canopy resulting in a decrease of light intensity, they survive in the forest interior forming a distinct forest layer. Absence of shrubs in managed forests is caused by human interference and sometimes also by grazing of domestic animals in forest stands.

4. ADAPTATIONS OF UNDERSTORY PLANTS TO FOREST INTERIOR ENVIRONMENT

Understory species growing in forest stands differ in their adaptive strategies. Adaptations to spatial and temporal variations in light conditions within the forest stand, including shade avoidance and shade tolerance, are remarkably manifested in species of temperate-zone deciduous forests. In the forests two basic synusia of understorey species occur:

i) spring synusium in the spring, stress -absent period with high light intensity and soil moisture content near field capacity. It consists of spring ephemerals, evergreen and semi-evergreen herbs and summer green species with early emergence

ii) summer synusium in the summer period with shade, soil drought and high temperature. It consists of summer and evergreen herbs, shrubs and young individuals of forest trees.

Plant components of the synusia differ distinctly in their main adaptive features and in adaptive strategies (Tab. 2). Stress

TABLE 2 . Adaptive features of understory plant species in deciduous forests

Features	Summer synusium	Spring synusium
A. Anatomical-morphological features		
Leaf-blade size	large	large
Thickness of leaf blade	very small	small
Mesophyll-cell size and length	large, short	large, short
Intercellular size	large	large
Leaf-nervature density	low	low
Stomata density and size	very low, large	low, large
Specific leaf area	very high	high
Chloroplast size	large	small
Chloroplast number per cell	low	very high
B. Physiological features		
Light-saturated photosynthetic rate	low	high
Light-intensity requirement for light saturation of photosynthesis	low	high
Light compensation point	low	high
Leaf dark respiration rate	low	high
Stomatal conductance for CO_2	low	high
Mesophyll conductance for CO_2	low	low
Leaf osmotic potential	high	very high
Subletal leaf-water deficit	high	low
Leaf water-holding capacity	low	very low
C. Biochemical or metabolic features		
Pigment content per dry mass	high	lower
Chlorophyll a:b ratio	high	high
Soluble protein content	low	high
Carboxylation enzyme activity	low	high
Capacity for electron transport	low	high

avoidance is typical for spring synusium species with high metabolic activity in April and May. The main adaptation of the spring synusium understory plants (geophytes and hemicryptophytes) consists of restricting the time of their phy-siological activity to the optimal ambient photoperiodic and water conditions. They exhibit several adaptive features characteristic for sun-tolerant species (Tab. 2). Vernal geophytes are shade and drought intolerant species. They survive the stress

162

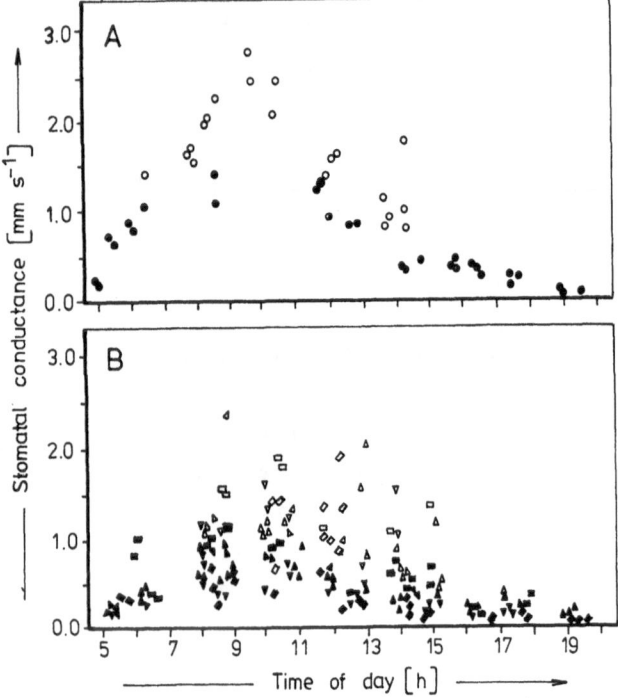

FIGURE 3. Daily variations in stomatal conductance of Pulmonaria officinalis (A) and other herbaceous species (B) in an oak-hornbeam forest. Open and closed symbols represent sunflecks and shaded conditions, respectively. 13 July 1977.

summer period individually since their bulbs, corms, rhizomes, or similar organs become dormant when stress develops. These spring ephemerals fulfill their complete individual life cycle during the light and wet spring period. They have evolved developmental mechanisms which allow them to complete their life cycle without having to face serious stress.

The development of shade in the forest interior in fall spring period leads to a wide range of responses by the understorey plants. The decrease in the compensation point of the spring-flowering geophytes is due to some internal rythm affecting the balance of photosynthesis and respiration rather than a progressive change in the environmental factors

Lieth, Asthon, 1961). Shade-induced changes in plant features are evident in summer herbaceous species with early emergence of shoots (Goryshina, 1979, 1981; Mitina, 1981). The rate of plant adaptation to new light conditions varies from some days to some weeks depending on species and environment. In evergreen herbaceous species the changes are also connected with forming of new leaves (aboveground organs).
Summer synusium species - summer and evergreen plants - are metabolically active throughout the growing period, including the whole shade and drought summer period. They are phanerophytes, chamaephytes and hemicryptophytes. They exhibit the ability to survive for time-limited periods in deep shade and usually tolerate simultaneous stresses of shade, soil drought, warm temperature and occasionally also mineral-nutrient deficiency.
The main adaptive features of summer synusium understorey species are summarized in Tab. 2. With very few exception the plants have a comparatively large surface in relation to their dry mass. Their leaves are large, belonging to the size class of mesophyllous and macrophyllous (20 to 100 and 100 to 500 mm^{-2}, respectively), thin , with large chloroplasts rich in chlorophylls. However, metabolic activity of the spe cies in summer period (with high frequency and severity of stresses) is low. Low relative growth rates of the species are genetically determined and, therefore, it was concluded that the natural selection in deeply-shaded habitats has been associated with the evolution of mechanisms of conserving energy (Grime, 1981).
Morphological and anatomical differences between spring and summer synusium species could be associated with adaptations of particular photosynthetic partial proces-

TABLE 3. Variations in selected ecophysiological features of main plant components of an oak-hornbeam forest at Báb, SW Slovakia

Features	TALL TREES (4*)		SHRUB LAYER (5)		GROUND LAYER (26)		
	Crown tops	Crown bottoms	Saplings	Shrubs	Vernal geophytes	Summer species	Tree seedlings
Specific leaf area ($mm^2 mg_{d.m.}^{-1}$)	7.6–12.8			28.6–38.5	30 – 55	30 – 64	30 – 37
Degree of leaf consistency ($mg_{d.m.}$ cm^{-2})	7.8–13.2	3.3–5.1	2.8–3.1	2.5–3.8	0.9–2.9	1.8–3.0	2.7–3.3
Stomatal density (number mm^{-2})	175–580	80–190	80–250	90–190	50–160	104–255	104–255
Maximum stomata conductance for H_2O ($cm\ s^{-1}$)	0.30–0.63			0.09–0.16	0.60–1.0	0.10–0.28	
Potential hydration of leaf tissue % of dry mass	90–140	140–200	170–210	200–350	400–910	300–630	140–260
Degree of leaf succulence (mg_{H_2O} mm^{-2})	7.8–15.6	4.0–7.3	4.6–6.7	ca 6	10–18 (44)	6.0–14.0	4.6–6.1
Subletal leaf-water deficit (% w.s.d.)	15–26	22–28	22–31	26–40		27–62	ca 35
Minimal xylem-water potential in natural environment (MPa)	–4.1to–3.3			–3.2to–2.6		–2.7to–1.2	ca –3.0
Leaf-blade chlorophyll content per dry mass (g kg^{-1})	4.3–7.8	9.2–13.0	10.3–13.1	10.5–17.4		14.3–27.3	12.2–16.7
Leaf blade chlorophyll content per area (g m^{-2})	0.3–0.7	0.3–0.6	0.3–0.5	0.3–0.6		0.3–0.6	0.3–0.45
Chlorophyll a:b ratio	2.9–4.2	2.6–3.2	2.6–3.0	2.5–3.4		2.5–3.1	2.4–3.2

* number of species observed

ses. However, Zelniker (1968) argued that physiological and metabolic adaptive features in summer synusium species are not adaptive but regressive changes resulted from insufficient development of photosynthetic apparatus. Goryshina (1979, 1980) supported this assumption by concluding that a decrease in the chloroplast number per unit of leaf area as well as shading may play an important role in summer decrease in photosynthetic rate of the forest herbs.

The role of reserves in the adaptation of summer synusium understorey species to stresses which occur in the period must be emphasized here. The reserves are formed in the more favourable conditions during fall spring, early summer, and autumn. The carbon balance of the plants exposed simultaneously to low light intensity, high soil drought and also high temperature remains positive (in balance) because of low respiration rates and occasional increases in the generally low rate of photosynthesis caused by sunflecks.(Fig. 3).

In sunflecks, higher transpiration rate, higher stomatal conductance and lower water potential as well as higher leaf temperatures were found, in comparison with shade areas (Pisek, Cartellieri, 1932; Rackham, 1975; Eliáš, 1979; Youngh, Smith, 1979; Woodward, 1980; Smith,1981). As Mitina (1981) found light-intensity requirements of summer herbs for maximum (= light-saturated) photosynthetic rate are from 10 to 20 klx in summer shade period; but light intensity in forest floor in summer is only 2 to 5 klx, in shade areas. Photosynthesis at maximum rates in summer species may therefore only be possible in sunflecks. The sunflecks can be therefore responsible for

an appreciable proportion of leaf carbon gain throughout a day, especially under light-limited conditions Gross, 1982 .

REFERENCE

Alekseiev VA (1975) The radiation regime of a forest. Leningrad, Izdat. Nauka.
Anderson RC, Loucks OL and Swain AM (1969) Herbaceous responses to canopy cover, light intensity, and throughfall precipitation in coniferous forests, Ecology 50, 255-263.
Barkman JJ (1979) The investigation of vegetation texture and structure. In Werger MJA, ed., The study of vegetation, pp. 123-160. The Hague, Dr. W. Junk publ.

Boardman NK (1977) Comparative photosynthesis of sun and shade plants, Ann. Rev. Plant Physiol. 28, 355-377.
Draskovits RM (1975) Light intensity studies in beechwoods of different ages, Acta Bot. Acad. Sci. Hung. 21, 9-23.
Eliáš P (1978) Water deficit of plants in an oak-hornbeam forest, Preslia 50, 173-188.
Eliáš P. (1979) Leaf diffusion resistance pattern in an oak-hornbeam forest. Biologia Plantarum 21, 1-8.
Evans GC (1956) An area survey method of investigating the distribution of light intensity in woodlands with particular reference to sunflecks, J. Ecol. 44, 391 -428.
Goryshina TK (1979) Some structural and functional features of leaf assimiliatory apparatus in plants of the forest-steppe oakwood II. Seasonal dynamics of plastid apparatus in the herbaceous cover, Bot. Zhurn. 64, 469-478.
Grime JP (1981) Plant strategies in shade. In Smith H., eds., Plant and the daylight spectrum. 1st Int. Symp. of the British Photobiology Soc., Leicester, pp.150-186. London, Academic Press.
Gross LJ (1982) Photosynthetic dynamics in varying light environments: a model and its application to whole leaf carbon gain. Ecology 63, 84-93.
Hutchison BA and Matt DR (1977) The distribution of solar radiation within a deciduous forest. Ecol. Monograph 47, 187-207.
Lieth H. and Ashton DH (1961) The light compensation points of some herbaceous plants inside and outside deciduous woods in Germany, Can. J. Bot. 39, 1255-1259.
Mitina MB (1981) Light curves of photosynthesis of the herbaceous plants of the oak-grove, Bot. Zhurn. 66, 1454-1464.
Pisek A and Cartellieri E (1932) Zur Kenntnis des Wasserhaushaltes der Pflanze II. Schattenpflanzen, Jahr. f. Wiss. Bota-

nik 75, 578-643.

Plašilová J (1970) A study of the root ecology of perennial herbs in the undergrowth of deciduous forests, Preslia 42, 136-152.

Rackham O (1975) Temperatures of plant communities as measured by pyrometric and other methods. In Evans GC, Bainbridge R, and Rackham O, eds., Light as an ecological factor: II, pp. 423-449.

Smith H (1982) Light quality, photoreception, and plant strategy, Ann. Rev. Plant Physiol. 33, 481-518.

Smith WK (1981) Temperature and water relation patterns in subalpine understory plants, Oecologia 48, 353-359.

Tužinský L (1977) Water regime of forest and agricultural soils on loess cover, Ved. Práce Výskum. Ústav. Lesn. Hospod. 25, 125-154.

Woodvord FI (1980) Shoot extension and water relation of Circea lutetiana in sunflecks. In Grace J, Ford ED and Jarvis PG, eds., Plants and their atmospheric environment, pp. 83-91. Oxford, Blackwell Sci. Pub.

Youngh DR and Smith WK (1979) Influence of sunflecks on the temperature and water relations of two subalpine understory congeners, Oecologia 43, 195-205.

Zellniker UL (1968) Adaptation of forest plants to shade, Bot. Zhurn. 53, 1478-1491.

Žuffa J (1975) Global radiation under the conditions of the oak-hornbeam stand in course of the vegetation period. In Biskupský V, ed., Research project Báb (IBP), Progress Report II, pp. 391-404, Bratislava, Veda.

GROWTH OF *Dactylis glomerata* L. AND *Bromus erectus* Huds. IN NATURAL HABITATS AND ALONG LIGHT AND WATER GRADIENTS

J. ROY (CEPE-CNRS, BP 5051, route de Mende 34033 MONTPELLIER CEDEX, France)

SUMMARY

In the Northern part of the French Mediterranean region, *Bromus erectus* is the dominant species of the herbaceous strata under *Quercus pubescens* canopies while *Dactylis glomerata* is found mainly in open swards corresponding to early stages of succession. Growth of the two grasses was studied in natural open and under canopy swards to determine how it is affected by the different habitats. Growth was also analysed along light and water experimental gradients to test if a differential response of the two species to light and water could explain their ecological distribution.

In spite of the light reduction under the oak canopy (35 and 75 % of the incident light with and without leaves respectively), wood and sward individuals had similar aerial biomass except in summer when the canopy improves water status and increases growth. Despite this similarity in aerial biomass, there were differences between the two species in tiller number, leaf number per tiller, and leaf surface area.

Compared to *D. glomerata*, *B. erectus* was less affected by light reduction (partially explaining their respective ecological distribution) and also by an increase in water stress. Ecotypic differentiations were found between wood and sward populations in both species but plasticity appeared to be the main factor that allows these species to grow in the two studied habitats.

1. INTRODUCTION

Dactylis glomerata and *Bromus erectus* are distributed over all the Northern and Central Europe. Like many other septentrional grasses their distribution probably extends to the Mediterranean region because of their high ecotypic differentiation (Roy, 1982).

In the Northern part of the French Mediterranean climate with cool winters, *D. glomerata* ssp. *hispanica* and *B. erectus* are found in several stages of the succession leading from abandoned vineyards to *Quercus pubescens* forest. *D. glomerata* is frequent in open swards but is progressively eliminated when the forest canopy becomes dense. *B. erectus* is the dominant species under these canopies and is less abundant in open swards unless, according to Escarré (1979) fire and grazing are frequent.

These contrasting ecological situations have been chosen to be the subject of a study on adaptative strategies. Several physiological, demographic and genetic aspects are under investigation in relation to environmental factors. The data presented here only concerns the growth of *D. glomerata* and *B. erectus* in two field sites (an open sward and a sward under a canopy) and along light and water experimental gradients.

Light and water availability differs in the open sward and under the canopy. The experiments with light and water gradients were conducted in order to determine if a differential response of the two species to these factors at least partially explains their ecological distributions. Intraspecific comparisons of open sward and wood populations were conducted to address the topic of the amount of ecotypic differentiation and plasticity that allows plants to occupy contrasting environments.

2. MATERIAL AND METHODS

The field site is located at St Martin de Londres, 25 km North of Montpellier. The open sward has vegetation which is similar to fields that are abandoned for 15 years (many species in the Labiaceae) and it is bordered by a plot with 50 year-old *Quercus pubescens* trees.

Light and water potential measurements were made with a linear pyranometer and a Scholander-type pressure chamber respectively. Rainfall is from the meteorological station ôf St Martin at a distance of 1 km from the field site.

For each of the four populations (*D. glomerata* and *B. erectus* in the open sward and the wood),

tillers were counted on 12 plants. The length
of mature leaves was measured on 6 of these
plants. Dry weight of green leaves per individual was calculated from the length of the leaves using a regression coefficient between length
and area and the ratio of lamina mass to lamina
area. Eighteen measurements were made from September 81 to September 82. The measurements of
the relationship between length-area-mass
were carried out on November 15th and April 28th.

In an other field site near and similar to
the first one, seeds of the two species were
harvested in the open sward and under the oaks
on August 81. They were sown on May 5th. Seedlings were transplanted to experimental boxes
two weeks later and were shaded and well watered
for two more weeks to promote their establishment. The experimental boxes consisted in
two soil moisture gradient boxes and two light
gradient boxes. The first ones were designed
after Pickett and Bazzaz (1976) and are drawn
in Fig. 1. The second ones had the same dimensions but were horizontal and were covered with
neutral filters of decreasing transmittance ;
soil was brought to saturation every evening.
Boxes were filled with and unleached red mediterranean soil. Seedlings were planted 11 cm
from each other.

Data were analysed by dividing the boxes in
50×50 cm^2 compartments each with 16 plants.
The extra row planted at each end of the gradient was not analysed. Measurements were taken
4 to 5 months after sowing. Tillers were counted
and the length of two of the longest leaves
of each individual was measured. Leaf number
per tiller was counted on one individual per
compartment. Dry weight of the aerial parts of
each compartment was measured after harvest and
divided by 16 to obtain the weight per individual.

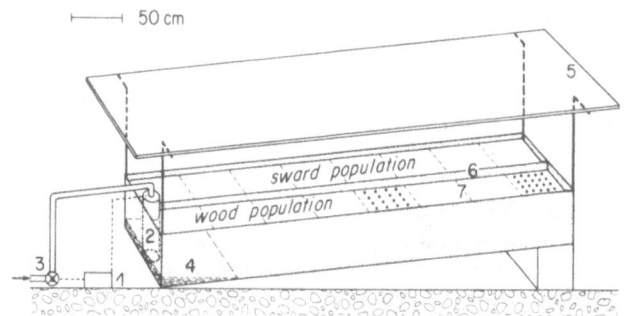

FIGURE 1. Inclined box providing gradient of soil
moisture. (An electronic device (1) opens or closes a valve (3) depending on the water level detector (2) so that the water table is held constant. A clear plastic sheet (5) protects the box
from rain. (6) is a true physical separation while
(7) indicates the analysis compartments).

3. RESULTS

3.1. In situ measurements

Under the *Q. pubescens* canopy, light energy
(300-3000 nm) from 10 a.m. to 18 p.m. is 75 % of
the incident energy in winter and 35 % when trees
bear leaves (Fig. 2). In fact plants receive almost full sun during one or two hours and about
10 % of the incident light the remaining part of
the day.

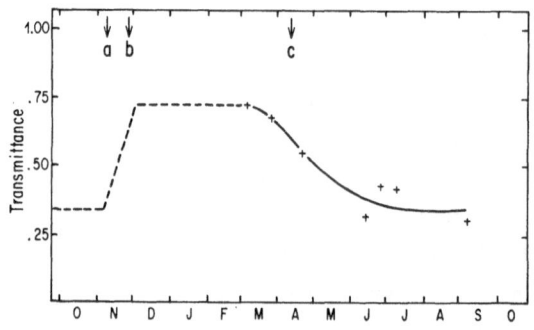

FIGURE 2. Seasonal transmittance of a *Quercus
pubescens* canopy. (Leaves turned yellow in the
beginning of november (a),fell at the end of
November (b) and appeared in April (c)). Dashed
line in the extrapolation from measurements with
or without leaves.

169

The seasonal evolution of minimum water potential (mid-afternoon value) shows how the tree canopy improves the water status of the herbaceous strata when trees bear leaves (Fig. 3). In both habitats *B. erectus* has water potentials that are higher than *D. glomerata* during summer drought.

In both habitats green leaf dry weight per individual of *B. erectus* compared to *D. glomerata* is higher and fluctuates more with the seasons (fig. 4). No drought induced dormancy is observed. Differences between sward and wood plants are not important until July when all *D. glomerata* and almost all *B. erectus* leaves become dry in the sward. The relative contribution of each component of the individual weight varies with the season and the habitat : Individuals under the canopy have less tillers (Fig. 5) but longer leaves (Fig. 6) and a slightly higher number of leaves per tiller (unpublished). shed).

5 which was demontrated by predawn water potential measurements (not shown).

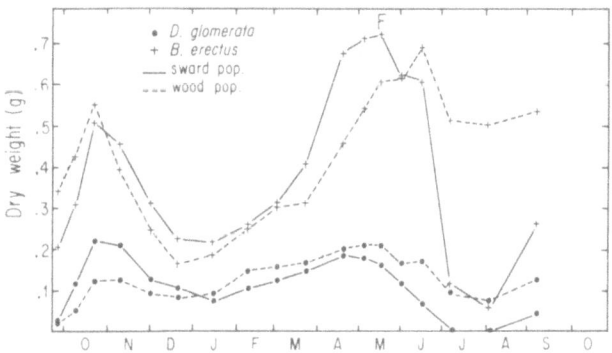

Figure 4. Seasonal dry weight of green laminas of sward and wood individuals of *D. glomerata* and *B. erectus*. (from Garnier and Roy, unpublished).

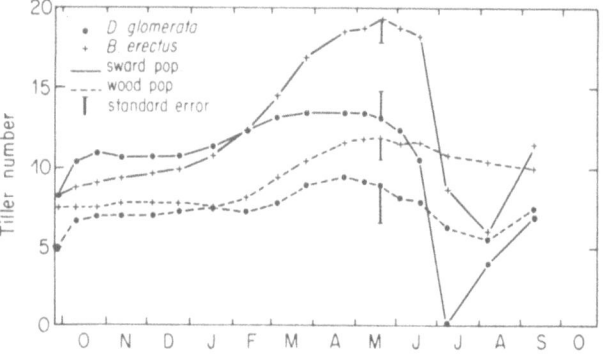

FIGURE 5. Seasonal number of tillers in sward and wood individuals of *D. glomerata* and *B. erectus*. (from Garnier and Roy, unpublished).

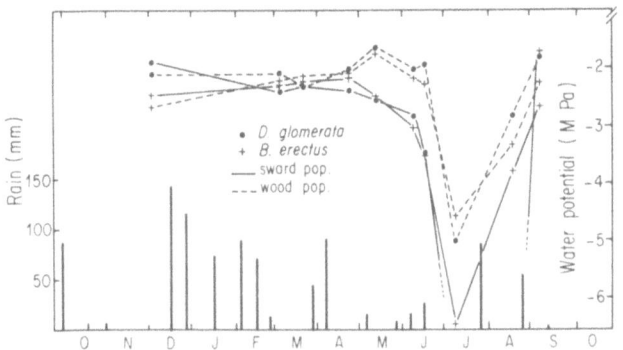

FIGURE 3. Rainfall and seasonal mid-afternoon leaf water potential of sward and wood individuals of *D. glomerata* and *B. erectus*.

3.2. Growth along light and water gradients

The magnitude of the light and water stress gradients prevailing in the boxes is shown in Fig. 7. Minimum water potential (mid-afternoon value) on a clear summer day is given in order to compare with the *in situ* measurements of Fig. 3, but it does not show clearly the increase in soil water stress from compartment 1 to

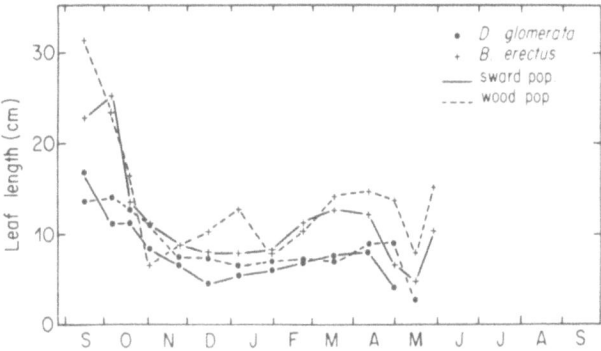

FIGURE 6. Adult leaf length as a function of leaf emergence date in sward and wood individuals of *D. glomerata* and *B. erectus*. (from Garnier and Roy, unpublished).

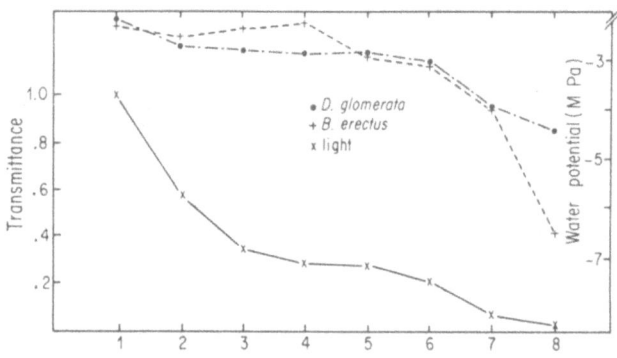

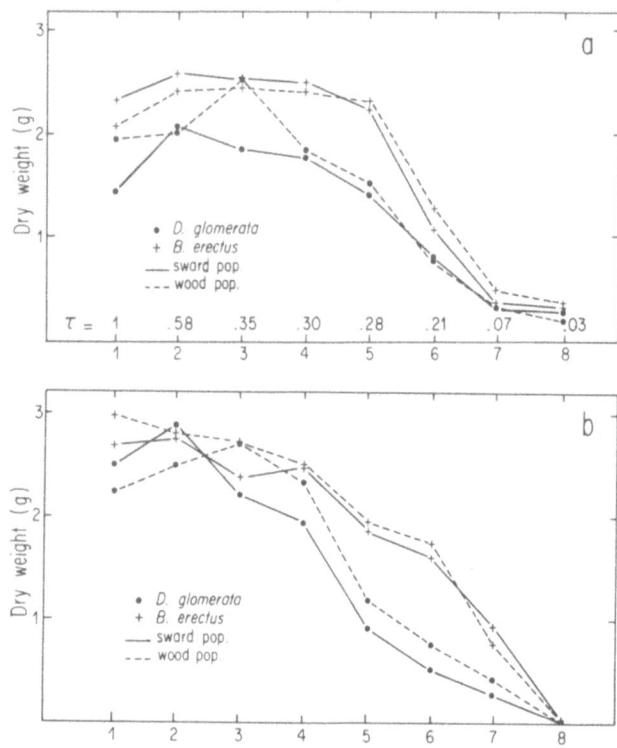

FIGURE 7. Light and water stress gradients in the experimental boxes measured through the screen transmittance and mid-afternoon water potential of *D. glomerata* and *B. erectus*. Numbers on the abscissa represent the analysis compartments and increase with increasing stress.

FIGURE 8. Dry weight per individual of sward and wood populations of *D. glomerata* and *B. erectus* cultivated along light (a) and water (b) gradients. (abscissa see Fig. 7).

Because of the mid-spring sowing, almost no plant flowered and the harvested dry weight relates mainly to green leaves. Fig. 8 shows that under optimal conditions of light and water, *D. glomerata* and *B. erectus* produced similar amounts of aerial biomass, but under both light and water moderate stress, *B. erectus* had a higher production than *D. glomerata*. In spite of the large decrease in light from compartment 1 to compartment 5 (100 % down to 30 %) both species but particularly *B. erectus* maintained high production. This was achieved through decreased tiller numbers (Fig. 9a) an increase in leaf length (Fig. 10a) and a constant leaf number per tiller. Fig. 9 and 10 show that the higher production of *B. erectus* under moderate stress mainly resulted from an increase of leaf length.

Differences between wood and sward populations occurred: wood populations performed better at lower hight intensity compared to sward populations as far as dry weight (*D. glomerata*) or tiller number (both species) were concerned. The wood population of *D. glomerata* had an optimum for dry weight production at a higher water deficit compared to the sward population.

4. DISCUSSION

4.1. Effects of light

The lower number of tillers in wood individuals in the field (Fig. 5) was probably the consequence of reduced light intensity since the experimental study showed a strong decrease in tiller number with shade (Fig. 9a). Such an effect of light is classical among grasses (Langer, 1963, Lawlor et al., 1981).

The increase in leaf length with shade (Fig. 10a) is also well documented (Taylor et al., 1968, Koller and Kigel, 1972). It results from an increase in cell number and cell length, but leaf width is only slightly affected (Forde, 1966).

4.2. Effects of drought

Drought decreases tiller number, leaf length

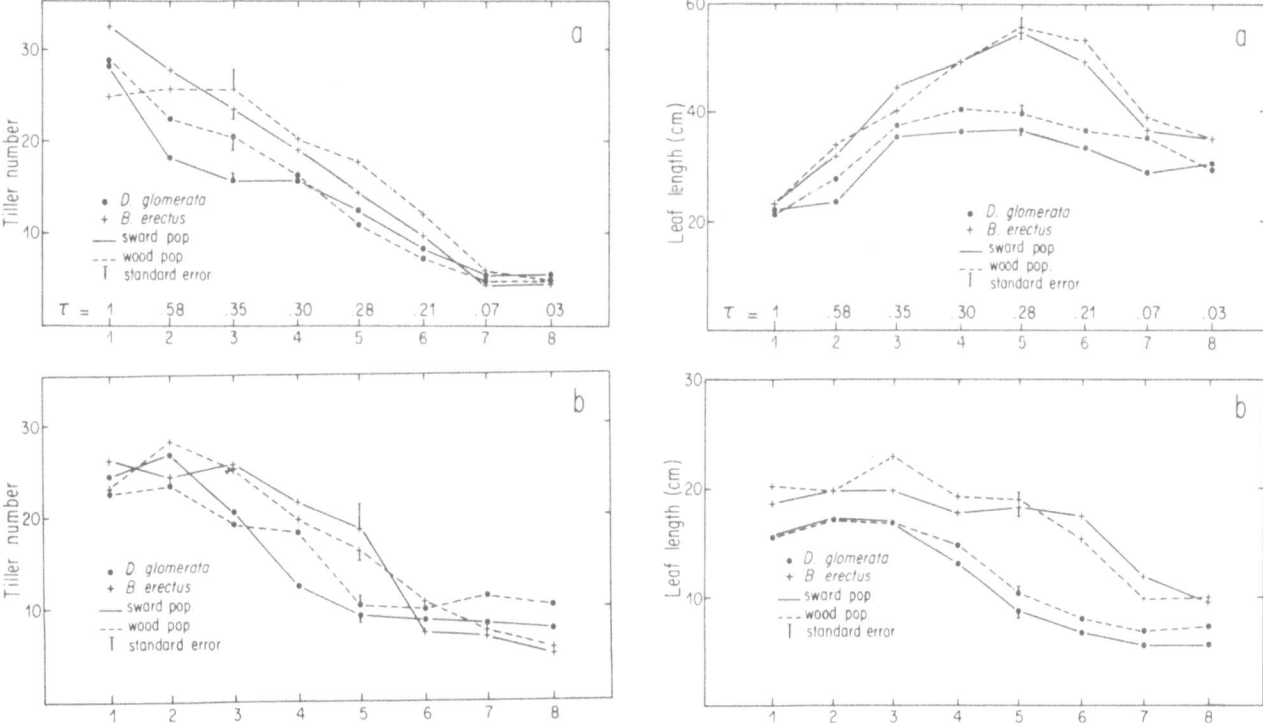

FIGURE 9. Tiller number per individual of sward and wood populations of *D. glomerata* and *B. erectus* cultivated along light (a) and water (b) gradients. (abscissa see Fig. 7).

FIGURE 10. Maximum leaf length of sward and wood populations of *D. glomerata* and *B. erectus* cultivated along light (a) and water (b) gradients. (abscissa see Fig. 7).

(Fig. 9b, 10b) and leaf number and it increases leaf senescence (Langer, 1963, Husain and Aspinall, 1970, Lawlor et al., 1981). Although leaf extension is very sensitive to drought through a decrease in cell turgor, production of leaf primordia stops at a higher water potential than does leaf extension (Husain and Aspinall, 1970).

4.3. Response to stress and ecological distribution

B. *erectus* dry weight is less affected by low light intensities than that of *D. glomerata* (Fig. 8a). This result partially explains its dominance under *Quercus pubescens* canopies. *D. glomerata*, though found mainly in open swards where water stress is higher, has a greater reduction in dry weight under drought conditions

as B. *erectus* (Fig. 8b).

In the field site *D. glomerata* has a dry weight per individual that is lower than B. *erectus* in the two swards and even at seasons when light and water are not limiting (Fig. 4). However under the optimal conditions of the experimental light and water gradients the two species had similar dry weights (Fig. 8). This suggests that growth of *D. glomerata* is limited in both the open sward and under the canopy. The factors that may be responsible could be soil nutrients and competition.

Relating our results to ecological distribution faces two problems : one is that the gradient experiments do not take factor interaction into account (Ernst, 1978). The second is that growth is only considered here while distribution might be related mainly to establisment. Seedlings of *D. glomerata* might be impeded by the thick oak

172

litter while vegetative reproduction of *B. erec-tus* might be lowered by the lack of photosyn-thate production in the open sward during summer.

4.2. Plasticity vs. ecotypic differentiation

Although some ecotypic differentiation is shown between sward and wood population of both (Fig. 8 and 9), plasticity appears to be the main mechanism allowing plant growth under different environmental conditions : In spite of a large light reduction, plants have been shown to maintain a high dry weight with tiller number reduction being compensated by an increase in leaf length. This parameter is particularly sensitive to environmental conditions : in *B. erectus* it varies from 30 cm to 7 cm with the seasons (Fig. 6) or from 20 cm to 55 cm with a reduction of light (Fig. 10). Data on photosynthetic rate and its plasticity would allow us to determine if increased leaf length compensates for light reduction as far as carbon acquisition is concerned.

ACKNOWLEDGMENTS

Thanks to the staff of the CEPE experimental field for building the boxes, to L. Sonié for numerous measurements, to G. Heim for comments on the manuscript and to L. Jackson for correcting the English.

REFERENCES

Escarre J (1979) Etude de successions post-culturales dans les hautes garrigues du Montpellierais, 3rd cycle thesis, Université des Sciences et Techniques du Languedoc, Montpellier, 171 p.

Ernst W (1978) Discrepancy between ecological and physiological optima of plant species. A reinterpretation, Oecol. Plant. 13, 175-188.

Forde BJ (1966) Effect of various environments on the anatomy and growth of perennial ryegrass and cocksfoot.1. Leaf growth, N.Z.J. Bot. 4, 455-468.

Husain I and Aspinall D (1970) Water stress and apical morphogenesis in barley, Ann. Bot. 34, 393-408.

Koller D. and Kigel J (1972) The growth of leaves and tillers in *Oryzopsis miliacea*. In Youngner VB and McKell CM, ed. The biology and utilization of grasses pp 115-134. New York, Academic Press.

Langer RHM (1963) Tillering in herbage grasses, Herbage Abstracts 33, 141-148.

Lawlor DW, Day W, Johnston AE, Legg BJ and Parkinson KJ (1981) Growth of spring barley under drought : crop development, photosynthesis, dry matter accumulation and nutrient content, J. Agric. Sci., Camb. 96, 167-186.

Pickett STA and BAZZAZ FA (1976) Divergence of two co-occurring successional annuals on a soil moisture gradient, Ecology 57, 169-176.

Roy J.(1982) Intraspecific variation in the physiological characteristics of perennial grasses of the Mediterranean region, Ecologia Mediterranea 8, 435-448.

Taylor TH, Cooper JP and Treharne KJ (1968) Growth response of orchard grass (*Dactylis glomerata* L.) to different light and temperature environments. 1. Leaf development and senescence, Crop Science 8, 437-440.

A MODEL OF GROWTH KINETIC THEORY OF ANACYSTIS NIDULANS IN LIGHT-LIMITED CULTURE

A.D. KARAGOUNI (University of Athens, Institute of General Botany, Panepistimiopolis, Athens 621, Greece)

1. INTRODUCTION

Consider a closed (batch) culture of any phototroph growing in an environment in which the supplied mineral resources, including carbon dioxide, are present in excess and therefore are present at concentrations which do not restrict the growth rate of the population. Furthermore, let us assume that all the physicochemical environmental parameters are optimal and remain constant throughout the growth period. The light intensity selected is chosen such that at some point during growth light, rather than a mineral resource, becomes the limiting factor.

Now, the rate of biomass production (or the rate of photosynthesis or the rate of carbon dioxide fication), dx/dt, is proportional to the initial biomass concentration, x,

Thus: $\frac{dx}{dt} \ \alpha \ x$ and so $\frac{dx}{dt} = \mu x$ (1)

where μ is a proportionality constant known as the specific growth rate.

The model developed for light-limited growth from the above equation (and this has been derived in a more complex and circuitous fashion elsewhere by Pipes, Koythoyannis (1962) depends on the relationship between the specific growth rate, μ, and the amount of light energy absorbed, I_a, by unit biomass (or per cell) such that:

$\mu \ \alpha \ I_a$ and so $\mu = A \ I_a$ (2)

Where A is a proportionality constant under the prescribed conditions for the conversion of radiant energy into chemical energy. Equation (2) is the analogue formula derived by Van Liere, Mur (1979) for *Oscillatoria agardhii* Gomont. Therefore the organism's specific growth rate is directly proportional to the absorbed light energy. At the beginning of a closed culture growth cycle when the initial biomass concentration, x,

is small, and there are no self shading effects restricting the quantity of light energy absorbed by each cell, the specific growth rate of the organism is proportional to the incident light intensity, I_i. However, as growth occurs and the biomass concentration increases, the absorbed light energy, I_a, must be a function of the incident light intensity and the biomass concentration since mutual shading causes a decrease in the amount of light energy absorbed by each organism.

Thus: $I_a \ \alpha \ \frac{I_i}{x}$ and so $I_a = B \ \frac{I_i}{x}$ (3)

Where B is proportionality constant.

Thus from equations (2) and (3):

$$\mu = C \ \frac{I_i}{x}$$ (4)

Where C is a composite proportionality constant of A and B. Since I_i has been defined as a fixed, chosen light intensity with a constant value during the course of the growth cycle, then:

$$\mu \ \alpha \ \frac{1}{x}$$ (5)

Thus this model predicts that for light-limited growth in batch culture the specific growth rate μ is ultimately proportional to $1/x$.

For light-limited growth in an open environment, the rate of change of the biomass concentration in the growth vessel, dx/dt, is given by the well known equation

$$\frac{dx}{dt} = \mu x - Dx$$

From equation (4) we have:

$$\frac{dx}{dt} = CI_i - Dx$$

and in a steady state culture when $\frac{dx}{dt} = 0$ by definition,

$$Dx = CI_i \quad \text{and so} \quad D = \frac{CI_i}{s} \quad \text{and}$$

$$D \: \alpha \: \frac{1}{x} \tag{6}$$

Thus data from steady state cultures should yield a straight line relationship when D is plotted against $\frac{1}{x}$ and the validity of this model is described and discussed in relation to the experimental results.

2. MATERIALS AND METHODS

The organism used in this study was *Anacystis nidulans* (strain number 625, Indiana University Culture Collection and obtained from N.G.Carr) and stock cultures were maintained as described by Pearce, Carr (1967). The details of the light limited cultures, growth media and growth conditions were as previously described (Slater, 1975; Karagouni, Slater, 1978; 1979).

3. RESULTS AND DISCUSSION

A number of closed culture growth curves were obtained as described by Karagouni (1979). *Anacystis nidulans* started growing under conditions where light (and other nutrients) were not growth limiting but ultimately the growth rate declined because light became the limiting factor before other nutrients were exhausted.

The plots showed typical growth curves and the data were used to calculate the maximum specific growth rate giving a value of $\mu_{max} = 0.16 \text{ h}^{-1}$ for absorbance data and $\mu_{max} = 0.27 \text{ h}^{-1}$ for cell number values. So the μ_{max} values obtained depended on what parameter was measured. However these results show that there was no stage in closed culture growth cycle where one could obtain a balanced exponential growth (Karagouni, 1979). This difficulty led to the use of conti-

nuous culture techniques which can be a powerful tool when studying the physiology of microorganisms with regard to problems related to energy metabolism in particular (Aiking et al, 1977).

Experiments were carried out in a light-limited chemostat. The growth rate was varied by altering the dilution rate and the incident irradiance was constant. The biomass concentration expressed in terms of absorbance or dry weight (μg dry weight ml^{-1}) or cell number (ml^{-1}) increased with decreasing growth rate (Fig. 1).

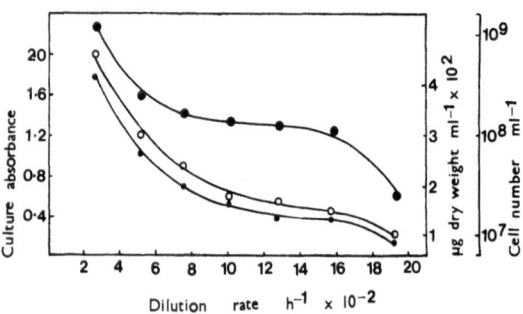

FIGURE 1. The influence of dilution rate on the steady state biomass concentration for *Anacystis nidulans* grown in light-limited chemostat culture. The biomass concentration was determined as absorbance (•), dry weight expressed as μg dry weight ml^{-1} (o) and cell number ml^{-1} (•).

Equation (6) indicates that a plot of the biomass concentration (x) against the reciprocal of the dilution rate (1/D) is a straight line. The results obtained from the continuous-flow chemostat culture designed to be light-limited showed that the above relationship between the biomass concentration (x) and the reciprocal of the dilution rate (1/D) was realistic at least at dilution rates higher than $D = 0.04 \text{ h}^{-1}$ (Fig. 2).

At dilution rates below $D = 0.04 \text{ h}^{-1}$ with the high biomass concentration obtained, a second nutrient began to limit growth and the above relationship was not valid. Below $D = 0.04 \text{ h}^{-1}$ the biomass concentration obtained was lower than the expected,

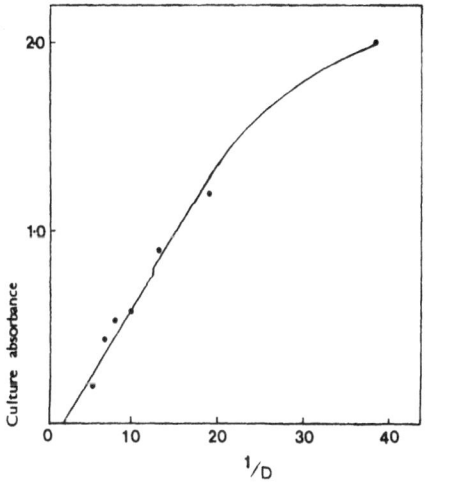

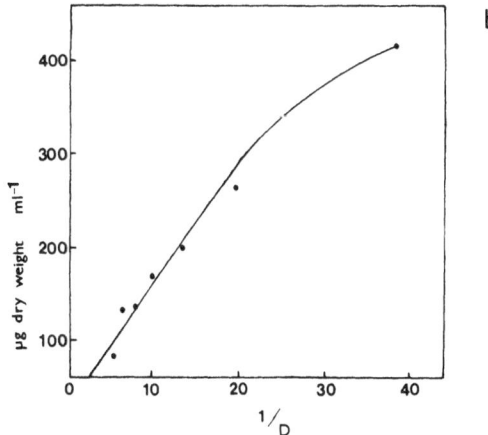

FIGURE 2. The relationship between the biomass concentration and the reciprocal of the dilution rate under light-limited conditions. Biomass concentration is expressed (a) in terms of absorbance and (b) in terms of dry weight (µg dry weight ml⁻¹).

the curve was not a straight line and in addition to light probably NO_3^- was a second limiting substrate, resulting in the lower stready state biomass concentration.

In addition other experiments indicate that light actually was the limiting factor since there was a rapid increase in culture biomass concentration in all cases when the incident light intensity was increased. Two experiments were undertaken in steady state light-limited chemostat cultures grown at $D = 0.03$ h⁻¹ and $D = 0.10$ h⁻¹, where at t = 0 the

incident light intensity increased 12 or 8-fold respectively. The culture absorbance started to increase within the first hour after the additional illumination (Fig. 3 and Fig. 4).

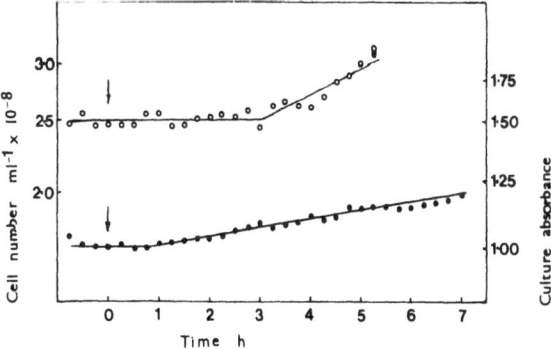

FIGURE 3. The increase in culture absorbance (o) and cell number (●) of steady state light-limited culture of *Anacystis nidulans* grown at a growth rate of 0.03 h⁻¹ during a "shift-up" induced by increasing the light-intensity, at t = 0.

The kinetic theory of a light-limited chemostat culture of *Anacystis nidulans* proposed here should be of some value in interpreting the above results obtained in laboratory cultures of *Anacystis nidulans*.

There are very few published works of algal or

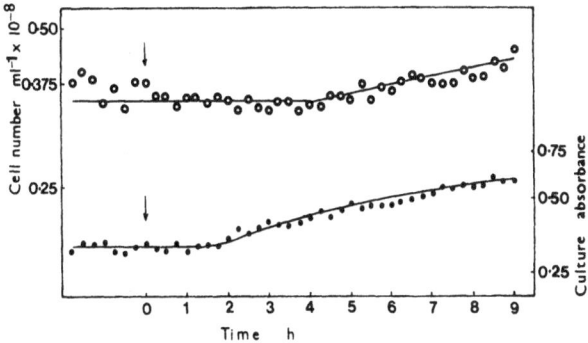

FIGURE 4. The increase in culture absorbance (o) and cell number (●) of steady state light-limited culture of *Anacystis nidulans* grown at a growth rate of 0.10 h⁻¹, during a "shift-up" induced by increasing the incident light intensity at t = 0.

cyanobacterial growth in continuous cultures under conditions of light limitation. Oswald (1970), Myers (1970) and Aiba, Ogawa (1977) measured the net yield on energy, the composite proportionality constant C. It seems that there will never be a 100% utilization of the limiting nutrient (be it light or water dissolved nutrient) (Herbert et al., 1956). Part of the light energy actually consumed is used for "maintenance" functions such as the maintenance of solute gradients across membranes, maintenance of cell integrity and turn-over of macromolecular components of the cell (Pirt, 1965). However, only Aiba and Ogawa (1977) measured the light energy escaping from the culture system. Gons (1977) and Van Liere, Mur (1979) measured the efficiency factor C_{growth} (which is identical to "true" yield value) and showed that it is dependent on incident irradiance. Finally whatever the realistic values of these constants may be, the function of μ versus I_a is mathematically a valid description of the growth of phototrophic organisms under conditions that are limited by the supply of light energy.

ACKNOWLEDGMENT

I wish to thank Prof.J.H.Slater for his valuable advice.

REFERENCES

Aiba S and Ogawa T (1977) Assessment of growth yield of a blue-green alga, *Spirulina platensis*, in axenic and continuous culture. JGM 102, 179-182.

Aiking H, Sterkenburg A and Tempest DW (1977) Influence of specific growth limitation and dilution rate on the phosphorylation efficiency and cytochromes content of mitochondria of *Candida utilis* NCYC 221. Arch.Microb. 113, 65-72.

Gons HJ (1977) On the light-limited growth of *Scenedesmus protuberans* Fritsch. Thesis, Univ. Amst. The Netherlands.

Herbert D, Elsworth R and Telling RC (1956). The continuous culture of bacteria; a theoretical and experimental study. JGM 14, 601-622.

Karagouni AD (1979) The physiology of the blue-green alga *Anacystis nidulans* grown in light- and carbon dioxide-limited continuous flow culture, Ph D Thesis Warwick Univ. England.

Karagouni AD and Slater JH (1978). Growth of the blue-green alga *Anacystis nidulans* during washout from light- and carbon dioxide-limited chemostats. FEMS Microb. Let. 4, 295-299.

Karagouni AD and Slater JH (1979). Enzymes of the Calvin cycle and intermediary metabolism in the cyanobacterium *Anacystis nidulans* grown in chemostat culture. JGM, 115, 369-376.

Myers J (1970). Genetic and adaptive characteristics in the *Chlorellas*. Prediction and Measurements of Photosynthetic Productivity. In Proc. of Int. Biol. Prof/Primary Productivity, pp 447-455, Wegeningen: Centre for Agric. Publ. and Documentation.

Oswald WJ (1970) Growth characteristics of microalgae cultures in domestic sewage: environmental effects of productivity. Pr. and Meas. of Phot. Prod. Proc. of Int. Biol. Prog/Primary Productivity, pp 473-488. Wegeningen: Centre for Agric. Publ. and Documentation.

Pearce J and Carr NG (1967). The metabolism of acetate by the blue-green algae, *Anabaena variabilis* and *Anacystis nidulans*. JGM 49, 301-313.

Pipes WO and Koutsoyannis SP (1962) Light-limited growth of *Chlorella* in continuous culture. App. Micr. 10, 1-5.

Pirt SJ (1965) The maintenance energy of bacteria in growing cultures. Proceedings of the Royal Soc. of London B163, 224-231.

Slater JH (1975) The control of carbon dioxide assimilation and ribulose 1,5-diphosphate carboxylase activity in *Anacystis nidulans* grown in a light-limited chemostat. Arch. of Microb. 103, 45-49.

Van Liere L and Mur LR (1979) Growth kinetics of *Oscillatoria agardhii* Gomont in continuous culture, limited in its growth by the light energy supply. JGM 115, 153-160.

PART V

NUTRITION

Fe, Co, Zn, Br, Rb, Cs, Se, Cr, Sb and Sc CONTENT AND GROWTH OF SOYBEAN NODULES AS AFFECTED BY NUTRIENT DEFICIENCY

L.A. TYANKOVA*and A.A. DAMYANOVA**(Institute of Plant Physiology*, Institute of Nuclear Energy**, Bulgarian Academy of Sciences, Sofia 1113, Bulgaria).

1. INTRODUCTION

With the increasing energy crisis as well as the rise in the price of fertilizers, the question of using biological fixation of nitrogen becomes urgent. The clarification of the parameters that limit or stimulate growth, development and functional activity of the nodules acquires peculiar significance (Tyankova et al., 1978; Tyankova, Dimitrova, 1981; Bothe, Trebst, 1981; Schilling, 1980, 1982).

The present studies are a part of investigations of the influence of deficiencies of macro- and microelements on the growth, development and productivity of the soybean plant, and the nodules, as well as the uptake, translocation, and distribution of macro-, micro- and trace elements in the organs of the plant.

Preliminary results are presented here on the uptake and accumulation of iron, cobalt, zink, bromine, rubidium, cesium, selenium, chromium, antimony and scandium in nodules of soybean plants grown under single deficiency of macro- and micronutrients.

2. MATERIALS AND METHODS

Soybean plants (*Glycine max* (L) Merr. cv.Beeson) with a vegetation period of about 140 days were grown as gravel culture in 10-liter plastic buckets in a greenhouse in May and June. Presowing treatment of the seeds was done with *Rhizobium japonicum*. There were four buckets for each variant of the experiment, and in each bucket there were 15 plants at the beginning of the experiment. The control plants were grown on full nutrient solution (FNS) or on water (H_2O), and the test plants grew for the whole period in absence of a particular element: nitrogen, sulfur, phosphorus, boron, potassium, calcium, magnesium, or iron. All variants received equal amounts of microelements from A^{-Z} with the exception of -Fe and

-B. For comparison, nodules of plants grown in soil under the same conditions were examined. The method of the experiment is described in detail elsewhere (Tyankova et al., 1979; Tyankova, Dimitrova, 1981). The samples for analysis were taken from 31- and 60-day-old plants.

A non-destructive method of neutron activation analysis was applied for determination of the element content of the nodules. This ensures simultaneous determination of a great number of elements in a small volume of samples for a short time at high sensitivity. The samples were homogenized and dried, under conditions that prevented additional contamination. The 80-to 100 mg samples so prepared were hermetically sealed in plastic ampoules. They were put for irradiation together with the coresponding standard material into the vertical channels of the research reactor IRT-2000 at a flux of $2.10^{13}n.cm^{-2}.sec^{-1}$. Kale powder Bowen was used as a standard material.

Gamma spectrometry of samples and standards has been carried out on the 4000-channel analyzer of the Didac type, with a 80 cm^2 Ge/Li detector. Processing of the digital information was done by the method of Covell (1959).

3. RESULTS AND DISCUSSION

The aboveground system, the roots, and the nodules react differently to the tested deficiencies (Fig. 1 and Fig. 2). In spite of the presence of a great number of nodules (Tyankova, Dimitrova, 1981) and the total nodule mass, the 31-day-old N-deficient plants retard first and to the greatest extend, compared to the tested deficiencies of anions (Fig. 1). The strongest growth-retarding influence is exerted by -K, followed by -Fe.

All older tested plants are considerably reduced in growth as compared to the FNS (Fig. 2). The highest negative effect was exerted from -K treatment followed by -N,-B,-Fe. A detailed analysis of the

180

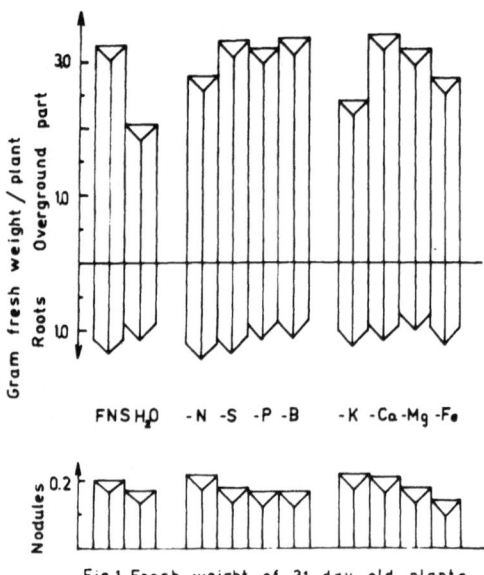

FNSH₂O -N -S -P -B -K -Ca-Mg -Fe

Fig.1.Fresh weight of 31 day old plants

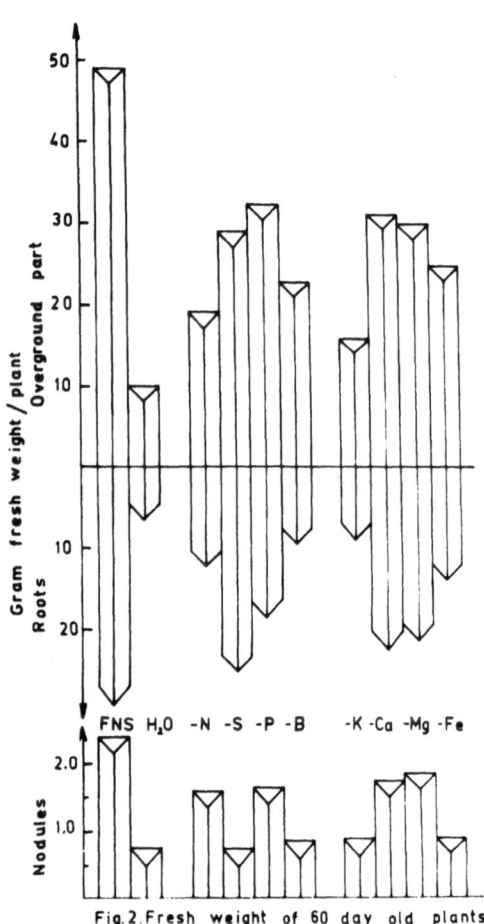

FNS H₂O -N -S -P -B -K -Ca -Mg -Fe

Fig.2.Fresh weight of 60 day old plants

formation and distribution of the organic mass in soybean plant organs, nodules included, has been made elsewhere (Tyankova, Dimitrova, 1981).
Of interest is the fact that the sulfur deficient plants have the least nodules and the least affected root system, and are comparatively well developed in the aboveground part (Fig. 2). The formation of nodule mass is considerably retarted in boron, potassium and iron deficiency.
Data of microelement and trace element analysis in nodules at the first trifoliate-leaf stage of the development (31-day-old-plants) and at the beginning of flowering (60-day-old-plants) are shown in Tables 1, 2, 3 and 4. The following statements can be made comparing the results received.

3.1. Iron

Depending on the age of the plants and the nutrition regime, the quantity of the nodules varies from 150 to 960.10^{-6} g.g^{-1} dw (Table 1). A considerable accumulation of Fe is observed in the nodules of 31-day-old plant in the absence of N and S; the lack of Ca and Mg retard its accumulation, and B and K deficiencies hardly change it. The differences are settled with aging. Most probably the comparatively high percent of Fe in the S-deficient nodules during flowering is due to the small nodule mass (Fig. 2). Obviously, the growth rate and amount conditions the Fe concentration in the nodule tissues during the later phase of development.

3.2. Cobalt

Cobalt varies in the ranges $0.21-0.98.10^{-6}$ g.g^{-1} dw (table 1). Its percentage in the plant nodules hardly changes with phase of development at the different deficiencies. Co content is lowest at -K and -S deficiency, absorption velocity increases at -P and in the later phase of development at -Mg and -Ca as well.

3.3. Zink

Zink varies comparatively slightly, $36-99.10^{-6}$g.g^{-1} dw (Table 1). Regardless of the development phase, the FNS treatment contains the largest quantities of Zn. Values are comparatively close in 31- and 60-day-old plants. Evidently -Ca suppresses its

TABLE 1. Content of iron, cobalt and zink in soybean nodules under variant mineral deficiency.

Variants	Age of plants and stage of development							
	31-day-old plants 1st trifoliate leaf				60-day-old plants Beginning of flowering			
	x 10^{-6} gram per gram dry weight (dw) or per plant							
	gram dw	% to FNS	plant	% to FNS	gram dw	% to FNS	plant	% to FNS
Iron								
FNS	370.8	100.0	12.2	100.0	479.2	100.0	207.0	100.0
H_2O	233.3	62.9	6.1	49.7	256.6	53.5	38.49	18.6
Soil	–	–	–	–	348.9	72.8	–	–
–N	958.8	258.6	35.5	290.8	189.9	39.6	76.9	37.1
–S	963.3	260.5	28.9	236.9	353.5	73.8	31.8	15.4
–P	262.5	70.7	7.3	60.2	224.6	46.9	68.5	33.1
–B	358.7	96.7	8.2	67.6	268.7	56.1	39.5	19.1
–K	394.1	106.3	13.4	109.8	234.6	48.9	30.7	14.8
–Ca	205.2	55.3	7.0	57.2	200.9	41.9	70.1	33.9
–Mg	319.2	86.1	9.6	78.5	224.2	46.8	79.4	38.3
–Fe	154.3	41.6	3.9	31.6	246.6	51.5	45.9	22.2
Cobalt								
FNS	0.7954	100.0	0.0262	100.0	0.7396	100.0	0.3195	100.0
H_2O	0.0798	10.0	0.0021	7.9	0.0723	9.8	0.0108	3.4
Soil	–	–	–	–	2.2598	305.5	–	–
–N	0.6907	86.8	0.0255	97.5	0.5578	75.4	0.2259	70.7
–S	0.3942	49.5	0.0047	17.7	0.2790	37.7	0.0251	7.9
–P	0.7908	99.4	0.0221	84.5	0.9783	132.3	0.2984	91.0
–B	0.9362	117.7	0.0215	82.1	0.7647	103.4	0.1124	35.2
–K	0.2115	26.6	0.0072	27.4	0.5353	72.4	0.0700	21.9
–Ca	0.7157	90.0	0.0243	92.7	0.8996	121.6	0.3140	98.3
–Mg	0.9248	116.3	0.0277	105.8	0.9441	127.6	0.3340	104.6
–Fe	0.6245	78.5	0.0156	59.5	0.6800	91.9	0.1265	39.5
Zink								
FNS	99.2	100.0	3.274	100.0	88.1	100.0	38.1	100.0
H_2O	65.2	65.7	1.695	51.8	42.8	48.6	6.4	16.9
Soil	–	–	–	–	70.3	79.8	–	–
–N	35.2	35.5	0.950	29.0	55.9	63.4	28.5	74.9
–S	79.2	79.8	2.376	72.6	67.0	76.0	6.0	15.8
–P	48.8	49.2	1.366	41.7	54.0	61.3	16.5	43.3
–B	40.5	40.8	0.931	28.5	55.3	62.8	8.1	21.4
–K	52.5	52.9	1.785	54.5	67.0	76.0	8.8	23.0
–Ca	36.4	36.7	1.237	37.8	30.3	34.4	10.6	27.8
–Mg	43.2	43.5	1.296	39.6	68.1	77.1	24.1	63.4
–Fe	65.5	66.0	1.637	50.0	65.8	74.7	12.2	32.2

absorption while -S has the opposite effect.

3.4. Bromine

Bromine varies in the ranges $0.7-4.7\times10^{-6}$ $g.g^{-1}$ dw (Table 2). The younger nodules of the tested plants contain on the whole smaller quantities of Br. Greater accumulation of Br in nodules is observed at anion, than at cation deficiencies. The lack of Fe increases considerably absorption of Br.

Little is known about the physiological role of Br in plants. In small quantities it can stimulate growth. The higher concentrations are not as toxic as iodine (Scharrer, Schropp, 1949), and within definite limits bromide can substitute for the chloride in plants (Broyer et al., 1954). Pedersen et al. (1974) report on peroxidases which bring about biosynthesis of bromine containing phenols by enzymatic halogening. Undoubtedly the established dependence between the content of Fe and Br in the nodules deserves attention.

3.5. Rubidium

Rubidium varies from 1.6 to 8.5×10^{-6} $g.g^{-1}$ dw (Table 2). During the whole test period the nodules of the deficient plant contained considerably less Rb from FNS. Mg-deficiency increases its concentration in comparison to the rest of the cation deficiencies, and -S and -P lead to its decrease. The nodules of plants grown at anion deficiencies contain more Rb than those lacking cations.

3.6. Cesium

Cesium varies from $0.21-0.86\times10^{-6}$ $g.g^{-1}$ dw (Table 2). There is almost three times greater content at -K, and increased concentration at -Fe and -P. The nodules' age does not change the pattern of the observed Cs content as a function of nutrient deficiency.

3.7. Selenium and Chromium

These elements belong to the mineral elements with limited or ambiguous importance (Baumeister, Ernst, 1978). In recent years these have been the object of numerous studies (Singh, Singh, 1978, 1979; Tyankova et al., 1981; Gujova, 1982). Selenium varies from 0.25 to 0.61×10^{-6} $g.g^{-1}$ dw (Table 3). It is contained in greater quantities in the test variants than in FNS. With aging, its concentration at anion deficiencies increases. The same is observed at -K, but this could be due, as for -S and -B, to the reduced growth of nodule mass. The considerable increase of Se on -P deserves special attention in view of the works of Sing and Sing (1978, 1979), who established an increase in the P-content of forage cowpea with Se treatment. They make use of high doses of P to remove the Se-toxicity. Resently, Gujova et al. (1982) reported that Se stimulated the uptake of K, P and Ca by maize seedlings. In our experiments it could be said, that all deficiencies are accompanied by an increase of Se concentration, although this effect was less pronounced with P and K deficiency.

Chromium varies from 0.17 to 0.6×10^{-6} $g.g^{-1}$ dw (Table 3). With aging, its concentration with anion-deficiencies decreases. With cations deficits, chromium remains almost unchanged. A strong increase is observed with -S and -B and a decrease is observed with -K, -Ca and -Mg treatment.

3.8. Antimony

Antimony varies from 0.12 to 0.88×10^{-6} $g.g^{-1}$ dw (Table 3). It diminishes with aging. There is considerable decrease of the concentration with -P and -K and an increase with -Mg, -Ca and -Fe.

3.9. Scandium

Scandium varies from 0.011 to 0.059×10^{-6} $g.g^{-1}$ dw (Table 4). It diminishes with aging. There are strongly increased quantities in comparison to the other deficiencies at -K.

In comparison to FNS the nodules of plants grown on soil contained smaller quantities of Fe, 72.0 % (Table 1), Zn, 70.2% (Table 1), and Rb, 58.2% (Table 1); almost equal quantities of Sb, 95.9%

TABLE 2. Content of bromine, rubidium, and cesium in soybean nodules under variant mineral deficiency.

Variants	Age of plants and stage of development							
	31-day-old plants 1st trifoliate leaf				60-day-old plants Beginning of flowering			
	x 10^{-6} gram per gram dry weight (dw) or per plant							
	gram dw	% to FNS	plant	% to FNS	gram dw	% to FNS	plant	% to FNS
B r o m i n e								
FNS	3.143	100.0	0.104	100.0	2.245	100.0	0.970	100.0
H_2O	0.997	31.7	0.026	24.9	0.873	38.9	0.131	13.4
Soil	-	-	-	-	4.757	211.9	-	-
-N	1.955	62.2	0.072	69.7	2.419	107.8	0.980	101.0
-S	1.335	42.5	0.040	38.6	2.648	118.0	0.238	24.5
-P	1.902	60.5	0.053	51.3	2.024	90.1	0.617	63.6
-B	2.804	89.2	0.065	62.2	2.183	97.2	0.321	33.1
-K	0.794	25.3	0.027	26.0	2.427	108.1	0.772	79.5
-Ca	1.179	37.5	0.040	38.6	1.786	79.5	0.623	64.2
-Mg	0.702	22.3	0.023	22.6	1.628	72.5	0.576	59.4
-Fe	1.855	59.0	0.046	44.7	3.347	149.1	0.623	64.1
R u b i d i u m								
FNS	8.571	100.0	0.283	100.0	8.515	100.0	3.678	100.0
H_2O	4.112	47.9	0.107	37.8	5.500	64.6	0.825	22.4
Soil	-	-	-	-	4.955	58.2	-	-
-N	6.807	79.4	0.252	89.0	6.482	76.1	2.625	71.4
-S	4.147	48.4	0.124	43.9	3.154	37.0	0.284	7.2
-P	3.131	36.5	0.088	30.9	2.361	27.7	0.720	19.6
-B	7.015	81.8	0.161	57.0	5.768	67.7	0.848	23.0
-K	2.901	33.8	0.099	34.8	3.208	37.7	0.420	11.4
-Ca	2.606	30.4	0.089	31.3	1.618	19.0	0.565	15.4
-Mg	6.324	73.8	0.190	67.0	3.787	44.4	1.341	36.4
-Fe	4.156	48.5	0.104	36.7	2.337	27.4	0.435	11.8
C e s i u m								
FNS	0.2836	100.0	0.0094	100.0	0.2926	100.0	0.1264	100.0
H_2O	0.0000	0.0	0.0000	0.0	0.0000	0.0	0.0000	0.0
Soil	-	-	-	-	0.8199	279.9	-	-
-N	0.2584	91.1	0.0096	102.2	0.2577	88.1	0.1044	82.5
-S	0.2519	88.8	0.0076	81.2	0.2622	89.6	0.0236	18.6
-P	0.3233	114.0	0.0090	96.2	0.3945	134.8	0.1203	95.1
-B	0.3192	112.6	0.0073	78.4	0.3192	109.1	0.0469	37.1
-K	0.7826	276.0	0.0267	284.3	0.8643	295.4	0.1132	89.5
-Ca	0.2159	76.1	0.0073	78.4	0.2869	98.1	0.1001	79.2
-Mg	0.2813	99.2	0.0084	90.2	0.2370	81.0	0.0839	66.3
-Fe	0.4253	150.0	0.0106	113.6	0.5842	199.7	0.1087	85.9

184

TABLE 3. Content of selenium, chromium and antimony in soybean nodules under variant mineral deficiency.

Variants	Age of plants and stage of development							
	31-day-old plants 1st trifoliate leaf				60-day-old plants Beginning of flowering			
	$\times 10^{-6}$ gram per gram dry weight (dw) or per plant							
	gram dw	% to FNS	plant	% to FNS	gram dw	% to FNS	plant	% to FNS
Selenium								
FNS	0.2823	100.0	0.0093	100.0	0.2776	100.0	0.1192	100.0
H₂O	0.0000	0.0	0.0000	0.0	0.1906	68.6	0.0286	25.5
Soil	–	–	–	–	2.1180	762.9	–	–
–N	0.3917	138.7	0.0145	155.9	0.4764	171.6	0.1929	161.8
–S	0.2471	87.5	0.0070	75.3	0.4638	167.1	0.0417	34.9
–P	0.4122	146.0	0.0115	123.7	0.6056	218.2	0.1840	154.9
–B	0.3742	132.6	0.0086	93.5	0.4553	146.0	0.0669	56.1
–K	0.5231	185.3	0.0178	191.3	0.6716	241.9	0.0880	73.8
–Ca	0.4824	170.9	0.0164	176.3	0.3377	121.6	0.1179	98.8
–Mg	0.3602	127.6	0.0108	116.2	0.3686	132.8	0.1305	109.4
–Fe	0.5296	187.6	0.0132	142.9	0.5527	199.1	0.1028	86.2
Chromium								
FNS	0.2925	100.0	0.0096	100.0	0.3040	100.0	0.1313	100.0
H₂O	0.0466	15.9	0.0012	12.6	0.1164	38.3	0.0175	13.3
Soil	–	–	–	–	0.5677	186.7	–	–
–N	0.4667	159.5	0.0173	179.8	0.2135	70.2	0.0865	65.8
–S	0.6615	226.1	0.0183	191.0	0.3812	125.4	0.0343	26.1
–P	0.0912	65.6	0.0060	61.6	0.3342	109.9	0.1019	77.6
–B	0.6024	205.9	0.0138	144.3	0.4489	147.7	0.0651	49.5
–K	0.1718	58.5	0.0058	60.6	0.1775	58.4	0.0232	17.7
–Ca	0.1785	61.0	0.0061	63.2	0.1802	59.3	0.0629	47.9
–Mg	0.1552	53.0	0.0047	48.5	0.1248	41.0	0.0441	33.6
–Fe	0.2771	94.7	0.0069	72.1	0.2962	97.4	0.0551	41.9
Antimony								
FNS	0.4146	100.0	0.0137	100.0	0.4681	100.0	0.2022	100.0
H₂O	0.1411	34.0	0.0037	26.7	0.1388	29.6	0.0208	10.2
Soil	–	–	–	–	0.4488	95.9	–	–
–N	0.1347	32.5	0.0050	36.3	0.4105	87.7	0.1612	82.2
–S	0.2735	66.0	0.0048	34.8	0.4572	97.7	0.0411	20.3
–P	0.3358	81.0	0.0094	68.6	0.1232	26.3	0.0376	18.5
–B	0.6569	158.4	0.0151	110.2	0.5822	124.4	0.0856	42.3
–K	0.4882	117.7	0.0166	121.1	0.2542	54.3	0.0333	16.4
–Ca	0.6311	152.2	0.0215	156.6	0.6381	136.3	0.2227	110.1
–Mg	0.7682	185.3	0.0231	168.2	0.8770	187.3	0.3105	153.5
–Fe	0.7374	177.9	0.0184	134.5	0.5690	121.6	0.1058	52.3

TABLE 4. Content of scandium in soybean nodules under variant mineral deficiency

V a r i a n t s	Age of plants and stage of development							
	31-day-old plants 1st trifoliate leaf				60-day-old plants Beginning of flowering			
	x 10^{-6} gram per gram dry weight (dw) or per plant							
	gram dw	% to FNS	plant	% to FNS	gram dw	% to FNS	plant	% to FNS
				S c a n d i u m				
FNS	0.0966	100.0	0.0032	100.0	0.0711	100.0	0.0307	100.0
H_2O	0.0105	10.9	0.0003	9.3	0.0194	26.7	0.0005	1.6
Soil	-	-	-	-	0.0652	91.7	-	-
-N	0.0185	19.2	0.0007	21.1	0.0162	22.8	0.0066	21.3
-S	0.0146	15.1	0.0004	13.6	0.0131	18.4	0.0012	3.8
-P	0.0185	19.2	0.0005	16.1	0.0111	15.6	0.0034	11.1
-B	0.0190	19.7	0.0004	13.6	0.0185	26.0	0.0027	8.9
-K	0.0542	56.1	0.0018	57.5	0.0590	83.0	0.0077	25.0
-Ca	0.0263	27.3	0.0009	27.3	0.0124	17.4	0.0043	14.1
-Mg	0.0371	38.4	0.0011	34.7	0.0246	34.6	0.0087	28.3
-Fe	0.0192	19.9	0.0035	110.6	0.0128	18.0	0.0024	7.7

(Table 3) and Se, 91.7% (Table 4); or greater quantities of Cr, 186.7%, Co, 305.5% (Table 1) and Se, 762.9% (Table 3). The number and the weight of the nodules from soil grown plants were considerably smaller than those of hydroponically grown plants.

4. CONCLUSIONS

The accumulation of nodule mass in soybean plants and the mineral content of the nodules are strongly influenced by the content of the nutrient media. When plants are grown at nutrient deficiency, the fresh weight is ordered as follows: nodules, S<Fe<H_2O<K<B<P<Ca<Mg<N <FNS; roots, H_2O<K<B<N<Fe<P<Mg<Ca<S<FNS; aboveground parts, H_2O<K<N<B<Fe<S<Mg<Ca<P<FNS. Depending on the nutrition regime and the age of the plants, the nodules contain varying amounts of minerals as follows (all valuesx10^{-6} g.g^{-1} dw substance): Fe, 150-960; Co, 0.21-0.98; Zn, 36-99; Br, 0.8-4.7; Rb, 1.6-8.5; Cs, 0.21-0.86; Se, 0.25-0.67; Sb, 0.12-0.88; and Sc, 0.011-0.059. Differences have been established in absorption

and accumulation of the test elements investigated in the nodules according to phase of soybean development and applied nutrient deficiency.

REFERENCES

Baumeister W and Ernst W (1978) Minerallstoffe und Pflanzenwachstum, Stuttgart-New York, Gustav Fischer Verlag.
Bothe H and Trebs A (1981) Biology of inorganic nitrogen and sulfur, Berlin, Springer Verlag.
Broyer TC, Carlton AB, Jonson CH and Stout PR (1954) Chlorine micronutrient element for higher plants, Plant Physiol. 29, 526.
Covell DA (1959) Determination of Gamma-Ray Abundance Directly from the total absorption Peak, Anal. Chem. 31, 1785
Gujova NV, Fatalieva SM and Kerimova ASh (1982) The effect of nutrition level and selenium on the ion uptake. In Genetic specificity of mineral nutrition of plants, p. 89. Belgrade, Sci. Assemblies, XIII.
Pedersen M, Saenger P and Fries L (1974) Simple brominate phenols in red algae, Phytochemistry 13, 2273.
Scharrer K and Schropp W (1949) Untersuchungen über die Wirkung von Chlor und Brom auf die Keimung und die erste Jugendentwicklung einiger Kulturpflanzen, Z. Pflanzenernähr., Düng. Bodenk. 46, 88.
Schilling G (1980) Zur Ernährung der Pflanzen mit Luftstickstoff, W. Wiss.u. Fortschriftt. 30, 467.
Schilling G (1982) Genetic specificity of nitrogen nutrition in leguminous plants. In Genetic specifi-

186

city of mineral nutrition of plants, p. 249. Belgrade, Sci. Assemblies, XIII.

Singh M and Singh N (1978) Selenium toxicity in plants and its detoxication by phosphorus, Soil Sci. 126, 255.

Singh M and Singh N (1979) The effect of forms selenium on the accumulation of selenium, sulfur, and forms of nitrogen and phosphorus in forage cowpea (*Vigna sinensis*), Soil Sci. 127, 264.

Tyankova L, Damyanova A and Dimitrova A (1981) Uptake and distribution of Se and Cr in soybean plants grown under nutrient deficiencies. Proc. II Int. Youth Symp. Regulation of Metabolism in Plants, Varna (in press).

Tyankova L and Dimitrova A (1981) Formation and distribution of dry matter in soybean plants, nodules included, as a function of nutrient deficiencies. Proc. II Int. Youth Symp. Regulation of Metabolism in Plants, Varna (in press).

Tyankova LA, Stoyanova ZP and Dimitrova AV (1978) Influence of the different supply of N,P,S,K,Mg and Ca on the formation of nodules and on the productivity of soya bean plants, C.R. Acad. Bulg. Sci. 31, 1447.

Tyankova L, Tsonev Ts, Dimitrova A and Kudrev T (1979) Nutrient deficiency induced changes in photosynthetic activities and growth of maize plants. Proc. I Int. Symp. Plant Nutr.,II, p. 148, Varna.

EFFECT OF MOLYBDENUM AND COBALT ON THE NITROGEN-FIXING ACTIVITY OF *RHIZOBIUM JAPONICUM* AND SOYBEAN YIELD

A. MARKOVA and D. CHANOVA (N. Poushkarov Institute of Soil Science and Yield Prediction, Sofia, Bulgaria)

1. INTRODUCTION

The studies of the effect of Mo and Co on the two components of the symbiotic system of leguminous plants, the nodule bacteria and the green plant, are still insufficient and most of the studies have been agrochemical in aspect (Gyulahmedov, Tagiev, 1980; Ivchenko, 1980; Shukla, Dwivedi, 1979; Tishenko, 1970; Yagodin, 1970; Yagodin, Sablina, 1981). Very few studies of this kind have been carried out in this country (Markova et al., 1981), while the use of trace elements in the nutrition of leguminous plants is becoming more and more widespread; a practice which calls for more investigation into trace elements nutrition of leguminous plants.

2. MATERIALS AND METHODS

We studied the effect of Mo and Co on the N-fixing activity and virulence of *Rhizobium japonicum*, strain 646, under the conditions of the leached chernozem occuring near the village of Gorni Dubnik.

The preliminary investigations showed that soil pH was 5.3, the humus content was 3.2 mg.kg^{-1}, total P was 0.161 mg.kg^{-1}, available N (determined by the method of Cornfield) was 41.3 mg.kg^{-1}. Available P was estimated at 2.6 mg per 100 g soil using the method of Egner-Riehm, available K was estimated at 38 mg per 100 g soil using 2N HCl, available zinc was estimated at 1.78 mg.kg^{-1} using an EDTA$^+$ (NH$_4$)$_2$CO$_3$ extract with pH 8.6 and available Mo was estimated at 0.18 mg.kg^{-1} in an oxalate extract. Therefore, this soil was characterized by a weakly acid reaction, fair N and Mo availability and good K and Zn availability.

An experiment was carried out with soybean in pots containing 1.200 kg air dry soil each, in 8 replications. The Beeson variety was tested. All the treatments included the application of 300 mg P$_2$O$_5$ and 148 mg K supplied as KH$_2$PO$_4$. The molybdenum was applied as (NH$_4$)$_6$Mo$_7$O$_{24}$.4H$_2$O at the rate of 1.5 mg Mo per kg soil and the cobalt was applied as CoSO$_4$.7H$_2$O at the rate of 3 mg Co per kg soil. The seeds were treated with *R. japonicum*. The investigation was carried out at two growth phases; beginning of pod formation (phase 1) and maturity (phase 2). The activity of nodule bacteria was determined during the first phase. Their N-fixing activity was estimated from the difference in the total N uptake between plants treated and plants untreated with the investigated strain, since in this soil there were no nodule bacteria of this kind. Total N was determined by the Kjeldahl procedure.

3. RESULTS

The data in Figure 1 show that the N-fixing activity of *R. japonicum*, strain 646, increases with respect to the control at both phases studied under the influence of Mo applied alone or in combination with Co by 31 and 13%, respectively, at first phase and 36 and 15%, respectively, at second phase. Therefore, Mo , as reported by other authors (Karaguishieva, Alibekova, 1978), exerts a positive effect on the N-fixing activity of nodule bacteria, while other studies have shown Co to have a positive effect on nitrogen fixation under the conditions of calcareous chernozem (Markova et al., 1981). The application of Co at the rate of 3 mg Co per Kg soil under the conditions of the experiment has little effect on the level of nitrogen fixation. The different behavior of Co in these two studied soils can be accounted for by the different pH values: 7.3 with calcareous chernozem and 5.3 with leached chernozem. The slight acidity has apparently results in an adequate supply of Co. These differences make the leached chernozem an important soil type for further study.

The molybdenum when applied alone has a positive effect on the level of nitrogen fixation (Fig. 2).

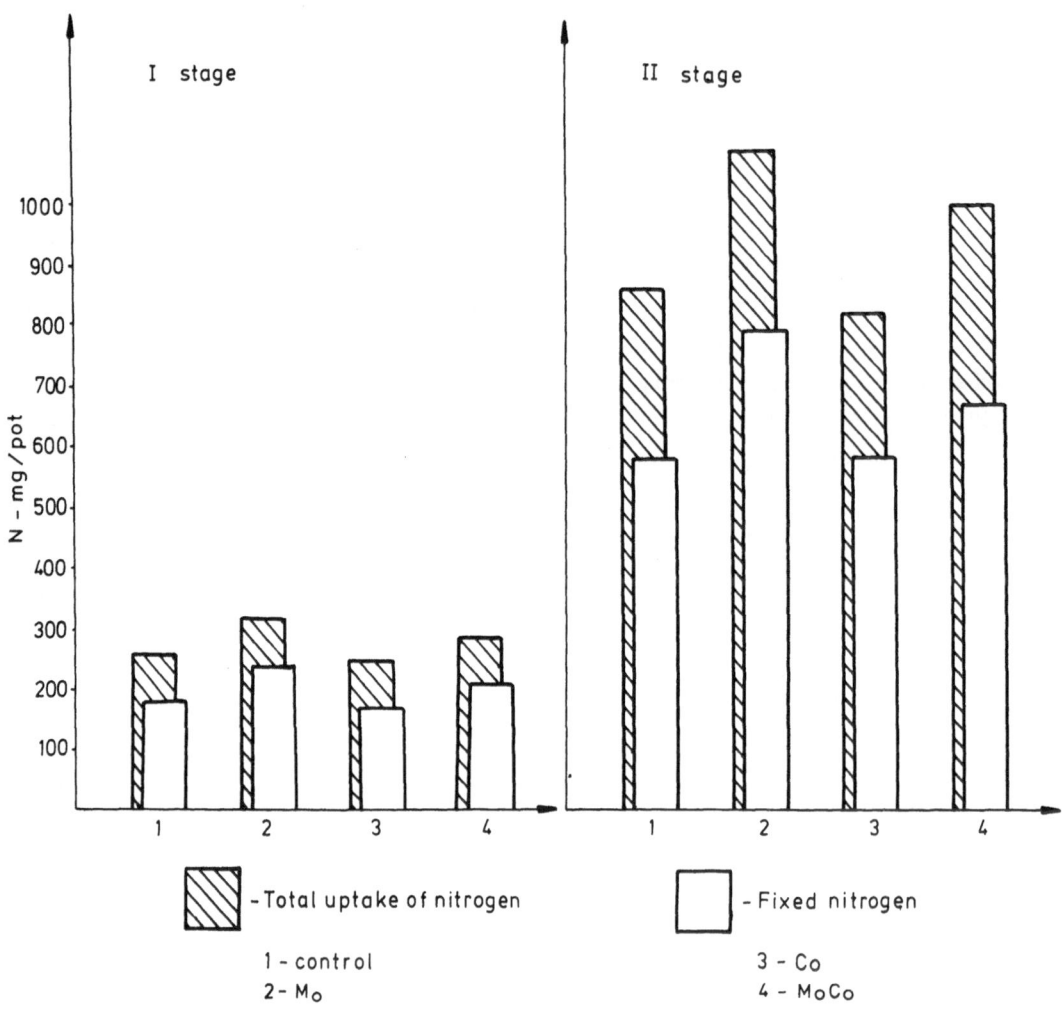

FIGURE 1. Effect of Mo and Co on N-fixing activity of *R. japonicum* in infected soybeans.

This effect is not due to an increase in the number of nodules and no significant differences were observed with respect to the control. However, the size of nodules increases up to two times and their weight by more than 1.5 times, which is considered by some authors (e.g. Raicheva, 1982) as an indication of higher N-fixing activity by the nodule bacteria.

The data given in Figure 3 indicate that the high N-fixing activity and nodule development in the presence of *R. japonicum* causes the yields of aboveground parts and grain to increase two to three times with respect to the control. This is accounted for by the insufficient supply of available N in soil as indicated by the preliminary agrochemical characteristics of soil. The application of Mo increases the yields of aboveground parts and grain by 22 and 24%, respectively, with respect to the infected control due to the improved conditions for nitrogen fixation established above. This increase is statistically significant. Patarinski and Mirchev (1976) report that at a soil molybdenum number (SMoN) $\leqslant 7.3$, some leguminous crops respond positively to the application of Mo fertilizers. In our study the SMoN is 7.1. The cobalt has little effect on the yield of soybean since it does not

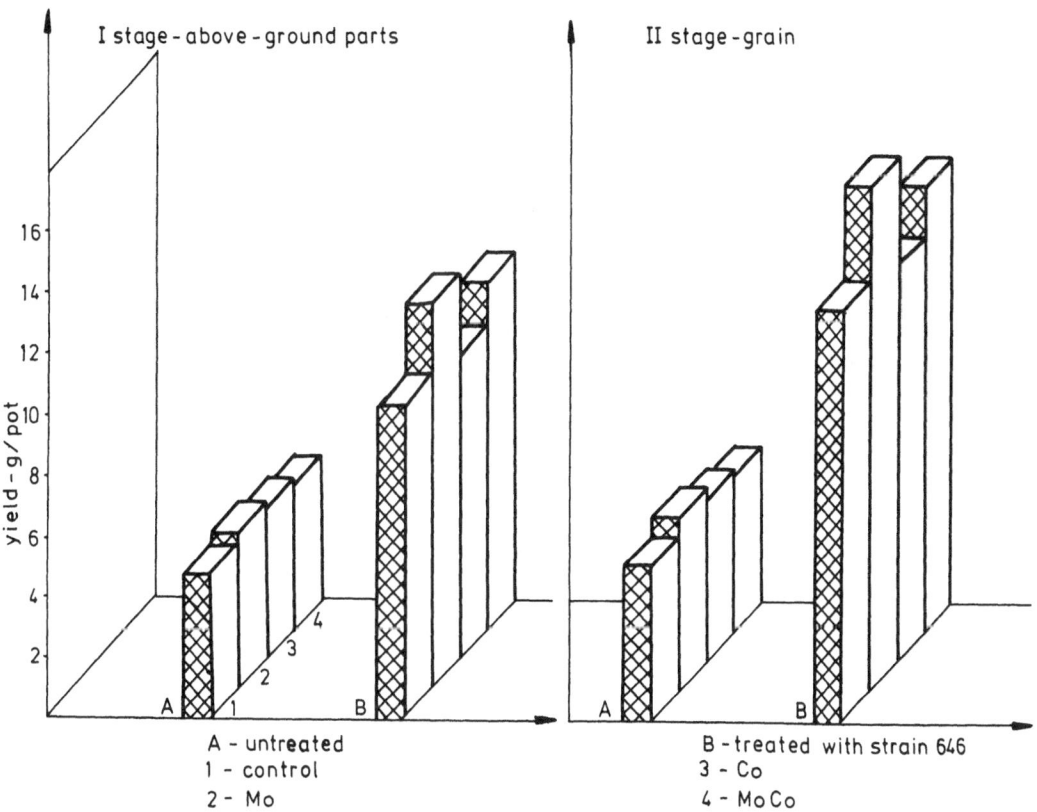

FIGURE 2. Effect of Mo and Co on the yield of soybean.

influence the N-fixing activity of nodule bacteria. When Mo and Co are applied together the data indicate an increase in the soybean yield during both phases investigated which, however, is not higher than the increase caused by the application of Mo only.

4. CONCLUSIONS

Treating seeds with the strain 646 of *R. japonicum* significantly increases soybean yield.

The application of Mo creates better conditions for symbiotic nitrogen fixation during both phases investigated: the N-fixing activity and virulence of *R. japonicum* (as measured by nodule size and weight) increase. The yields of aboveground parts and grain also increase.

Supplemental cobalt does not significantly affect

these indices. The combined application of Mo and Co is not expedient with this soil type, since the aggregated effect on the N-fixing activity and soybean yield is not higher than the effect caused by Mo only application.

REFERENCES

Gyulahmedov AN and Tagiev VD (1980) Vlijanie mikroelementov na symbioz bobovih s klubenkovimi bakteriami, Dokl. ANASSR 36(12), 76-79.

Ivchenko VI (1980) Fisiologicheskoe znachenie molibdena dija rastenii. Mikroelementi v okruza- justei srede. Vlasjuk PA, ed., pp. 89-92. Kiev, Naukova dumka.

Karaguishieva D and Alibekova Sh (1978) Vlijanie mikroelementov na simbioticheskuju azotfiksaziju soei, Vestnik selskohosyaistvenoi nauki,Kazahstana 2, 28-31.

Markova A, Chanova D and Raicheva L (1981) Vlijanie na zinka i cobalta varhu azotfiksirastata aktivnost na *R. japonicum* i varhu dobiva na sojata. Peti kongres po mikrobiologia, Nedyalkov St, ed.,

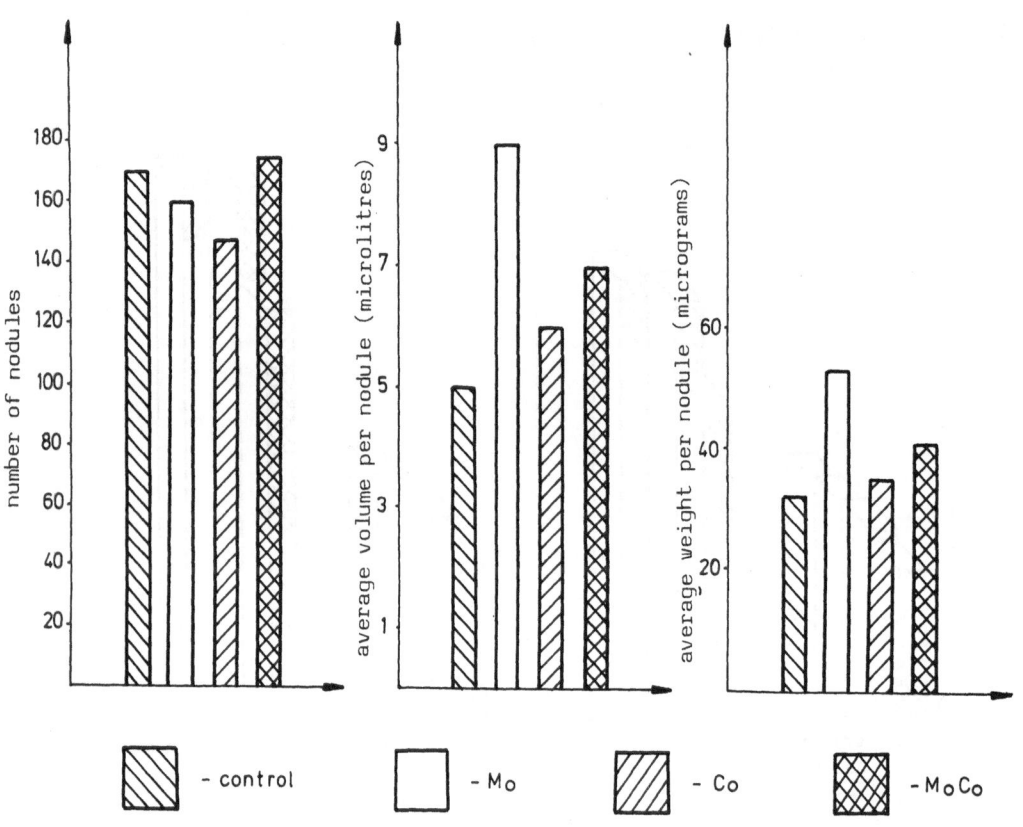

FIGURE 3. Effect of Mo and Co on the virulence of *R. japonicum* as defined by nodule number, size and weight.

pp. 415-419. Sofia, Sojuz na nauchnite rabotnizi.

Patarinski N and Mirchev S (1976) Molibden, toreneto pri intensivno zemedelie. Enikov K, ed., pp. 107-112. Sofia, Zemizdat.

Raicheva L (1982) Doktorska disertazia. Institut "N. Poushkarov". Sofia.

Samzevich SA and Karbanovich AI (1970) Vlijanie molibdena i klubenkovih bakterii na rasvitie rastenii goroha i fiksazii imi azota. In Samzevich SA, ed. Fisiologia i biohimia mikroorganismov, pp. 187-191. Minsk, Nauka i tehnika.

Shukla SW and Dwivedi RS (1979) Influence of trace elements on growth of Rhizobia and on nodulation in *Trifolium alexandrium*. Proc. Indian Nat. Sci. Acad., 45(6), 639-642.

Tishenko IV (1970) Autoreferat doktorskaja disertazia. Moskva.

Jagodin BA (1970) Kobalt i productivnost bobovih rastenii. In Peive YV, ed. Kobalt v zisni rastenii, pp. 8-39. Moskwa, Nauka.

Jagodin BA and Sablina SM (1981) Deistvie kobalta na urozai grechihi i soderjanie v njem elementov mineralnogo pitanija i rutina, Izv. TSHA 6, 68-72.

PART VI

INTERACTIONS

THE MEASUREMENT OF ADAPTATION IN CULTIVATED PLANTS

P.J. KALTSIKES (The Athens School of Agricultural Science, Iera Odos 75, Athens, Greece, T.T. 301)

1. INTRODUCTION

Cultivated plants differ in their reaction to the various environmental factors (temperature, rainfall, soil type etc) prevailing in the areas where they are grown. Some perform well under a wide range of environmental conditions while others perform poorly under adverse conditions and considerably better when conditions improve. Many studies have shown that plants usually respond linearly to changes in the environment when the latter is measured by its effect on the plant character being studied (Finlay, Wilkinson, 1963; Eberhart, Russell, 1966; Perkins, Jinks, 1968). When this linearity exists it provides two measures of a genotype's ability to cope with changes in the environment. These are: the linear regression coefficient of, and the deviations from, the regression line of a plant's attribute on some environmental index. These two measures can be used in the comparison of the sensitivity, or lack thereof, of various genotypes to changes in the environment. The third criterion, desirable mean performance (e.g. high yield, low height etc), is necessary for the successful commercialization of a genotype. The ability of a genotype to withstand changes in the environment without undue impairment to its mean performance is termed stability or adaptability. The purpose of this paper is to discuss the statistical methods used for the measurement of stability, or its converse, sensitivity, of a particular genotype or a group of genotypes.

2. METHODOLOGY

2.1. The Models

Stability can be studied (Fripp, Catten, 1971): Purely statistically (Finlay, Wilkinson, 1963; Eberhart, Russell, 1966) or on the basis of a biometrical-genetical model originally proposed by Perkins, Jinks (1968) and later extended by Freeman, Perkins (1971).

For these studies a group of t genotypes is grown in s different environments and the observations obtained are described by one of the following linear models (Fripp, Catten, 1971):

$$Y_{ij} = \mu + d_i + \varepsilon_j + g_{ij} \qquad (1)$$

where

Y_{ij} = mean phenotypic expression of the ith genotype in the jth environment

μ = grand mean of the experiment

d_i = the effect of the ith genotype

ε_j = the effect of the jth environment

g_{ij} = the effect of the interaction effect of the ith genotype with the jth environment.

On the basis of the above model the analysis of variance would take the form given in Table 1. Beginning with model 1 one can proceed to the regression analysis by using model 2 as follows

$$Y_{ij} = x_i + \beta_i \varepsilon_j + \delta_{ij} \qquad (2)$$

where

x_i = the grand mean, over all environments, of the ith genotype

β_i = the regression coefficient corresponding to the ith genotype when Y_{ij} is regressed on ε_j

δ_{ij} = the deviation, in the jth environment, of the ith genotype from the regression line of Y_{ij} on ε_j

TABLE 1. The analysis of variance when t genotypes are grown in s environments

$$Y_{ij} = \mu + d_i + \varepsilon_j + g_{ij}$$

Source of variation	Degrees of freedom
Genotypes (G)	$t - 1$
Environments (E)	$s - 1$
Genotypes X Environments (G X E)	$(t-1)(s-1)$
Residual error (from replication)	$ts(r - 1)$
Total	$ts(r - 1)$

Model 2 provides two measures of stability, namely β_i, the regression coefficient and $(\sum_j \delta^2_{ij})/(s - 2)$, the mean square for the deviations from the regression line.

If g_{ij} is regressed on ε_j, then one obtains two other, but equivalent, measures of stability namely β_{di} and $(\sum_j \delta^2_{ij})/(s - 2)$. The linear model for this regression is given in equation 3

$$Y = \mu + d_i + \varepsilon_j + \delta_{ij} \qquad (3)$$

or

$$= \mu + d_i + \varepsilon_j(1 + \beta_{di}) + \delta_{ij}$$

where

β_{di} is the regression coefficient corresponding to the ith genotype when g_{ij} is regressed on ε_j. The following connect models 2 and 3

$$x_i = \mu + d_i \qquad (3a)$$
$$\beta_i = 1 + \beta_{di} \qquad (3b)$$
$$\delta_{ij} = \delta_{ij} \qquad (3c)$$

Equation 3b holds because when model 2 is used both ε_j and g_{ij} are regressed on ε_j, the regression coefficient of ε_j in itself being unity. When model 2 is used, the sum of squares due to regression is equal to $(1 + \beta_{di})^2 \sum_j (\varepsilon_j)^2$ and encompasses variation due to the environment and

its interaction with the genotype, whereas when model 3 is used the corresponding sum of squares equals $(\beta_{di})^2 \sum_j (\varepsilon_j)^2$ (Fripp, Catten, 1971; Perkings, Jinks, 1968).

Up to this point we have not delved into the nature of ε_j. The environmental index should be independent of the Y_{ij}'s but in practice this has not been adhered to. Usually the environmental component ε_j has been estimated as the grand mean of all genotypes in the jth environment. As this contains the particular Y_{ij} being regressed on it, the Y and the X variables of this regression analysis are not independent. This becomes clearer if we examine the various sums of squares resulting from the two analyses (Table 2).

It was noted by Freeman, Perkins (1971) that the sum of squares corresponding to the joint regression and having 1 degree of freedom when model 2 is used, equals the sum of squares due to environments with $(s-1)$ degrees of freedom when model 3 is used. It can be stated, therefore, that the use of non-independent environmental indices invalidates this kind of analysis, unless progeny phenotypes are regressed on ε_j's obtained from the analysis of their parents (Breese, 1969). The problem of the lack of independence of ε_j and Y_{ij} can be overcome by using an independent environmental index z_j (Freeman, Perkins, 1971) according to the following model

TABLE 2. The partitioning of the available degrees of freedom when t genotypes are
studied in s environments according to the regression model being used
(based on Fripp, Catten, 1971)

$Y = x_i + \beta_i \epsilon_j + \delta_{ij}$

Source of variation	Degrees of freedom
1. Among Genotypes (G)	t - 1
Within Genotypes (WG)	t (s-1)
2. Joint Regression *	1
3. Heterogeneity of regressions } WG	t - 1
4. Deviations	t (s-2)
Total	ts - 1

$Y_{ij} = \mu + d_i + \epsilon_j + \beta d_i \epsilon_j + \delta_{ij}$

Source of Variation	Degrees of freedom
1. Among Genotypes (G)	t - 1
2. Among Environments (E)*	s - 1
Genotypes X Environments (GXE)	(t-1) (s-1)
3. Heterogeneity of regressions	t - 1
4. Deviations } G X E	(t-1) (s-2)
Total	ts - 1

* If ϵ_j is not independent of Y_{ij}, then the same sum of squares (Joint Regression
and Among Environments) is assigned two different degrees of freedom.

$$Y_{ij} = \mu + d_i + \bar{\beta} z_j + \bar{\delta}_j + \beta_{di} z_j + \delta_{dij} \quad (4)$$

or

$$Y_{ij} = \mu + d_i + \beta_i z_j + \delta_{ij} \quad (5)$$

Where:

β_i = regression coefficient of the ith genotype when Y_{ij} is regressed on to z_j

$\bar{\beta}$ = combined regression coefficient. It is
equal to the means of all β_i

β_{di} = $\beta_i - \bar{\beta}$ or the coefficient from regres-
sing g_{ij} on to z_j

δ_{ij} = the deviation of the ith genotype, in
the jth environment, from its regres-
sion on to z_j

$\bar{\delta}_j$ = deviation of the mean of all genotypes
in the jth environment from the combined
regression line ie $\epsilon_j - \bar{\beta} z_j$

δ_{dij} = deviation of the ith genotype from its
linear regression on z_j in the jth envi-
ronment less $\bar{\delta}_j$ ie $\delta_{ij} - \bar{\delta}_j$

The above models also circumvent another problem in-
herent in the use of models (2) and (3), namely the
partitioning of the G X E sum of squares into parts
corresponding to the genotypes under study which is
not valid from the statistical point of view due to
the low number of degrees of freedom that are available (Freeman, Perkins 1971). If the purpose of

the analysis is the comparison of the stabilities of the various genotypes, then the sum of squares to be partitioned should be the total within genotypes sum of squares which encompasses both the among environments and the G X E variation.

The estimates of z_j can be obtained in several ways (Freeman, Perkins, 1971):

1. The replicates of the experiment are divided into two groups. One is used for the measurement of the genotype X environment interactions and the other for the assessment of the environment.

2. A group of related genotypes is included in the experiments to provide an assessement of the environment.

3. A group of genotypes or a single genotype is used to assess the environment. These need not be related to the genotypes under study.

4. Some physical measurements of the environment i.e. temperature, nutritional substrate etc., are used as an index.

All these methods have been tried. The general conclusion is : if the number of genotypes under study is large, then the mean of all genotypes per environment provides a sufficiently independent assessment of the environment (Freeman, 1973).

The analysis of variance corresponding to models 4 and 5 is given in Table 3. The various mean squares of this table are tested for significance by the F test, the denominator being the error mean square obtained from replication. If the heterogeneity mean square is significant, then some regressions are heterogeneous. They can be tested in pairs by means of a t - test. If the mean square for the environmental residual is not significant, then z_j describes adequately, in a statistical sense, the changes in ε_j (the additive environmental component of the genotypes under study) that follow changes in the environment. If at the same time $\bar{\beta}$ does not differ from unity, then z_j estimates ε_j adequately, reducing model 4 to model 2 (Freeman, Perkins,

1971). The item "combined regression" represents the variation due to fitting a single line to all genotypes, while the item "heterogeneity of regressions" represents the variation due to the deviations of the individual regression lines from the combined regression line.

2.2. Stability parameters

2.2.1. Regression Studies

On several occasions the term stability of a genotype was used throughout the text. There is, however, disagreement as to how stability is measured. It is to this problem that we now turn.

Finlay, Wilkinson (1963) employing regression techniques characterized the environment by the mean yield of all varieties grown in it. To obtain better linearity they used the logarithmic scale. According to them a regression coefficient of unity indicates a variety of average stability. If the regression coefficient is greater than unity, then the corresponding variety is of below average stability while if it is smaller than unity the variety has above average stability. If the regression coefficient is zero, then the variety has absolute phenotypic stability (The values of the regression coefficients would be smaller by one, if model 3 notation is used). On the basis of the above, however, a stable cultivar would perform relatively well in poor environments and relatively poorly in "good" environments. To get around this Finlay, Wilkinson (1963), proposed that an ideal variety should have maximum phenotypic stability plus maximum yield in the most favourable environment. These ideas are shown in Fig. 1. Unfortunately from the 277 barley varieties used by Finlay, Wilkinson (1963), no one combined both attributes of the ideal variety. As usual a compromise was suggested.

A refinement to the regression technique for the purposes of comparing the stability of various varieties was proposed by Eberhart, Russell (1966). They performed their analysis on the basis of a model similar to model 2, given earlier, but using as ε_j the mean of all varieties at the jth environment minus the grand mean. The analysis of variance according

TABLE 3. Joint regression analysis for t genotypes grown in s environments when an independent assessment of the environment is used.

$$Y_{ij} = \mu + d_i + \bar{\beta}z_j + \bar{\delta}_j + \beta d_i z_j \; + \delta_{dij}$$

Source of variation			Degrees of Freedom
1. Among genotypes (G)			t - 1
Among environments (E)			s - 1
2. Combined Regression	} E	$[\bar{\beta}]$	1
3. Environmental Residual		$[\bar{\delta}_j]$	s - 2
Genotypes X Environments (G X E)			(t-1)(s-1)
4. Heterogeneity of regressions	} G X E	$[\beta\delta_i]$	t - 1
5. G X E residual		$[\delta_{dij}]$	(t-1)(s-2)
Total			ts - 1

The error sum of squares can be computed from the replicates and has ts(r-1) degrees of freedom. In this case, of course, the total degrees of freedom are no longer ts - 1 but tsr - 1.

All items in the above table are orthogonal to each other.

---

to Eberhart, Russell (1966), which even though it was reported earlier can be considered a modification of the analysis of Freeman and Perkins (1971), is given in Table 4. A stable variety according to Eberhart, Russell (1966), has a regression coefficient of unity and deviations from the regression line as small as possible. A high performance is also a prerequisite. The Eberhart-Russell analysis (Table 4) differs from that of Freeman and Perkins (Table 2) in that the two residual sums of squares of the latter are combined to give the "pooled deviations" item in the former and in that the E and G X E items are combined. This in turn is partitioned to give the "environments linear" conponent sum of squares which is allocated one degree of freedom when in effect it is the same with the total sum of squares for environments which has (s-1) degrees of freedom.
The tests of significance in the Eberhart-Rus-

sell approach assume that the deviations for each genotype are homogeneous. The procedure outlined above provides estimates of phenotypic stability statistics. Tai (1971) was concerned that when relatively few genotypes are tested over a restricted set of environments, the phenotypic stability parameter provided by the Eberhart - Russell approach may differ from the corresponding genotypic estimates. Tai (1971), therefore, proposed a method which, by removing the error deviate term from the stability parameters, provides estimates of genotypic stability.
Tai uses two measures of stability: α_i, which measures the linear response to the environment and λ_i which measures the deviations from this linearity in terms of the magnitude of the error variance i.e.

$$\lambda_i = \frac{\sigma^2_{\delta i} + \sigma^2_e}{\sigma^2_e}$$

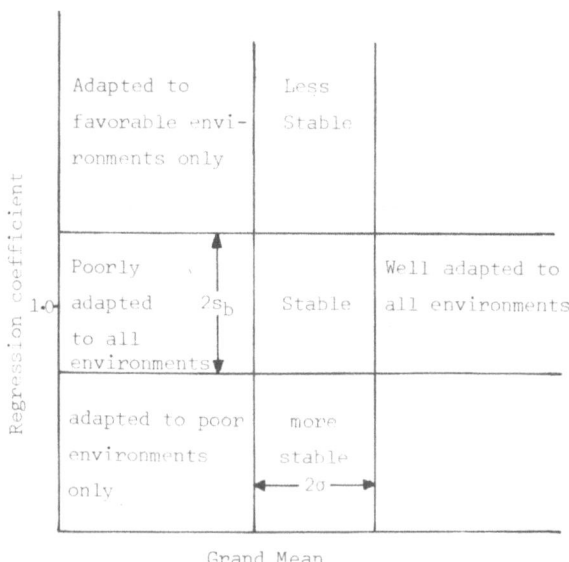

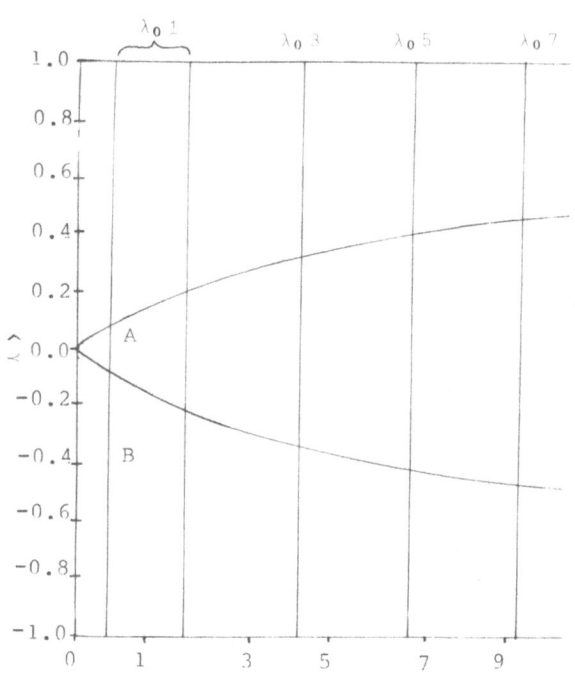

FIGURE 1. Graphical representation of the method used by Eberhart, Russell (1966) to select stable (b $\cong$ 1) and high yielding genotypes. Stable genotypes should fall in the rectangle formed by the four lines. s_b, s standard deviation of the regression coefficient and the grand mean, respectively (Adapted from Eberhart, Russell (1966) and Finlay, Wilkinson (1963).

FIGURE 2. The two regions of stability according to Tai (1971). See text for explanation. The varieties falling in area A have average stability while those which fall in area B possess above average stability.

TABLE 4. The analysis of variance of t genotypes in s environments (r replications per environment) for estimation of stability parameters according to Eberhart and Russell (1966)

Source of Variation	Degrees of freedom
Genotypes (G)	t - 1
Environments (E)	
Genotypes X Environments G X E }	t(s - 1)
Environments (linear)	1
G X E · (linear)	t - 1
Pooled deviations	t(s - 2)
genotype 1	s - 2
genotype 2	s - 2
genotype t	s - 2
Pooled error	s(t-1) (r-1)

A perfectly stable variety has $\alpha = -1$ and $\lambda = 1$ whereas a variety of average stability has $\alpha = 0$ and $\lambda = 1$. Tai's method is given graphically in Fig. 2 where α and λ are represented by two orthogonal axes. The hyperbola contains 95% of the predictions for α when its true value is zero. The first two vertical lines give the 90% confidence region for $\lambda = 1$, while the remaining lines depict the upper limits of the intervals when $\lambda > 1$. Thus the area determined by α and λ is divided into two stability regions. The varieties falling in the area marked A have $\alpha = 0$, $\lambda = 1$ and average stability while those in B have $\alpha < 0$ and $\lambda = 0$ and are characterized by above average stability. Yet another stability measure was proposed by Hanson (1970) which has the value

$$D_i^2 = \sum_j [\, g_{ij} + (1-\alpha)\, \varepsilon_j]^2$$

where α is not the same as in Tai's (1971) notation but is the minimum observed value of $(1+\beta_i)$. D_i combines the regression and deviations from regression into a single value.

2.2.2. G X E analysis

In addition to the stability measures described above, all of which come from regression analysis, others, emanating from analysis of variance have been proposed. Thus Wricke (1962) uses the term *ecovalence*, which is the contribution of a given genotype to the G X E interaction sum of squares i.e. $\sum_j g_{ij}^2$. The smaller this is the more stable the variety.

Shukla (1972) partitioned the sum of squares for G X E into components corresponding to each of the genotypes under study and proposed the stability variance σ_i^2 of a genotype as the measure of stability. This quantity is defined as the variance over environments of $(g_{ij} + \bar{e}_{ij})$, where $\bar{e}_{ij}$ is the mean over replicates of the error component (e_{ijk}) of an observation described by the model $Y_{ijk} = \mu + d_i + \varepsilon_j + g_{ij} + e_{ijk}$. Shukla

(1972) obtained unbiased estimates of σ_i^2 and an approximate test for deciding whether a genotype was unstable or not.

Jowett (1972) and Easton, Clements (1973) compared the various stability parameters on the same sets of data (Fig. 3). No general conclusions can be drawn from their work. It appears, however that Wricke's ecovalence is not as informative as the others (Jowett, 1972) while the Finlay-Wilkinson method cannot identify genotypes which are adapted to intermediate environments (Easton, Clements 1973). Becker (1981) argued that stability of performance in cultivated plants has two concepts: the biological concept, under which the genotype should have minimal variance under different environments and the agronomic concept, under which a genotype should have minimal interactions i.e. small values of Wricke's ecovalence. He found that for five crops there were positive correlations between coefficients of regression and variances and deviations from regression with ecovalences, respectively. The latter correlation was confirmed by Heine, Weber (1982) indicating that either the ecovalence or the regression coefficient can be used. The same authors, however, also found that neither the ecovalence nor the regression parameters (coefficient and deviations) were constant over years or over the different trials in any one year and argued that these parameters could not be used for the usual evaluation experiments which only last three years. From the foregoing it is clear that there is no lack of methods for the evaluation of the stability of a group of genotypes. Most of them are interelated and provide overlapping information. From the point of view of the plant breeder the need exists for a method that allows predictions of a genotype's stability. Such methods do exist but are beyond the scope of this paper.

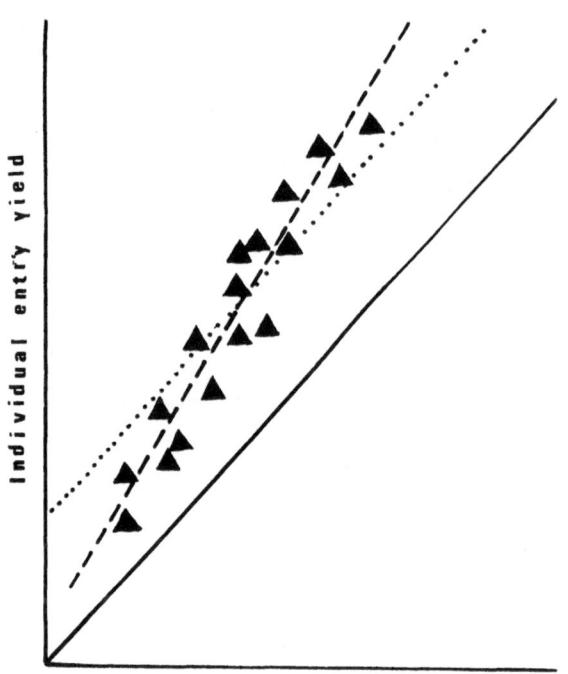

Environment mean yield

FIGURE 3. The relationships among some of the stability parameters in use. Solid triangles represent the observations of an hypothetical genotype. The solid line describes the average response to the environmental index. The dashed line gives the Finlay-Wilkinson line, while the dotted line (b=1) corresponds to a particular genotype. The Eberhart-Russell stability parameter is a function of the deviations from the dashed line while the parameter of Wricke depends on the deviations from the dotted line (From Easton, Clements, 1973).

REFERENCES

Becker HC (1981) Correlations among some statistical measures of phenotypic stability, Euphytica 30, 835-840.

Breese EL (1969) The measurement and significance of genotype-environment interactions in grasses, Heredity 24, 27-44.

Easton HS and Clements RG (1973) The interaction of wheat genotypes with a specific factor of the environment, J. agric. Sci.Camb. 80, 43-52.

Eberhart SA and Russell WA (1966) Stability parameters for comparing varieties, Crop Sci. 6,
36-40.

Finlay KW and Wilkinson GN (1963) The analysis of adaptation in plant breeding, Aust. J. Agr.Res. 14, 742-754.

Freeman GH and Perkins JM (1971) Environmental and genotype-environmental components of variability, VIII, Relations between genotypes grown in different environments and measures of these environments, Heredity 27, 15-23.

Freeman GH (1973) Statistical methods for the analysis of genotype-environment interactions, Heredity 31, 339-354.

Fripp YJ and Caten CE (1971) Genotype-environmental interactions in *Schizophylum commune*, Heredity 27, 393-407.

Hanson WD (1970) Genotypic stability, Theoret. Appl. Genet. 40, 226-231.

Jowett D (1972) Yield stability parameters for sorghum in East Africa, Crop Sci. 12, 314-317.

Heine H and Weber WE (1982) Die Aussagekraft statistisches Masszahlen fuer die phaenotypische Stabilitaet in amtlichen Sortenpruefungen bei Winterweizen und Koernermais, Z. Pflanzenzuechtg. 89, 89-99.

Perkins JM and Jinks JL (1968) Environmental and genotype-environmental components of variability, III, Multiple lines and crosses, Heredity 23, 339-356.

Perkins JM and Jinks JL (1973) The assessment and specificity of environmental and genotype-environmental components of variability, Heredity 30, 111-126.

Shukla GK (1972) Some statistical aspects of partitioning genotype-environmental components of variability, Heredity 29, 237-245.

Tai GCC (1971) Genotypic stability analysis and its application to potato regional trials, Crop Sci. 11, 184-190.

Wricke G (1962) Über eine Methode zur Erfassung der ökologischen Streubreite in Feldversuchen, Z. Pflanzenzuechtg. 47, 92-96.

BRANCHING PATTERNS IN COLUMNAR CACTI

M. L. CODY (Department of Biology, University of California, Los Angeles, California 90024, USA)

1. INTRODUCTION

In recent years they study of vegetation structure has developed along novel and substantially different lines of investigation, following innovative approaches by Halle and Oldeman (1970) and Horn (1971). It has become fashionable to talk of the "architecture" of vegetation in general and of tree and shrub individuals in particular. A majority of this work has been conducted on tropical trees (see recent syntheses by Tomlinson and Zimmerman, 1978; Halle, Oldeman and Tomlinson, 1978). It includes attempts by various authors to show that branching patterns of certain tree species conforms to one of a limited number of standard forms (or "models"), and raises the possibility that certain vegetation types might be dominated by trees or shrubs that follow a particular branching mode.

There are at least three areas of potential difficulties that might hamper future studies of the adaptive geometry of branching patterns: i) trees and shrubs have complex three-dimensional branching patterns that are difficult to quantify; ii) the potential genetic control of branching during growth and development is likely to be modified and perhaps overridden by variable environmental factors amongst sites; and iii) although gross morphological features, such as crown shape or size, might be readily interpreted as adaptations to a particular environment (e.g. Horn, 1971), the selective advantage of different branching patterns amongst species and between habitats is at best obscure.

A good deal of attention has been paid to the branching pattern of the tropical tree *Terminalia catappa*, because of the relative ease with which its unusually regular form of branching (i.e. Aubreville's model) can be quantified (Fisher, 1977; Halle et al., 1978). This advantage is even more pronounced in the columnar cacti, those tree- and shrub-like members of the tribe Pachycereeae of the family Cactaceae.(For nomenclature I follow Gibson's (1982) revision of this section of the family). See Shreve and Wiggins (1964) and Wiggins (1980) for flora and description of the vegetation within the area of concern, the Sonoran Desert. Plant species-area curves for most of the sites mentioned in this paper are given in Biehl and Cody (1983). In this paper I describe a simple method for the quantification of branching patterns in the columnar cacti. I show how branching patterns change: i) with ontogeny within species, ii) amongst species that coexist at a certain desert site and iii) within species between different desert sites. I attempt to isolate particular influences on cactus branching patterns and pay particular attention to the possible effects of both the physical environment (slope, aspect, substrate) and the biotic environment (particularly co-occurring species of cacti). An advantage to working with branching patterns in columnar cacti is that these plants are leafless; the branches are of considerable thickness and their surfaces are photosynthetic. Further, there is relatively little variation in branch thickness within individuals or species; they do not show the graded series of diameters from trunk through decreasing branch sizes to twigs characteristic of most shrubs and trees. Columnar cacti are almost geometric in their structure, and can be modeled or constructed with reasonable accuracy by taking a cylinder of the appropriate (constant) diameter, chopping it up into segments, and arranging these segments in appropriate fashion to imitate the natural branching form of the species in question.

Despite the obvious convenience of columnar cacti for studies of branching patterns, very little research has been published on the subject. Felger and Lowe (1967) showed that branch characters of senita (*Lophocereus schottii*) changed geographically

from southern Sonora (Mexico) to southern Arizona (USA) by increasing stem (branch) radius and decreasing rib number (1.5X), concomitantly decreasing overall surface-to-volume ratio (2X) from south to north. They interpreted these changes of stem width and ribbing with latitude as a result of selection for larger stem radii to the north (to better resist freezing temperatures) and also as a response (form unspecified) in increased total and especially summer rain to the south. In fact there are many possible interpretations to stem size and to branching pattern in general, some of which are discussed next.
a) A plant might be selected to maximize cover (areal spread of the vertical projection of its branches per unit biomass or per cost in terms of structural material in the plant). This is the argument used by Honda and Fisher (1977) in their studies of *Terminalia* branching. b) A plant might evolve growth and branching patterns such that it overtops and shades out other plant individuals. This might be accomplished by a rapid vertical growth followed by a more horizontal branching form. Both a) and b) are forms of space pre-emption, with a vertical component in the latter absent from the former. Neither a) nor b) are likely to be important selective effects in desert habitats where the larger plant individuals are widely spaced and the vertical projections of their branches form well separated map areas. Space pre-emption may, however, be a factor in the branching of cacti in the more mesic and closed canopy thorn scrub habitats that border Sonoran Desert further up the moisture gradient (see, e.g., Gentry, 1943). c) Cacti might be under selective pressure to maximize photosynthetic area (branch surface area) per unit volume, with resultant advantages in more rapid carbon fixing and growth. Given a certain biomass or volume, in the form of a cylinder of radius r and length l, volume is equal to $(pi)r^2l$ and its surface area to $2(pi)r.l$. Thus the ratio of surface area to volume is $2/r$, which is obviously greater at smaller values of r. However, because

the shear strength of a beam (or stem) is proportional to its radius r, plants with small branches or stems will lack the structural strength to support a large biomass high off the ground, and thus should be confined to a more horizontally spreading habit. d) On the other hand branch radius might be affected by selection for increased water storage capacity in the plant. Imagine that a cylinder of overall radius r, longitudinally pleated with n ribs (as are columnar cacti), represents a cactus. At maximal water depletion the ribs are deeply indented; when the cactus is replete with water the ribs will barely show and the perimeter of its cross-sectional area, directly proportional to its water storage capacity, will be about 2rn. Thus selection for water storage, besides favoring more ribs, will increase stem radius r, and oppose selection for increased surface area-to-volume ratio which favors smaller stem radius. e) Cacti might branch in such a way that, for a given volume (biomass), maximal surface area is exposed to the sun, and self-shading by branches is minimized. This argument might gain weight if the findings of Nobel (1980a, 1981) and Woodhouse et al. (1980), that certain leaf- and stem-succulents in the Colorado Desert are light-limited in the sense that if more light were available growth could be more rapid, were found to be broadly applicable. f) More generally, the positioning of branches in space might have a good deal to do with regulating the microclimate around these branches. Particular branch orientations might, for example, affect local air movements and potentially decrease heat load and the moisture lost through transpiration (Yeaton, 1983a,b), even though cacti operate on the CAM system and their stomate open at night. Alternatively, many otherwise healthy-looking cacti die after being blown over by the wind, and a branch orientation that minimizes bending moments due to winds might be a useful attribute (see, e.g. Alexander, 1971; Lennon, 1979). Detailed microclimatological work of the sort described by Miller and Stoner (1979) is necessary to elucidate such possible effects of and on branching pattern.
There are still further potential selective influences

on cactus branching pattern. g) Pollinators are one candidate; bats and moths are known to pollinate some columnar cacti (Gibson and Horak, 1978), and it is conceivable that branching pattern has been influenced by selection to display flowers, often terminal on the branches, at maximal height above the ground or in certain attractive arrays for increased pollinator visits and maximal seed set. In addition, herbivores such as woodrats (*Neotoma*) might influence branching pattern directly by eating off growing tips (Blom and Bratz, 1976; Patterson, 1980), and indirectly by selecting for cacti that orient their branches to minimize the availability of food for such herbivores.

In this paper no firm conclusions will be made about a hierarchy amongst selective factors such as those just discussed. Where relevant, comments will be made on what appear to be the dominant forces that determine cactus branching patterns, but clearly the observed patterns are products of several, possibly many and sometimes conflicting, selective forces. The likelihood or resolving the role of each factor from observational data alone is slight, yet some of the stronger influences on branching patterns might still become apparent.

2. METHODS

I have measured the branching patterns of cacti in the genera *Pachycereus*, *Carnegiea*, *Lophocereus*, and *Stenocereus* (which includes *Lemaireocereus* and *Machaerocereus* - Gibson, 1982). These measurements have been taken during various intervals since 1977, when the approach was first attempted in Sonoran Desert north of La Paz, Baja California, Mexico. Further details of field sites and visits to them are given in the following section. Branching patterns in columnar cacti may be quantified in a very simple way. In Fig. 1 several Sonoran Desert species are depicted in a schematic fashion; since branches by and large ascend vertically with minimal horizontal growth (except in *Stenocereus gummosus*), information on branching pattern is accumulated vertically through the plant. I record the number of branches that intersect a horizontal plane at different heights above the ground, with no distinction made between branches and a main axial trunk. (Main trunks occur only in the larger tree-like species of *Pachycereus* and *Carnegiea*). Usually branches can be counted at successive heights from a single vantage point, but occasionally in individuals with very dense branching, a succession of vantage points is adopted by the observer moving around the plant. Heights above ground level were initially measured with poles marked off in height intervals and subsequently a clinometer. After the first few hundred plants were measured in that fashion, I verified that my ability to estimate heights was sufficiently accurate to dispense with direct measurement, although the clinometer was employed for many taller plant individuals. I used height intervals in feet rather than metres, since I had more confidence and practice in estimating intervals in the former units. Height intervals were scaled non-linearly, on a logarithm base around 1.3, such that measurements are made at increasingly wider intervals at increasing height above the ground (in accordance with the Weber-Fechner Law of estimation of just-noticeable differences -e.g. Stevens, 1957). Cactus individuals were measured for branching pattern at each of a number of sites selected for given characteristics, such as uniform flat bajada, slopes of constant aspect, cover of particular values. At each site the branching pattern of each species of columnar cactus was quantified in the above manner by recording each cactus individual as it was encountered. Thus the sample of individuals from which measurements were taken is representative of the population as a whole, and no effort was made to avoid or include certain individuals with e.g. specific morphological traits(extra large or small, unusually copious branching, etc.). Sampling continued, within each cactus species, until the measurements accumulated were adjudged representative of the species in question. In practice, this amounted to around 100 individuals of the commoner species at the site, rather fewer of the less common species.

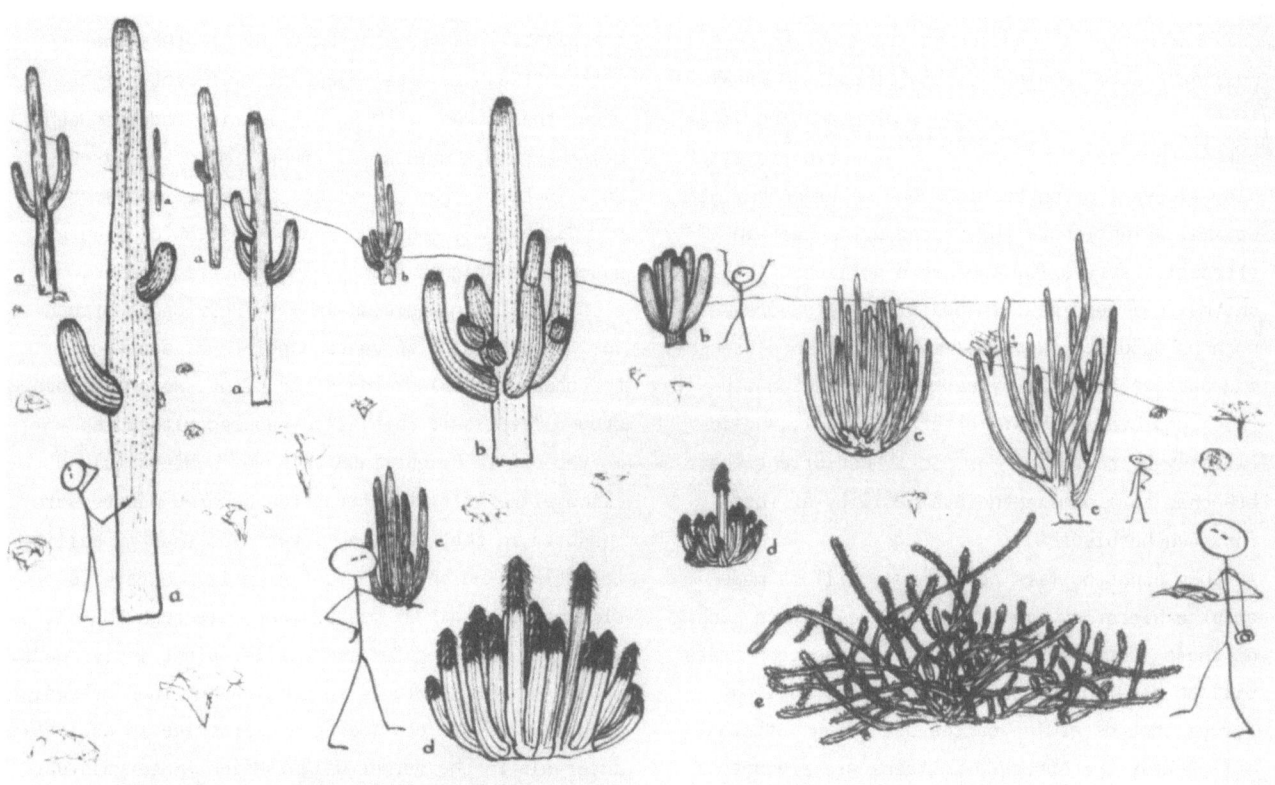

FIGURE 1. Schematic presentation of columnar cacti and their diverse branching patterns. The species represented are: a. saguaro *Carnegiea gigantea* (Cg), b. carbon *Pachycereus pringlei* (Ppr), c. organpipe *Stenocereus thurberi* (St), d. senita *Lophocereus schottii* (Ls) and e. agria *Stenocereus gummosus* (Sg).

The sample of individuals for each species was then divided into size classes, either by biomass or by height of individuals (height of the tallest branch). Usually some 6-8 size classes were differentiated, with 6-12 individuals in each size class. In this way the branching pattern of a species at a site is represented by a family of size-specific curves, plotting average numbers of branches against height above the ground. Branching patterns may be compared within and between species simply by inspection, since each branching "fingerprint" is usually quite distinct, being both species- and site-specific. However, reference is made initially to statistical tests that are employed to distinguish the branching patterns of size classes and different species, but in general I have avoided a repetitive and redundant use of such statistics, referring the reader instead to a comparison of the graphical data. A more profitable line of enquiry, especially appropriate to the testing of certain hypotheses about the adaptive significance of the branching patterns, employs the calculus of variations (see, e.g., Weinstock, 1974). In this way a particular branching pattern might be drawn a priori that, for example, maximizes the height above the ground of cactus biomass, subject to constraints of perhaps minimal wind resistance given a radius r, but this topic will be developed in a later paper (Cody, in prep.).

The branching pattern of a species can be generalized in a simple way by drawing an envelope of the size-specific branching curves. In order to draw a representative envelope, I use the curve for either the largest 10 individuals or the largest 15% of the individuals in the sample population, which almost invariably include all other curves of the smaller size classes within it.

3. FILED SITES

A map of the geographic distribution of Sonoran Desert vegetation and of adjacent thorn scrub is given in Fig. 2, which also shows field sites where measurements of cactus branching patterns were made are labeled also. A brief description of these sites is given in this section.

1) La Paz. This site is located in rich and diverse Sonoran Desert at km 29 northwest of the city of La Paz in southern Baja California (see Fig. 2). Four species of columnar cacti are common at the site (*Pachycereus pringlei*, *Stenocereus thurberi*, *Stenocereus gummosus* and *Lophocereus schottii*, species which will be abbreviated henceforth as

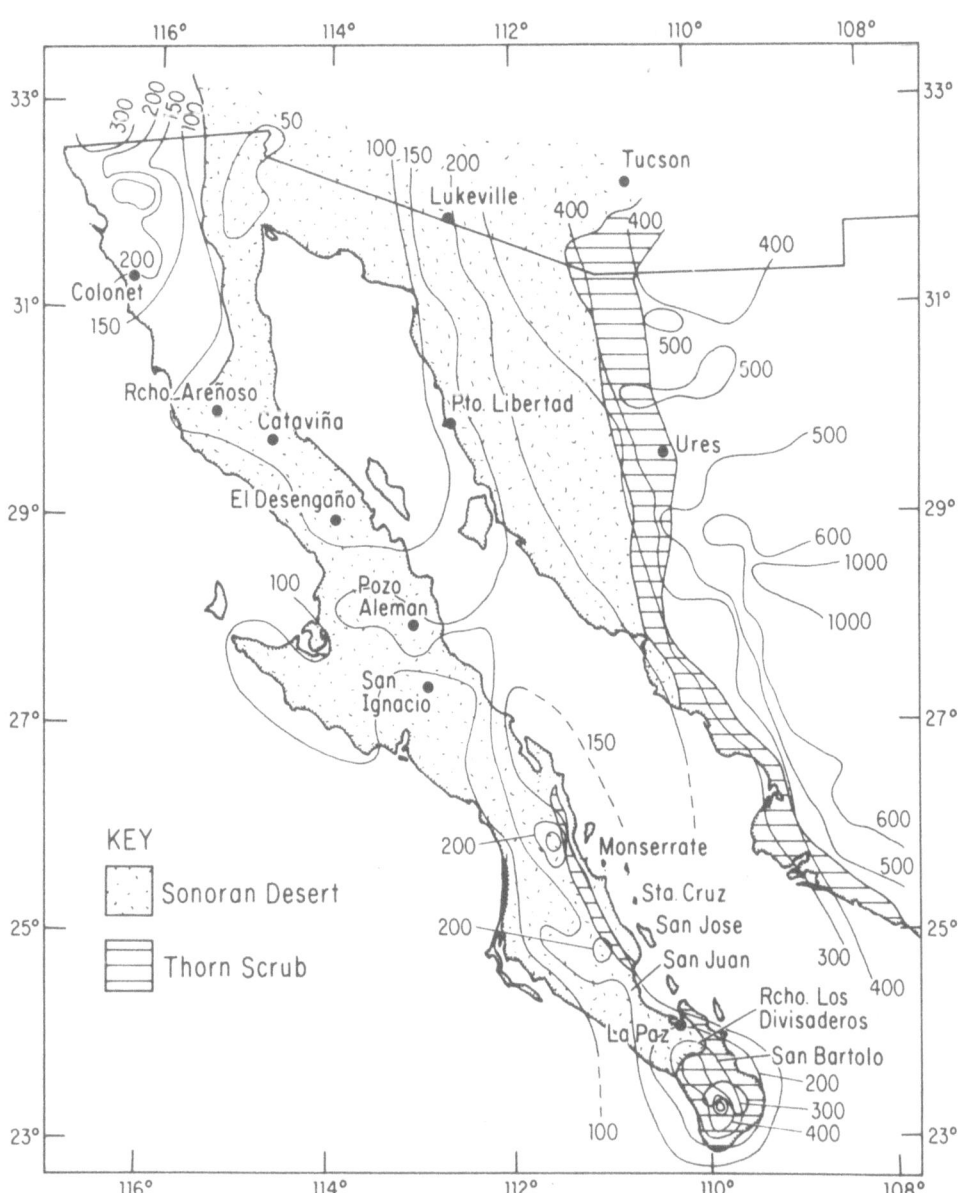

FIGURE 2. Map of the Sonoran Desert and adjacent thorn scrub vegetation. Field sites used in this study are marked on the map, and isohyets of precipitation, in mm, are also included.

Ppr, St, Sg and Ls). This site will serve as a standard of reference to which other sites will be compared. Besides columnar cacti, small trees of *Olneya, Cercidium, Prosopis, Bursera* and *Jatropha* spp. are common as are numerous shrub species such as *Larrea, Ambrosia* and *Opuntia*. Data were collected at this site in September 1977.

2) San Ignacio. Branching patterns were measured at several additional sites in Sonoran Desert to the north on the peninsula, and next at a site 20 km west of San Ignacio at an elevation of 165m adjacent to the transpeninsular highway. The vegetation here is dominated by Ppr, St and Sg, together with *Yucca vallida, Bursera, Jatropha, Fouquieria diguetii* and *Pedilanthus*. Three species of shrubby cylindro-opuntias are present, but Ls is rare at the site. Data were collected in March 1982.

3) Pozo Aleman. Further north, 27 km east of El Arco on the road joining that town to San Francisquito on the Gulf coast at elev. 410m near Pozo Aleman, the columnar cacti Ppr, Sg and Ls are common, but St is rare. Here the larger dominant plants, besides cacti, are *Cercidium, Fouquieria columnaris, Larrea, Viscainoa* and *Yucca vallida*, with the smaller *Agave cerulata, Ambrosia chenopodifolia* and *Acalypha* also common. The site was studied in January 1981.

4) El Desegnano. This site is 17 km east of the transpeninsular highway on the Bahia de los Angeles road, and as at the preceding site Ppr, Sg and Ls are common but St is too rare to measure. The larger common plants are *Yucca vallida, Fouquieria columnaris, Cercidium* and *Viscainoa*, with *Ambrosia magdelana* and *A. chenopodifolia, Agave deserti* and *A. shawii, Simmondsia, Solanum* and *Eriogonum* all common. Data were collected in February 1979.

5) Catavina. Amongst the spectacular granite boulders in the shallow arroyos about 5 km north of the El Presidente Hotel at Catavina and 2 km east of the highway, Ppr and Ls are both common, but St and Sg are both absent. In fact the northern limit of St on the peninsula is about half-way between sites 4 and 5, but Sg ranges north almost to the U.S. border. The elevation is 670m, and prominent plants are *Fouquieria columnaris, Pachycormus discolor,* and *Bursera microphylla* with various smaller shrubs of *Ambrosia, Larrea, Encelia, Trixis* and *Viguiera*. This site was studied in August 1979, April 1980 and April 1981.

6) Rancho Arenoso. This site is located 3 km east of the transpeninsular highway some 48 km south and east of El Rosario, at elevation 340m. In striking contrast to nearby site 5, the substrate is basalt rather than granite. Here Ppr, Ls and Sg are joined by *Myrtillocactus cochal* (Mc), a striking candelabra-like species that is particularly abundant in the coastal region of northwestern Baja California but occurs sporadically down the peninsula with relatives in central Mexico (see Cody, Moran and Thompson, 1983). The common larger plants here are *Prosopis, Malosma laurina, Prunus fremontii, Fouquieria columnaris* and *F. splendens, Bursera microphylla* and *B. hindsiana, Ptelea* and *Larrea*, with numerous species of the smaller *Eriogonum, Viguiera, Encelia, Ambrosia, Jatropha, Euphorbia* and *Haplopappus* shrubs. This site was visited in April and May, 1982.

7) Colonet. At a site 5 km east of the transpeninsular highway, on a road leading east to the Observatorio in the Sierra San Pedro Martir from just south of Colonet, the vegetation is dominated by plant species typical of the Mediterranean-climate zone in which this coastal site lies. However, many desert plant species extend north into this vegetation, including Sg and especially Mc, which occur together with *Bergerocactus emoryi* (Be), which has a distribution restricted to the coastal areas. The elevation is around 30m, and the vegetation has typical chaparral and coastal species such as *Malosma laurina, Rhamnus, Artemisia californica* and *Eriogonum fasciculatum*, as well as transition species such as *Harfordia, Haplopappus rosaricus* and *Aesculus parryi*, and more desertic species such as *Agave shawii, Simmondsia, Jatropha* and many cactus species.

8) Isla Monserrate. Branching patterns were measured on several islands in the Gulf of California, the

northernmost of which is Monserrate, a moderate-sized island (17km^2) of moderate isolation (some 13km from the peninsular coast; see Case and Cody, 1983, plus chapters and appendices therein, for physical and biological characteristics of the Gulf islands, including plant species lists). Monserrate is an island of somewhat uncertain status with respect to a possible Pleistocene land connection to the peninsula, but the weight of the evidence suggests that is was so connected. Its flora is relatively complete, with the small trees *Olneya*, *Cercidium* and *Bursera* common, and a long list of cactus species including Pppr, St, Sg, and Ls. Data were collected on Monserrate in March 1977.

9) Isla Santa Cruz. South of Monserrate, Isla Santa Cruz is a smaller (14km^2) oceanic island (with no recent landbridge connections) and a much abbreviated flora in comparison with Monserrate (72 species compared to 101; see Cody, Moran and Thompson, 1983). Although all four columnar cacti at La Paz occur on Santa Cruz, only Ppr is common, and both Ls and Sg are extremely rare.

10) Isla San Jose. This island is both large (194km^2) and close (5km) to the peninsula, and is a landbridge island. Its flora is very much like that of the adjacent peninsula, and includes all of the local peninsular columnar cacti. Data on cactus branching patterns were collected from all over the islands of Monserrate and Sta. Cruz, and from the hillsides and bajadas of the northern one-third of San Jose.

11) San Juan Canyon. This canyon adjoins the Gulf coast at the end of a road that runs east to the Gulf from El Cien across the southern end of the Sierra de la Giganta. The site was selected around 15 km inland as a comparable peninsular site to that on the nearby island of San Jose, and is around 50m elevation. Data were collected from Sta. Cruz, San Jose and San Juan in June 1977.

12) Tucson Mountains. Three field sites are located in the Sonoran Desert across the Gulf of California in Sonora, Mexico, and Arizona, U.S.A. The first is 25km southwest of Tucson in the Tucson Mountains at 790m elevation, and is representative of this desert towards its northern limits where just one columnar cactus, *Carnegiea gigantea* (Cg), occurs. Its commoner associates are *Prosopis*, *Cercidium*, *Larrea*, *Lycium*, *Ambrosia* and *Gutierrezia*, together with other cacti such as *Opuntia fulgida*, *O. acanthocarpa* and *O. spinosior*. This site was employed in March 1981.

13) Lukeville. As one proceeds south from southern Arizona into northern Sonora, additional species of columnar cacti are encountered. The first additional species is *S. thurberi*, which occurs together with Cg in Organpipe Cactus National Monument north of Lukeville, Arizona. The site is at 700m elevation, just west of Alamo Canyon near to the Ajo Mountains, on bajadas where *Cercidium*, *Olneya*, *Simmondsia*, *Larrea*, *Ambrosia deltoidea* and some half-dozen *Opuntia* species are common.

14) Puerto Libertad. Although Ls just crosses the border into Arizona at Lukeville, it does not become common until one moves south into northern Sonora. Here also Ppr reaches its northern limits, and near Puerto Libertad one finds the four species Cg, Ppr, St and Ls together in typical and undisturbed Sonoran Desert vegetation. These species were measured where they occur together at 85m elevation 23 km inland from the Gulf coast on the Pto. Libertad road. In this vicinity Sg occurs also, but it is very restricted in range and forms isolated coastal populations that are the only examples of the species in Sonora. Data were obtained from this site and from Lukeville during November 1977.

15) Ures thorn scrub. In central and eastern Sonora, towards the foothills of the Sierra Madre Occidental, the Sonoran Desert gives way to thorn scrub vegetation (see, e.g. Gentry, 1943) at higher elevations with greater precipitation. The thorn scrub is dominated by *Acacia* spp., *Lysiloma*, *Bursera* and other short trees and tall shrubs that form a dense and thorny thicket woodland with a more or less closed canopy between 3-6m. This habitat is predominant around Ures, central Sonora, and in it

columnar cacti are quite common. In particular St and a congener of Ppr, *Pacycereus pecten-aboriginum* (Ppa), are both common in thorn scrub. A site 6 km west of Ures at elv. 410m was studied in July 1979.

16) San Bartolo. Thorn scrub vegetation occurs also in southern Baja California, in the Cape region to the south of La Paz, where precipitation is higher than in the central peninsular deserts and in particular a greater proportion of the rainfall total falls in summer (see Fig. 2). On the road to Cabo San Lucas south of La Paz, one sees a vegetational gradient corresponding to increasing precipitation, from typical Sonoran Desert in the vicinity of La Paz and its southern outskirts, to typical thorn scrub near San Bartolo some 140 km south of La Paz. At a site 12 km north of San Bartolo both Ppa and St are common in vegetation dominated by *Lysiloma divaricata* and *L. candida, Bursera* and several *Acacia* spp. There are a very few individuals of Sg to be found here. This site was visited in September 1977 and March 1982.

17) Rancho Los Divisaderos. About half-way between La Paz and San Bartolo, at km 178 on the transpeninsular highway near San Pedro, a site was selected that represents the transition between the Sonoran Desert to the north and thorn scrub to the south. At Rcho. Los Divisaderos, elev. 340m, both congeners Ppr and Ppa are present, the congeners St and Sg are also common, and Ls is likewise present although quite rare. Here shrub stature and density, as well as canopy contiguity, reach values intermediate between those at the La Paz and San Bartolo sites. Large *Bursera* and *Jatropha* are common, but *Acacia* and *Lysiloma* are still scarce. This site was surveyed in March 1982.

4. THE BRANCHING PATTERNS

4.1. The ontogeny of branching

4.1.1. La Paz site. The first question I approach is how different cactus species add branches or biomass as they grow; that is, how does the branching configuration of individuals change as size increases? Beginning with Ppr at the La Paz site, the population sample of ca. 60 individuals is subdivided into size-specific subsets, and for each size category the average number of branches is plotted against height above the ground. The results are shown in Fig. 3. In Fig. 3a, the population is subdivided according to height classes of individuals; the curves for 7 height classes are shown, each an average of the number of individuals indicated next to the curves. These curves form a set of family showing how individuals have added branches in moving from one height class to the next. The tallest 10 individuals are averaged to form the outside curve of the family. In Fig. 3b individuals are grouped by biomass, which is assumed proportional to the total length of the branches plus trunk. In order to compute estimated biomass, the number of branches in each height interval is multiplied by the length of the height interval, and these figures are then summed over height intervals for each individual to give a total length or biomass figure. The cactus individuals in a population are then subdivided into biomass (total length) classes, and individuals in each class averaged as before. Again a family of curves results, showing how branches are added in moving up to the next size category. Note that each axis is similarly scaled, logarithmically with a base around 1.3. A good deal of information is summarized in Fig. 3 for the species Ppr. The curves show that branching begins when individuals reach about 3m or 10ft height, when branches are initiated around 6ft or 2m, and that the maximum height of individuals is around 10m or 30+ft. The vertical separation of adjacent curves in the family shows how branches are added at particular heights, and their horizontal separation on the graph shows how a given number of branches occurs at greated heights (to the right) between size classes. Thus the relative proximity of the curves in the family at 6-8ft shows the limited number of branches added in this height interval; subsequent size is added by this limited number of branches elongating vertically, and by

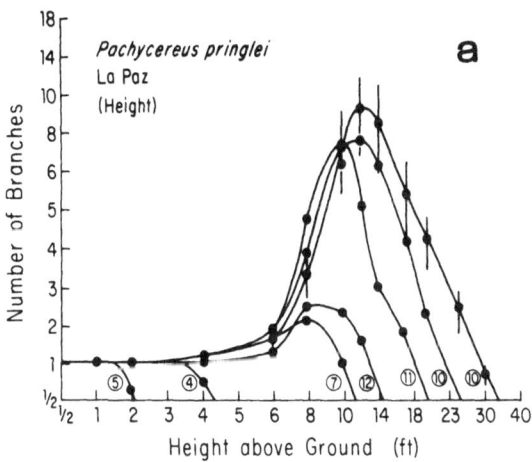

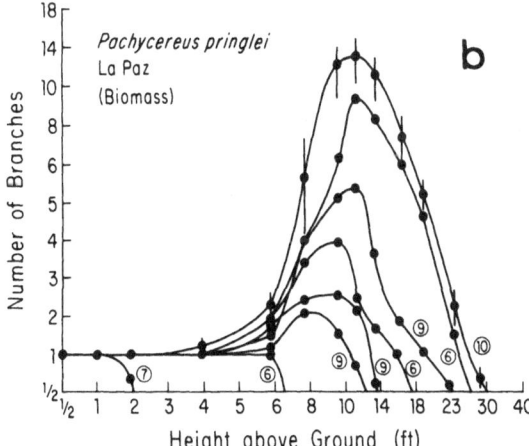

peak around 12ft or 3 1/2m in this species.

In Fig. 4(a and b) the branching patterns of St are portrayed in a similar fashion, again with size classes averaged by height and by biomass respectively. In contrast with Ppr, this species adds many branches at just above ground level early in its growth, but increasing size later in developement is achieved by adding branches higher off the ground

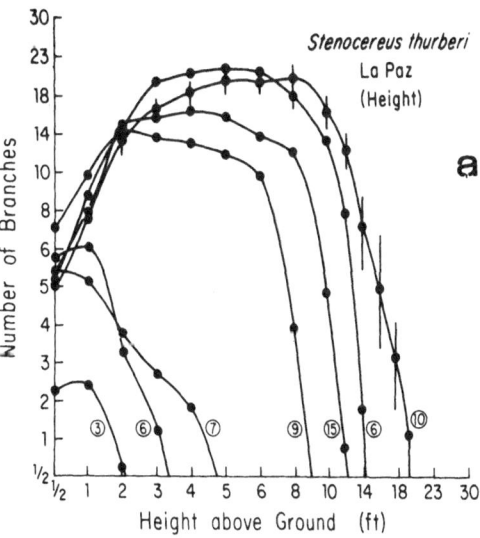

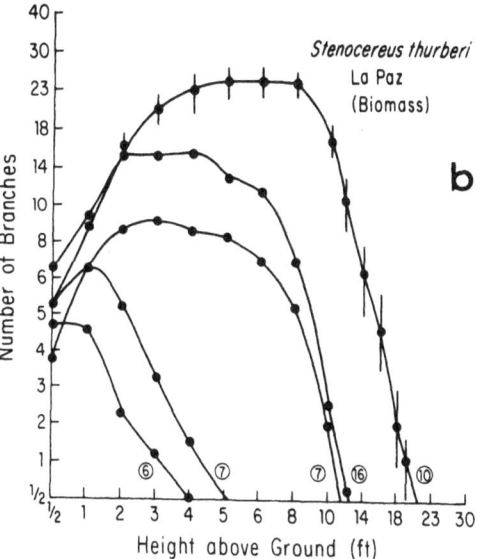

FIGURE 3. The branching pattern of *Pachycereus pringlei* at the La Paz site is shown by plotting the number of branches intersected by a horizontal plane at various heights above the ground, versus height above the ground. In Fig. 3a individuals are grouped into size categories according to their heights; in Fig. 3b according to their total stem length. The curves show how branches are added to individuals as the plants grow, and provide a family of size-specific curves that describe the ontogeny of branching in the species. Error bars are given for the largest size categories, and the numbers of individuals in each size class are shown adjacent to the curves.

secondary branching on them between 8-10ft above ground level. Subtracting the areas under adjacent curves shows where branches or biomass is added between adjacent size classes. Branch numbers

FIGURE 4. Branching patterns of *Stenocereus thurberi* at La Paz, showing individuals grouped by height (a) and biomass or total stem length (b). Error bars are given for the curves of the largest size classes.

210

and by branch elongation especially. St achieves higher branch numbers than Ppr (ca. 20 vs. ca. 12), branches earlier and reaches peak branching earlier (at 6ft rather that 12ft), but reaches more modest heights (20ft rather than 35ft). Standard error bars are included on the outer curves of Figs. 3 and 4, and reflect the variability of individual branching numbers within the largest size classes. In general this variability is more pronounced when averaging is by height than by biomass, and henceforth I use biomass (total length) averaging to generate size-specific branching curves for populations. The branching pattern of the third species at the La Paz site, Ls, is shown in Fig. 5a. Again grouping by biomass classes, the species gives a family of size-specific curves in which each curve can be imagined to be generated from its precedent by a particular but relatively straight-forward extrapolation. In Ls the trend observed from Ppr to St is continued: Ls has more branches at and just above ground level, reaches higher peak branch numbers (around 100) at lower heights (around 4ft), but the species grows to overall lower heights. This pattern is continued with the fourth species at the site, Sg (Fig. 5b). The family of curves that depicts Sg's branching pattern is different again from the preceding species, in that more of its ontogenetic development is channeled into a lateral spreading at ground level (where curves are well separated between 1/2-2ft) and less to growing taller (since curves are relatively close in height - abscissa- at low branch numbers). In fact individuals of this species form dense patches many meters across, even though they reach heights of no more than 3m or so. It is impossible to penetrate and count branches in these patches, formed evidently by successive long-term vegetative growth, however an outer curve that more accurately reflects their existence would be much taller (to hundreds of branches) but would spread very little further to the right (individuals limited in height). In fact these

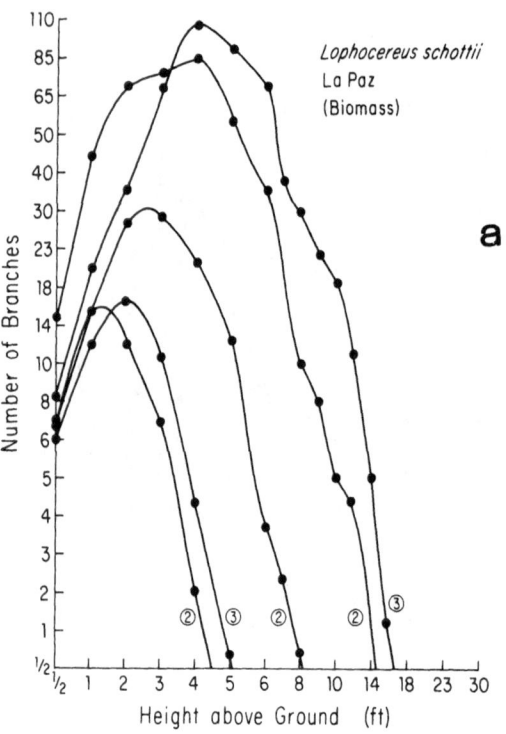

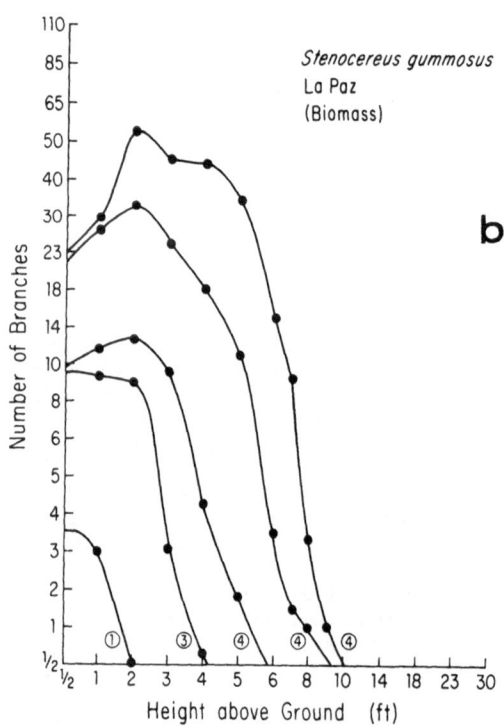

FIGURE 5. (a) Branching pattern of *Lophocereus schottii* at La Paz, and (b) that of *Stenocereus gummosus* at the same site.

very large Sg individuals are rooted at many
places, as branches spread laterally, root distal-
ly and die off proximally, and contribute to the
individual's spread and cover. By convention,
Fig. 5b includes individuals with a single root-
ing point.

4.2. The cactus community

4.2.1. La Paz.
The four columnar cactus species
at the La Paz site form a sequence in branching
pattern from Ppr at one extreme to Sg at the
other. Ppr is tall and trunked, with limited
branching high off the ground. Sg is low and
spreading, with a great deal of branching at
and just above ground level. St and Ls are intermediate in the sequence, filling out the trend
from the taller and sparsely branched to the
shorter and more copiously branched. The sequence
is illustrated in Fig. 6, which shows the outer-
most curves or envelopes for the largest plants
of each of the four species.

All four species of columnar cacti are common at
the La Paz site, although not equally so. Their
relative abundances there are Ppr:0.37, Sg:0.32,
St:0.23, Ls:0.08; all four coexist scattered
throughout the homogeneous-appearing vegetation
on a flat and uniform sandy substrate with no
apparent microhabitat preferences. With reference
to the branching envelopes of Fig. 6, it is
possible that coexistence is favored by species
differences in these curves. Indeed, they are
reminiscent of e.g. species distributions over
habitat gradients, such as *Solidago* species over
a gradient of moisture availability (Werner, 1979),
or the distribution of *Haplopappus* species over
slope and substrate gradients on Mojave Desert
hills (Cody, 1978). But these are examples of
habitat subdivision, and the segregation of
cactus branching patterns suggests something more
in the line of resource partitioning within a
habitat. The community branching patterns invite
the question as to what resource gradient is
being differentially exploited by the cacti, or
how might differences in resource exploitation

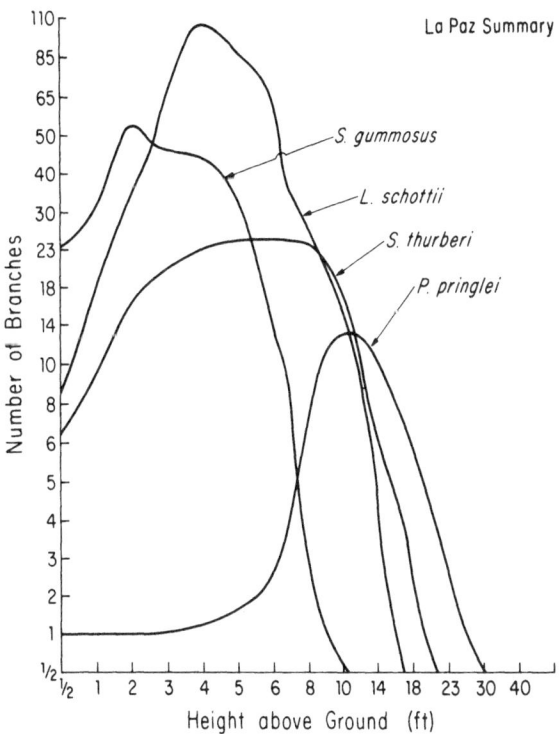

FIGURE 6. The envelopes of the branching patterns
of the four species of columnar cacti at the La Paz
site are drawn to illustrate species-specific differences.

strategies be used as a framework to interpret the
morphological series illustrated by the four species
in Fig. 6. To obtain further clues to the possible
influences on branching patterns, we next incorporate
data collected from other desert sites.

4.3. Central peninsular sites

4.3.1. Reduced species diversity.
To the north of
La Paz, in the central and north-central parts of
the Baja California peninsula, I surveyed branching
patterns in several other Sonoran Desert sites. From
the south, these are San Ignacio, Pozo Aleman, El
Desengano and Catavina. The first three sites represent level sandy flats with no topographic relief
in sight (as at the La Paz station), but rather than
four columnar cactus species as in the south, each
supports three species. At San Ignacio these are Ppr,
Sg and St, with the first two common, the third rare

and Ls absent. Their branching patterns are summarized in Fig. 7a. At both Pozo Aleman and El Desengano the three species present are Ppr, Sg and Ls, with St extremely rare at the former and absent from the latter site. The relative abundance of the three commoner species are 6:7:1 at Pozo Aleman and 7:5:2 at El Desegnano. Branching patterns for these two sites are shown in Fig. 7b and 7c. Further north at Catavina the study site differs in two respects: the substrate is pure granite unlike the blackish basaltic sands to the south, and there is topographic relief in the jumbled granite boulders of the area. Data were collected at Catavina from the sandy bajadas amongst the boulders (not from rocky slopes), and here just two species occur, Ppr and Ls, both commonly. The site

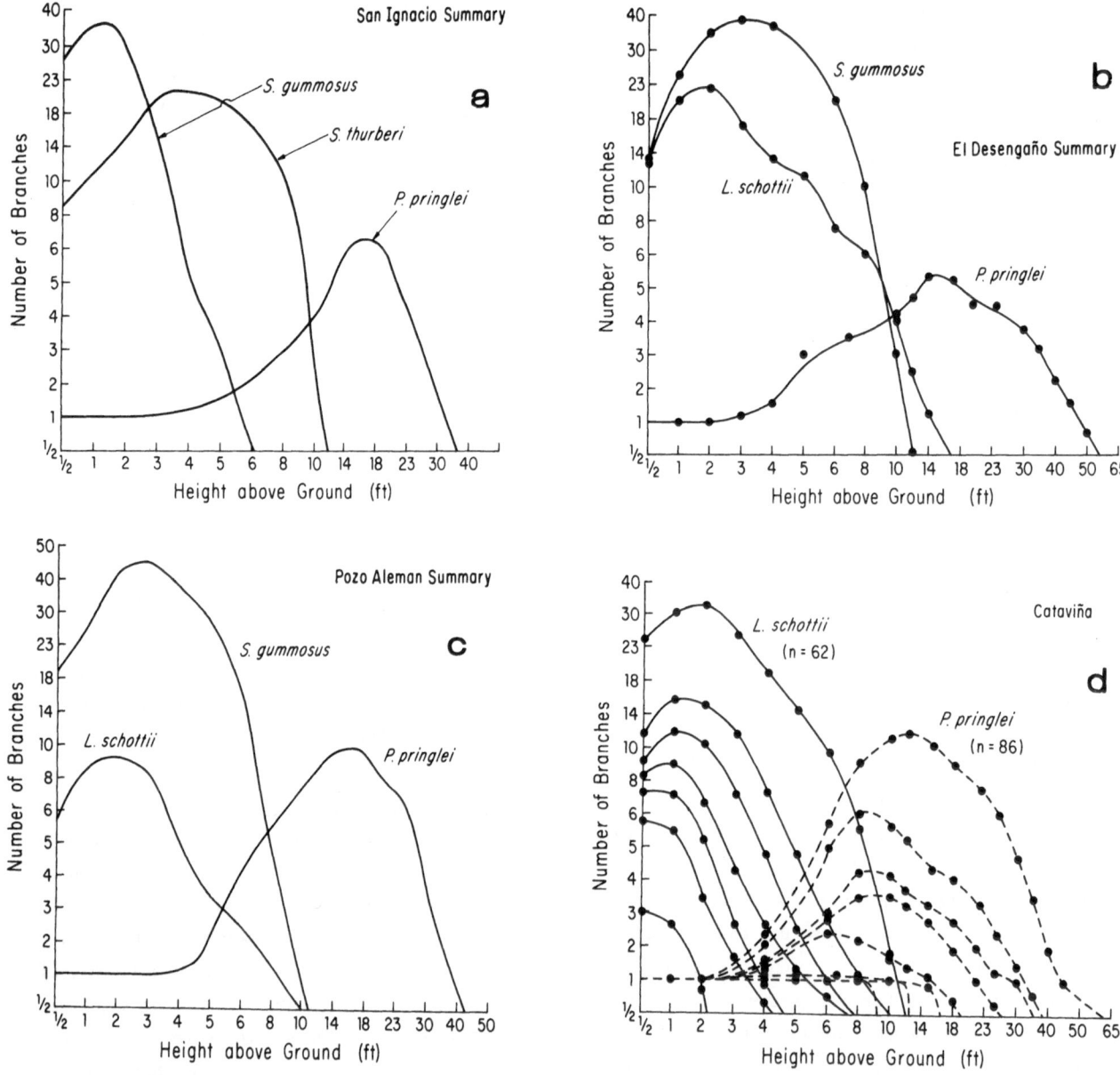

FIGURE 7. Summary curves for cactus branching patterns at four desert sites in the middle and northern parts of the Baja California peninsula. Three sites are three-species communities, but in only two of these are the three species the same; the fourth site is a two-species site, Catavina, for which size-specific curves are drawn as well as envelopes.

is north of the peninsular range of St, but Sg occurs outside the Catavina granite both to the south and to the north. The Catavina data are summarized in Fig. 7d.

The data summarized in Fig. 7 can be used to test the effects on branching patterns of the presence of different species combinations, although of course not independently of other variables that vary between sites, amongst which climatic changes with latitude are perhaps the more obvious. Inspection of Fig. 7 shows that branching patterns change within species amongst sites, and it seems possible that some of these changes might be attributable to the different species combinations at different sites. For example, Ppr reaches peak branching lower down where it coexists with a single additional species at Catavina, compared to the other three sites where it is found with two additional species. Peak branching height (H) at Catavina is 10.83 ±0.49 ft, significantly lower (by Student's t-test for the difference between the means, $p < .05$) than at Pozo Aleman (H = 15.5 ±1.24) or San Ignacio (H = 15.45 ±0.78), and lower but not significantly so than at El Desegnano (H = 13.14 ± 1.75). The two most similarly branched species at La Paz, St and Ls, do not occur together in the Fig. 7 sites; St occurs at the San Ignacio site, where Sg is low (<6ft) and branch number peaks at H = 1.33 ±0.16 off the ground. At Pozo Aleman and El Desengano where Sg coexists with Ls rather than St, its branching peaks at H = 3.04 ±0.26 and 3.23 ±0.47 ft respectively, and is a taller plant; all these shifts are statistically significant. For a summary of the means and their confidence limits of peak branch number (B), peak branch height (H), total plant stem length (L), which is proportionaly to biomass, stem radius (Rd) and rib number (Rb), for all cactus species at various sites, consult Appendix A.

We note finally the very different branching pattern in Ls amongst the northern three sites represented in Fig. 7. Plant size and branch number increases for Pozo Aleman to El Desengano,

and at Catavina where Ls coexists with only Ppr, it is taller and more branched higher off the ground. The three sites represent a gradient from St being very rare to absent entirely. Plant size L, for the largest 10 individuals in the population sample, increases from 40 to 114 to 145, and B increases from 9.2 to 22.0 to 32.4. However, peak branch height H is lowest at Catavina where Sg is absent (1.61 +0.26, vs. 2.73 +0.92 and 2.15 +0.25).

4.4. The northeastern desert sites

4.4.1. Addition of saguaro. To look further at the effects on branching patterns of changing species composition, I next consider the results from Sonoran Desert sites across the Gulf of California to the northeast. These sites support saguaro (Cg), which does not range into Baja California, and exclude Sg, with its very limited range east of the Gulf. Starting in the north and working south, in the Tucson Mountains only Cg is present; it is in fact the most northerly ranging columnar cactus. The branching of a large sample of individuals is summarized in Fig. 8a, from which it is seen that branching begins in individuals over 16ft in height and that branches are initiated at around 10ft off the ground. Branches are limited, on average, to two plus the main trunk, even in the largest class of individuals which reach 35ft (10m) in overall height.

Further south in Organ Pipe Cactus National Monument, the saguaros are taller and more branched. Here they are joined by St, which as before fills the eco-morphological niche of lower and prolifically branched plants that show little change in branch number over most of their height. This is shown in Fig. 8b, which may be compared to the two-species system at Catavina, Fig. 7d; despite the fact that both species are different between the two sites, there is an obvious similarity between sites in the way species segregate in branching within sites. Note also that St at Lukeville and Ls at Catavina are both more copiously branched close to the ground than is either of these two species at La Paz where they occur with Sg, which occupies the role of lowest, more spreading and most branched species at the southern end of the

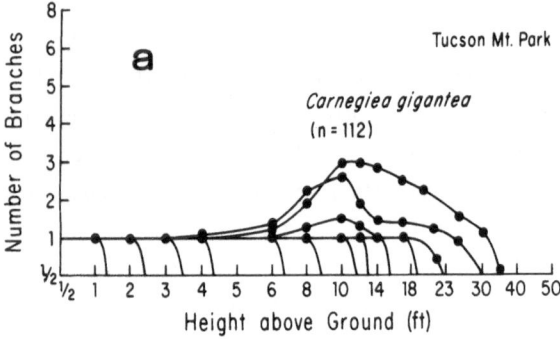

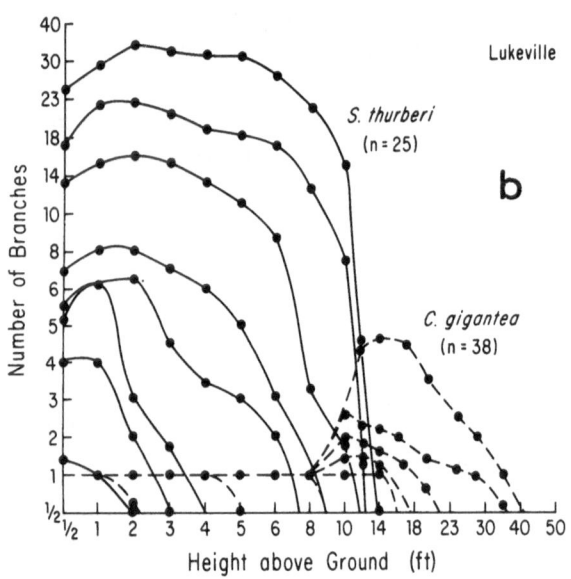

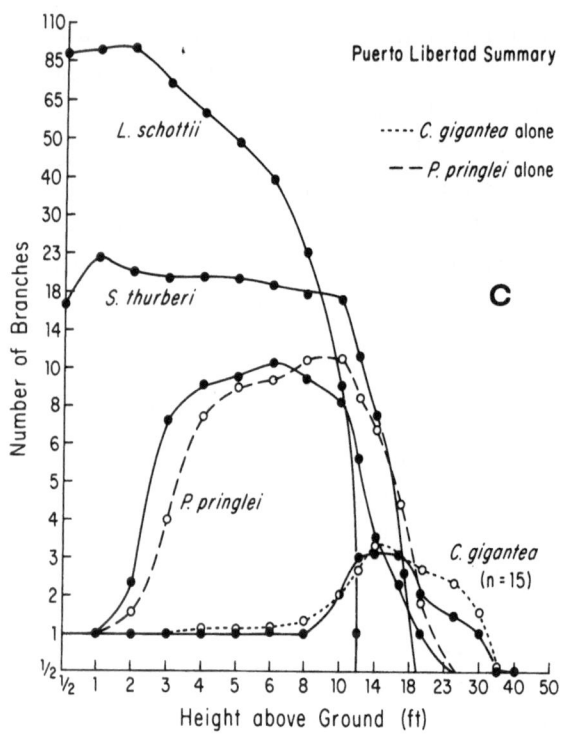

FIGURE 8. (a) Branching pattern of *Carnegiea gigantea* in Tucson Mt. Park. (b) In Lukeville St is added to Cg, to form a two species system, and at Puerto Libertad, (c), four species are present. (c) Includes two envelopes for Cg and Ppr; the (four) solid lines are those for species where they occur together, whereas the dashed line for Ppr and the dotted line for Cg are the branching patterns of the two species where they occur just outside the zone of coexistence.

peninsula. At height 1ft at Catavina the largest 10 Ls individuals average 30.1 +1.8 branches, and at 1ft at Lukeville St averages 29.5 +1.8 branches; these figures are not significantly different from each other, but both are different from the 20.3 and 16.7 that the two species average respectively at La Paz.

The third northeastern site is further south, in northern Sonora near Puerto Libertad. Here both Ppr and Ls are added to Cg and St, to make a four-species system like that at La Paz, with the substantial difference that Cg, not Ppr, occupies the role of the tallest and least branched species at Pto. Libertad and Sg, not Ls, the role of the

shortest and most-branched species at La Paz; the other three species of the four are in common between the two sites.

In the absense of Sg and the presence of Cg at Pto. Libertad, the branching patterns of the three species Ppr, St and Ls are substantially different from what they are at La Paz. The peak branching in each is shifted to the left, which is exactly what would be expected if branching patterns were influenced by the other members of the cactus community; the addition of Cg at the right-hand extreme and the removal of Sg from the left-hand extreme is correlated with a shift in Ppr, St and Ls away from the new Cg and towards the upper left of the graphs vacated by

Sg. For the three species in common between the northeast and southwest sites, Ppr, St and Ls, H = 10.67 +0.99, 1.9 +0.31 and 1.33 +0.28 respectively at Pto. Libertad, and 12.5 +1.57, 5.47 +0.45 and 3.71 +0.3 respectively at La Paz. These differences are statistically significant for St and Ls, but not for Ppr.

There is a further opportunity at Pto. Libertad to examine the potential effect of one cactus species on the branching of another. For the first 10 km inland from Pto. Libertad and the Gulf coast the vegetation is low and sparse, with only Ls common amongst the columnar cacti. As the road rises inland, Ppr and St are successively added to the increasingly richer-looking vegetation, presumably in response to increasing precipitation. By 23 km inland Cg joins the community, and at the census site just described the cactus relative abundances are St:0.65; Ppr:0.25; Ls: 0.07 and Cg:0.03. Further inland and higher in elevation, both St and Ls persist, but Ppr becomes scarse and then absent as Cg becomes common. 33 km inland the ratios of Cg:Ppr are 43:8, and by a point 37km from Pto. Libertad there were no more Ppr in a census, although scattered individuals may be found here and there to Caborca and some 60 km north of that town (Gibson, pers. comm.). I measured branching patterns of Ppr 14.5km inland from Pto. Libertad and of Cg 37km inland, both sites outside the zone where the two species co-occur commonly. The envelopes of these two sets of curves are included in Fig. 8c, as dashed lines. There is a weak tendency for Cg to be more branched both higher and lower off the ground where it occurs without Ppr, but these differences are not statistically significant. In Ppr, however, individuals outside the overlap zone with Cg are distinctly taller and more branched higher off the ground; that is, its branching patterns shifts towards that of the (absent) Cg. The maximum B is 10.9±0.37, at H=8.4±0.85ft off the ground outside the overlap zone with Cg, and inside this zone B=10.7±0.99 at H=5.8±0.83ft; the differences between the B values are not significant, but for

H they are (t = 2.17, d.f = 18; p<0.005). Ppr is significantly smaller in biomass within the overlap zones (L = 103 ±4.9 inside, L = 127 ±6.3 outside, t = 3.01, d.f. = 18, p<0.005), by virtue of the fact that branch numbers are significantly higher between 103 +4.9 inside, L = 127 +6.3 outside, t = 3.01, d.f. = 18, p<0.005), by virtue of the fact that branch numbers are significantly higher between 10-20ft, although significantly reduced at 3ft and lower but not significantly so at 3-6ft. Thus the branching patterns of Ppr shift towards Cg in its absence; note specifically that this shift in Ppr to larger stature and increased branching occurs on the low-productivity side of the overlap zone, the coastal side towards which overall plant stature, species diversity and plant density are much decreased. The implication that it is the absence of Cg that has mitigated these shifts in Ppr branching becomes therefore all the stronger.

4.5. Branching patterns on Gulf islands

The columnar cacti as a group are excellent island colonists (Cody, Moran and Thompson, 1983). Ls occurs on most and St and Sg on all major islands of the southern Gulf, and Ppr especially is adapt at maintaining populations on even the tiniest of the Gulf islands. With decreasing size and isolation, cactus species drop out in the order St, Ls, Sg and lastly Ppr. Yet some smaller islands with a full complement of peninsular species often show dramatic differences in cactus abundances. We begin here by discussing Monserrate, on which Ppr, St, Ls and Sg are all present, with relative abundances on the island of 0.23, 0.15, 0.13 and 0.49, the same rank order as at La Paz except for the inversion of Ppr and Sg. The branching envelopes of these four species on Monserrate are shown in Fig. 9a. In Ppr peak branching height is not significantly different from that at La Paz, but the plants are much smaller in biomass (L = 67 ±10.2 vs. 157 ±14.7) due to less copious branching (B = 4.6 ±0.72 vs. 12.5 ±1.57). The St branching pattern appears similar to that at La Paz, and in fact there are no significant differences in B, H or L between island and peninsula. It is in Ls

216

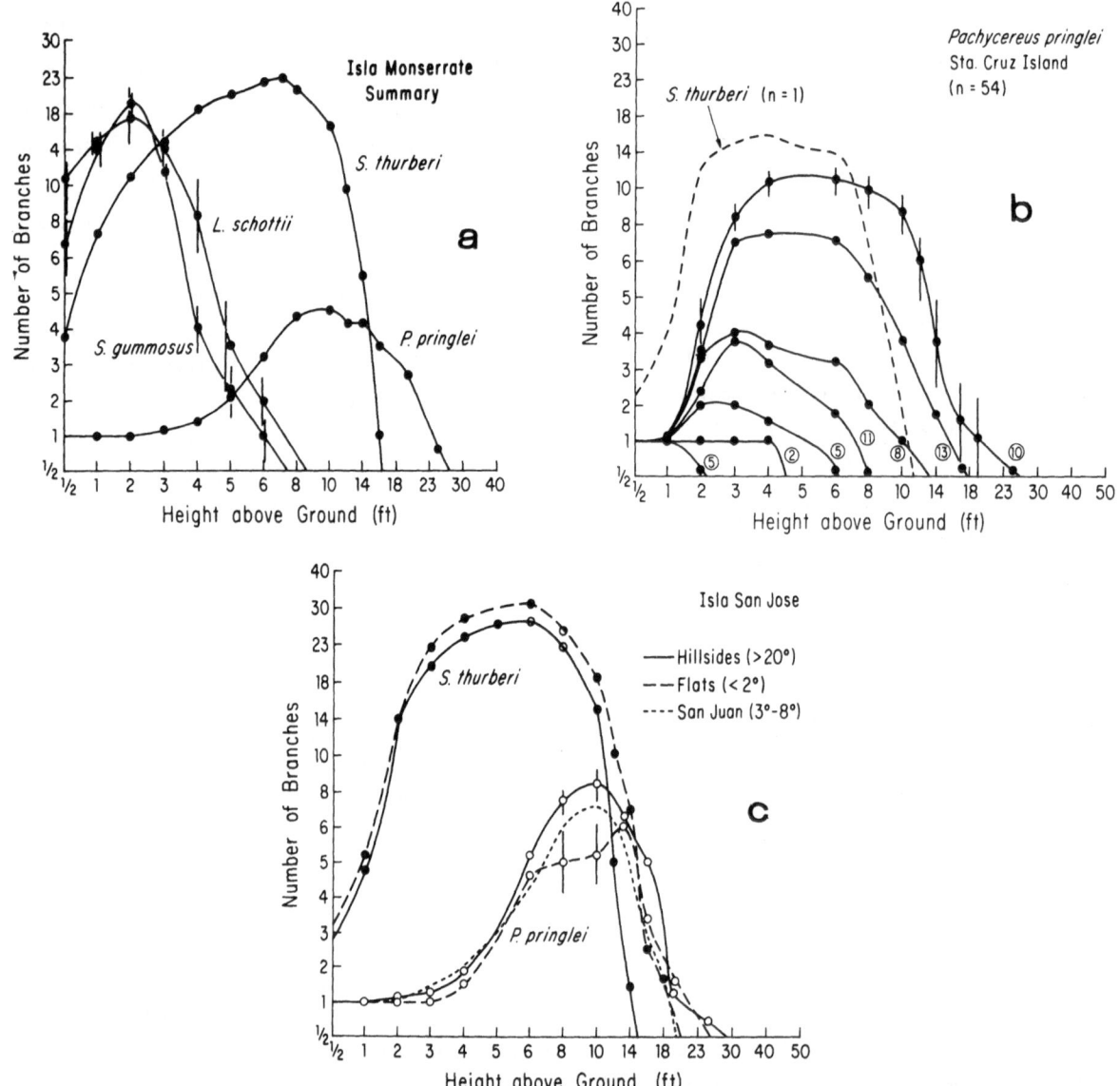

FIGURE 9. Branching patterns of cactus species on Gulf islands. (a) Monserrate is a smaller and isolated landbridge island, with four species of columnar cacti, two of which (Ls and Sg) are convergently similar in branching pattern. (b) Santa Cruz island is dominated by Ppr, which outnumbers St by around 30:1, and Ls and Sg are much rarer still. (c) San Jose is a large landbridge island close to the peninsula with a full complement of columnar cacti. Here the influences of slope on branching pattern were tested, with significant effects on Ppr but not St. San Juan Canyon provided additional data on Ppr, and is on the nearby mainland.

and Sg that branching differences are most apparent. Whereas these two species show clear differences at the peninsular site, on Monserrate the two appear to have converged in branching pattern. Both have peak branching B reduced significantly below peninsular figures, but on the island they do not differ significantly from each other (Ls:

B = 17.5 ±0.34: Sg: B = 19.5 ±1.15; t: 0.57, p≫0.05). Similarly, their peak branching heights are different at La Paz (H = 3.71 ±0.30ft and 2.88 ±0.42ft respectively, t = 1.61 and p = 0.07), whereas on Monserrate H = 1.78 and 1.80 respectively, t = 0.06, p≫0.05. Finally the two are of the same overall size on Monserrate (L = 58 ±9.0 and 49 ±1.3

respectively), compared to peninsular figures that
differ interspecifically and are both far greater
than on the island (see Appendix A for all such
figures). Thus on Monserrate, while there are four
taxonomic columnar cacti, morphologically in terms
of their branching patterns there are just three
distinct entities: Ppr, St and the convergently-
similar Ls + Sg. The island community appears to
show a reduction in species diversity that is
usual in island situations, but reduced "morphological" or "ecological" species diversity rather
than taxonomic diversity in the group.
To the south of Monserrate lies Santa Cruz, which
is smaller and generally steeper than Monserrate,
with a reduced flora (see Field Sites above). On
first approach the island appears to support only
Ppr, which is abundant everywhere, but on closer
inspection in fact St, Sg and Ls are present on
the island. Cencuses conducted on various parts
of the island produced 543 Ppr, 18 St, 1 Ls and
3 Sg. Thus for ecological purposes the Sta. Cruz
island community is a one-species system of
columnar cacti, with the other three species
occupying at best very minor roles. The branching
pattern of Ppr on Sta. Cruz is shown in Fig. 9b,
and the most cursory inspection reveals substantial
changes from the peninsular Ppr. Neither peak
branch number B nor biomass L differ significantly in the largest size classes between Sta. Cruz
and the peninsula, but in peak branching height
they are very different, with H = 12.3 +0.44 at
La Paz and H = 4.8 +0.50 on Sta. Cruz. Thus Ppr
has shifted left on the graph in a location where
it dominates the vegetation and the other columnar
cacti are very rare.
A sample of 10 individuals of St was measured on
Sta. Cruz, almost the entire island population I
encountered, and the largest individual is graphed
on Fig. 9b. Its branching curve appears like those
of St at La Paz, although smaller, and it serves
to illustrate how much Ppr has shifted relative
to its usual peninsular pattern. This shift in
Ppr, in circumstances of reduced ecological
species diversity, parallels the many known cases

of niche shifts and niche expansions in island
populations in general (see Carlquist, 1965 for
an excellent review of island plant populations
and their differences relative to mainland ancestors), and may constitute another example of such
competitive release.

4.6. Effects of topography on branching pattern
At all of the mainland and peninsula sites so far
discussed, data were collected at level sites with
a minimum of topographical relief. In this section
I give the results of measurements taken specifically to evaluate the effects of hillsides versus
flat bajadas on individual branching patterns.
These data were collected on San Jose island, a
large landbridge island in the southern Gulf and
one with a flora and vegetation very similar to that
of the adjacent peninsula. The island has the same
four species of columnar cacti as the La Paz site,
and from samples of 30 individuals of Ppr on flat
bajadas with slopes $< 2^{\circ}$ and of 52 individuals on
hillsides with slopes $> 20^{\circ}$, I represent the branching curves of the largest 15% in Fig. 9c. Although
neither biomass nor branch height H differ significantly between hillsides and flats, peak branch
number does at 8ft and at 10ft off the ground.
Comparable samples of St from flats and hillsides
fail to reveal any significant differences in
branching between these two extremes (see Fig. 9c).
To check on whether there is any island effect per
se on San Jose, and to further test the effects of
slopes on branching patterns, I collected an additional set of data on Ppr from San Juan Canyon on
the nearby peninsula, on slopes between 3° and 8°
in the bottom of the canyon. These are summarized
in Fig. 9c, and show a branching envelope intermediate between those of the flats and the hillsides on San Jose. Thus there appears to be an
unequivocal effect of slope on the branching
patterns of Ppr in this area, such that individuals are more branched on steeper slopes. Note
that this effect cannot explain the differences
in branching pattern of Ppr on Sta. Cruz island,
where H rather than B is the variable that is

different. Further, the magnitude of the difference
in branching pattern is slight between individuals
of different environmental extremes on San Jose, a
far less dramatic shift that observed on Sta. Cruz
where the cactus community composition is changed.
I also attempted to ascertain the effect of slope
on branching pattern in Cg in the Tucson Mts. Here
I divided the total sample of 112 individuals into
two sets, one growing on slopes of 30° and above
and the other on lower angle slopes. Here, however, there were no statistically significant
differences, and Fig. 8a gives the averages of
the combined data sets.

4.7. Branching patterns in thorn scrub

Next I investigated the effects on cactus branching
patterns of a substantial shift in vegetation,
from the Sonoran Deserts treated so far to thorn
scrub, at San Bartolo in the Cape region of Baja
California south of La Paz. Here both Ppa and St
are present, in about equal abundance, and Sg is
extremely rare. Fig. 10a gives the branching
pattern of Ppa at this site, and Fig. 10b that of
St. The summary data are presented in Fig. 10c,
from which it can be seen that the two species
are similar in overall height, are similarly unbranched low to the ground, and reach peak branching numbers at the same height above the ground,
but St is overall more branched than is Ppa. The
two species can be compared to Ppr and St at the
La Paz site. Compared to its congener to the north,
Ppa is shorter, significantly smaller in biomass
and branches significantly lower, but maximum
branch numbers do not differ significantly. St is
the same in overall size, peak branch number and
total biomass, but again branches at a significantly different height, higher at San Bartolo (H =
9.8 ±0.59ft) that at La Paz (H = 5.47 ±0.45). The
heights at peak branching do not differ between St
and Ppa at San Bartolo (t=1.13, d.f.=18, p>0.05).
Thus it appears that the two species converge in
peak branching height where they grow in the
closed-canopy thorn scrub, and differ in this
habitat only by the greater extent of branching

in St over Ppa.
To check on whether this result might be region-
specific rather that habitat-specific, I compare
the San Bartolo data with figures collected at a
second thorn scrub site, at Ures in central Sonora,
across the Gulf some 655km due north. In the Ures
thorn scrub again Ppa and St are the only columnar
cacti, and the branching patterns of the two are
given in Fig. 10d. Here peak branch number is not
significantly different between species, but peak
branch height is. Compared with the same two
species at San Bartolo, Ppa, is larger in biomass
at Ures and branches higher off the ground, but
branch number is not significantly different; St
is not significantly different in biomass, peak
branch number or peak branch height.
One difference between the thorn scrub at Ures
and that at San Bartolo is the taller vegetation
at the former site, around 1 1/2 m taller than at
San Bartolo where the canopy is some 3m high. This
difference might account for the higher peak
branching of Ppa at Ures. In all, it seems that
the branching patterns of the two species in thorn
scrub are affected by their growing in more continuous and thicker vegetation than conspecifics
and congeners in the open desert sites. Their
branching patterns are far more similar to each
other, and specifically branching height appears
to be affected by the higher vegetation cover. It
is of particular interest to note, in Fig. 10d,
how St initially grows upwards through the vegetation rather than branching extensively at lower
heights as it does in the desert; this would appear
to be obvious response of the species' growing in
a more shaded habitat.
On the vegetational gradient south from La Paz to
San Bartolo, Ppr is replaced by Ppa, Ls and Sg
drop out and St extends throughout but with a
changed branching pattern. I selected a site about
halfway along this gradient, at Rancho Los Divisaderos, to better calibrate the changes in branching of species restricted to either end of the
gradient and of species that extend across the
whole range of desert to thorn scrub. At this site

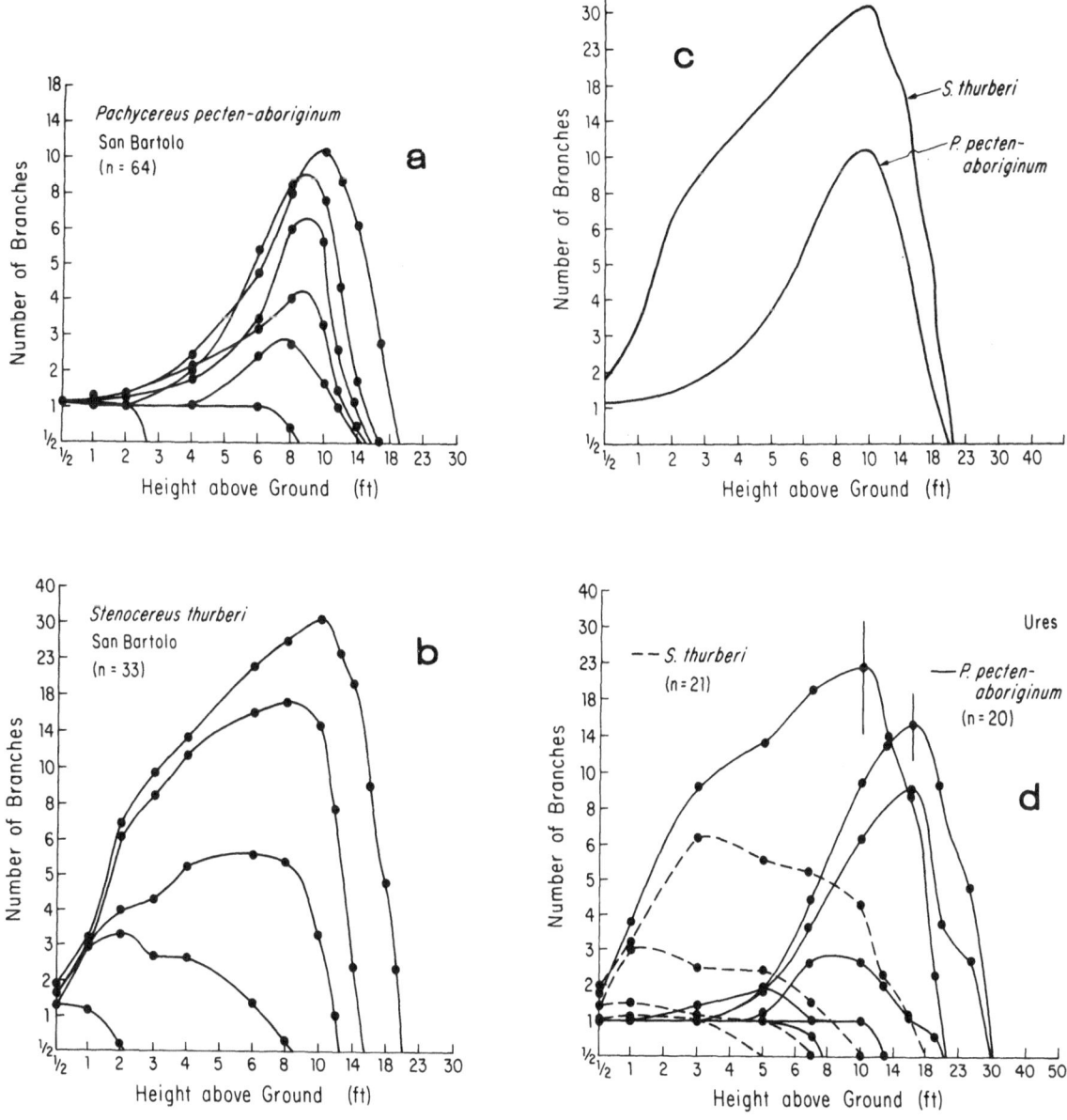

FIGURE 10. Cactus branching patterns in thorn scrub. (a) *Pachycereus pecten-aboriginum* at San Bartolo, in dense thorn scrub with a canopy around 3 1/2m, and (b) St at the same site. (c) Summary data for the San Bartolo two-species system, and (d) data from Ures in eastern Sonora, a similar thorn scrub habitat with the same two cactus species present, canopy about 4 1/2m.

the relative abundances of Ppr, Sg, St and Ppa are 0.39, 0.22, 0.22 and 0.17; it is flat, uniform vegetation with the cacti scattered throughout. Ls is much rarer than the other four species, about I individual per 240 of the first four combined.

Considering first the two *Pachycereus* species here where they coexist, they are clearly similarly branched. Peak branch numbers do not differ significantly between the two, but Ppr, is taller, with significantly greater biomass and is branched significantly higher off the ground. Ppa is significantly more branched between 4ft and 10ft off the ground.These differences appear in Fig. 11a and b. Relative to its branching at the La Paz site, Ppr is larger in biomass and reaches peak branching

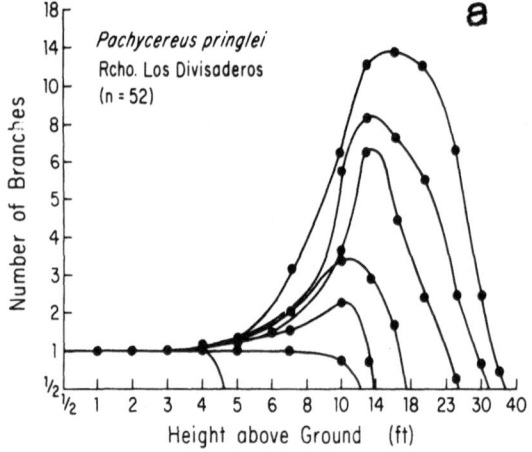

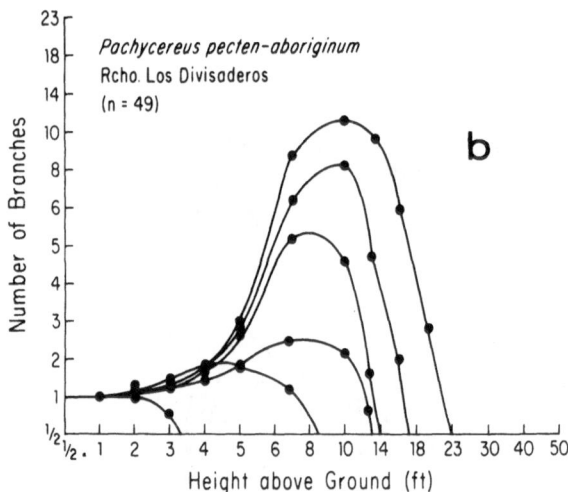

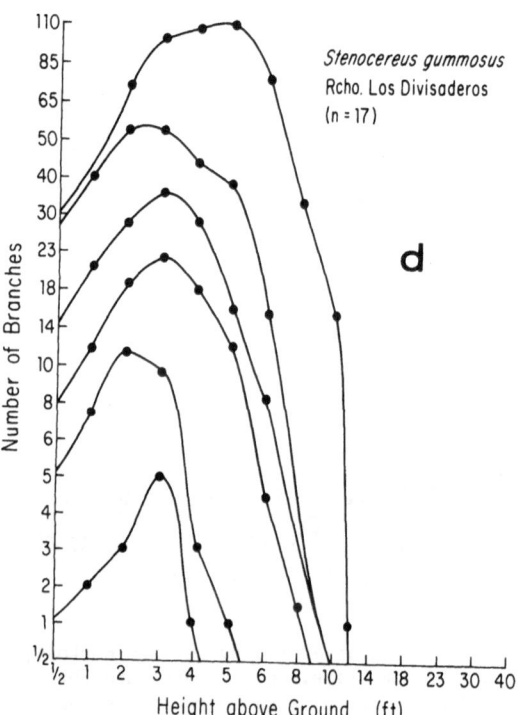

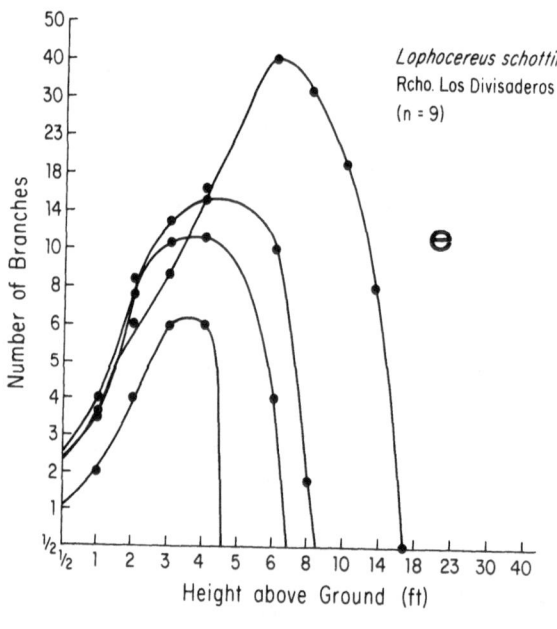

FIGURE 11. Rancho Los Divisaderos is a site intermediate between the desert at La Paz and the thorn scrub Ppr, Ppa, St are all quite common, but Ls is rarer and Sg rarer still. Note in the summary (f) that and Ppa are relatively similar also. To the south, one species of each pair will drop out, as will Sg too.

higher off the ground, at H = 16.3 ±0.64, vs. 12.3 ±0.44 at La Paz.

Branching in the two *Stenocereus* species is shown in Fig. 11c and d. Sg. is shorter, and more highly branched at lower heights than is St, differences of the same sort as observed at La Paz. The small sample size of Sg at Rcho. Los Divisaderos precludes statistical comparison, but the larger

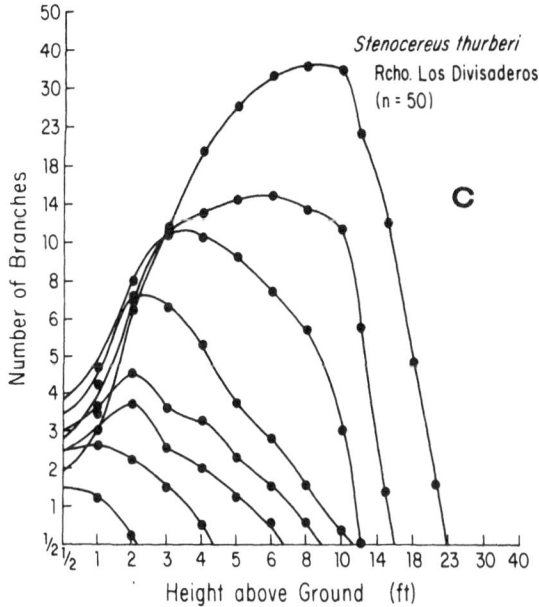

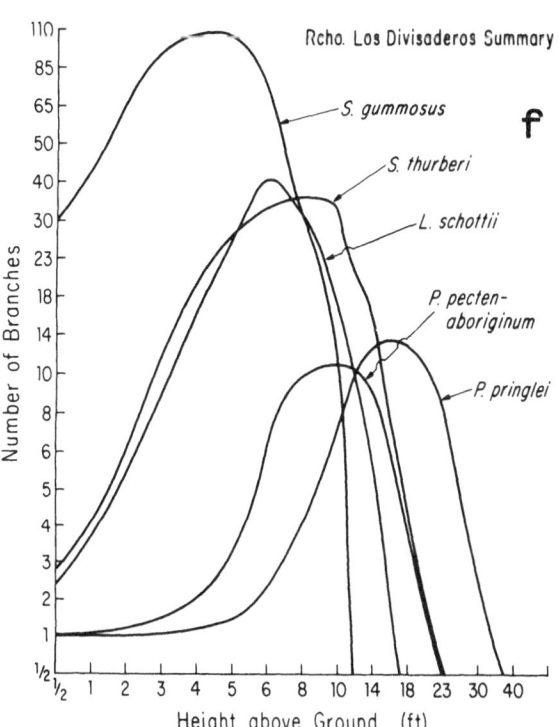

at San Bartolo. Five columnar cacti are present. St and Ls are similar in branching pattern ,

individuals there branch higher off the ground than at La Paz and appear greater in biomass. St is likewise larger in biomass at Rcho. Los Divisaderos, and it has larger branch numbers signifi-

cantly higher off the ground than at La Paz. The branching fingerprints of Sg are relatively similar between Rcho. Los Divisaderos and La Paz (cf. Figs. 5b and 11d), with the qualification that Sg is more truncate or arborescent and less branched close to the ground at the former site. The differences in St between the two sites are more pronounced, with initial growth adding more to the vertical and less to the horizontal spread of biomass in individuals at the edge of the thorn scrub.

Lastly, the small sample of Ls from Rcho. Los Divisaderos is shown in Fig. 11e. Like both St and Sg, it is more arborescent here (with a higher H and reduced branching below H), but its marginal status in this habitat might be reflected in the much smaller biomass of individuals at the site. Furthermore, Ls is particularly similar to St in branching at Rcho. Los Divisaderos, and can be seen for the site's summary graph in Fig. 11f. The two species do not differ significantly in either B or H, have very similar overall heights and are similar branched throughout their branching envelopes. It would thus appear that the two converge via the constraints of the denser and more closed vegetation to the same branching pattern, and that of the two just St will persist as the vegetation continues to increase in density and height to the south.

St can be compared in branching pattern among the three sites from La Paz south to San Bartolo. The species tends to have rather more branches and a somewhat larger biomass at the two southern sites, but the greatest difference amongst the three St populations is the height at which branch number peaks. This steadily increases from the desert into the thorn scrub, from 5.5 ±0.45ft to 8.4 ±0.53ft to 9.8 ±0.59ft, all shifts significantly different. On the other hand, Ppa shows no significant differences in B or H between San Bartolo and Rcho. Los Divisaderos, but does shift to smaller biomass at the southern site. If it can be considered a replacement for Ppr, as indeed it must be, the genus shows a trend to smaller biomass and decreased branching in the thorn scrub, and to a branching lower off the ground than in the open desert. The

222

thorn scrub habitat, with its dense vegetation and and closed canopy, apparently selects against branching low to the ground and for a uniform peak branching height that corresponds to level of the canopy in which the cacti grow, and selectively eliminates species such as Ls and Sg which typically branch and spread widely. It favors in addition Ppa over Ppr, perhaps because of the higher radius and reduced surface area/unit volume of the latter (see Discussion below).

4.8. Columnar cacti in the Mediterranean-climatic zone

In the same way that Sonoran Desert gives way to thorn scrub as summer precipitation increases to the east in Sonora and to the south in the Cape region of Baja California, the desert merges into chaparral and coastal sage scrub in the northwestern part of the Baja California peninsula where the Mediterranean-climate region of greater winter rainfall and increased coastal humidity takes over. A site representative of these changes is Colonet, some 140 km north of the area regarded as transitional between desert and coastal sage/chaparral, at El Rosario. Of the four columnar cactus species at La Paz, only Sg is present, but in addition two other species occur there. One of these is *Myrtillocactus cochal*, Mc, which is particularly abundant in this coastal region although it extends sporadically south through the peninsula. The third species at Colonet is *Bergerocactus emoryi*, a species that is restricted to the coastal areas of northwestern Baja California. The branching patterns of species at Colonet are shown in Fig. 12. Individuals of both Sg and Be grow largely though adding branches close to ground level; larger individuals are those that have spread to a greater extent horizontally. The growth form of Mc, however, is very different, since individuals of different size classes have close to the same number of branches at any given height above the ground, and grow by ramifying more, higher off the ground. The community pattern, Fig. 12d, shows some similarity to that of 3- or

4-species communities to the south, since there is the same sequence of shorter and more horizontally spreading species of lower radius, Be, through the intermediate Sg to the taller Mc branched higher off the ground with the largest stem radius. Selection may have segregated branching strategies in a similar fashion in the coastal scrub as in the desert, with at one extreme a species that is a superior space pre-emptor and at the other a species that reaches above the bulk of the vegetation to ramify higher off the ground.

By selecting a site to the south of Colonet and closer to the transition near El Rosario of desert to coastal scrub vegetation, a different combination of cactus species can be found. At Rcho. Arenoso the vegetation is more desertic, a mixture of predominantly Sonoran Desert species with a few of Mediterranean-climate coastal scrub, just the reverse of what it is at Colonet. At Rcho. Arenoso Ppr, Ls and Sg are present together with Mc, but Be does not extend this far inland. A comparison of their branching patterns with those at Colonet will show the effects, if any, of the drier and more seasonal climate at Rcho. Arenoso.

The terrain around Rcho. Arenoso is steep valleys and arroyos cut into an uplifted basaltic plain, with little in the way of sandy bajadas. Branching data summarized in Fig. 13a were collected from slopes facing $\pm5^o$ of due south, and of slope values $18^o \pm2^o$, with the exception of the Mc measurements, which were taken from individuals located on slopes of the same angle but facing $+5^o$ of north, west and east in addition to south.

Compared to their branching patterns at Colonet, Mc is significantly more branched and Sg significantly less branched at Rcho. Arenoso. Mc is not significantly different in biomass or peak branching height, but Sg branches significantly higher at Rcho. Arenoso. The shift in Sg, to more branching lower down at the more mesic site, is in accordance with stronger selection for space pre-emption in the thicker vegetation of the coastal scrub. At the desert Rcho. Arenoso site, Mc is no longer the tallest columnar cactus, for Ppr occupies this slot;

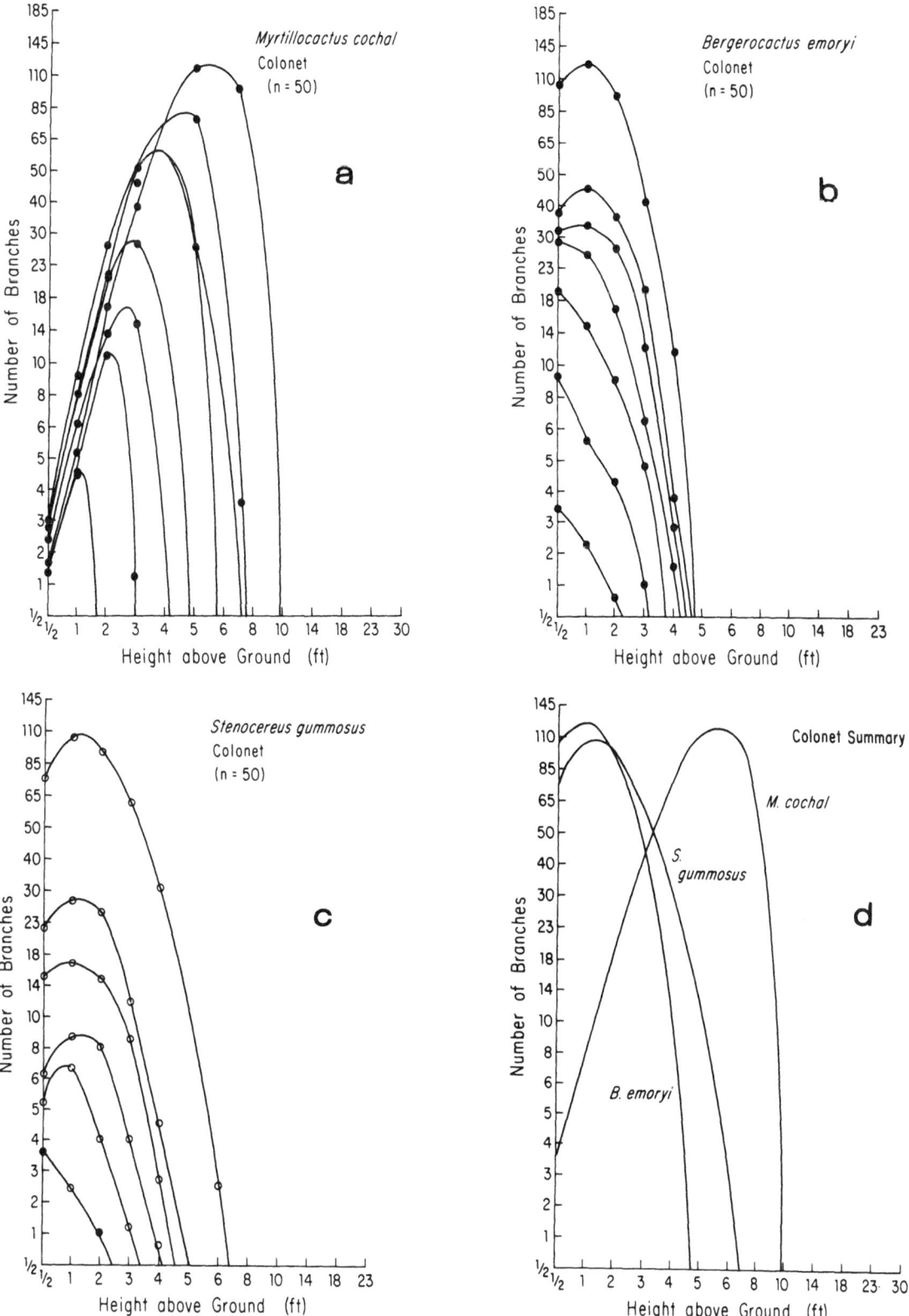

FIGURE 12. Colonet is a site within the Mediterranean-climate zone in the northwestern part of Baja California. Three cacti are present, all relatively thin-stemmed and well-branched. Note the emphasis in Sg and Be of growing horizontally rather than vertically, and in Mc of steadily increasing branch number with increasing height off the ground.

224

Mc appears to be "pushed" to branch somewhat lower, and rather more prolifically.

In terms of habitat similarity, the closest site to the Rcho. Arenoso hillsides studied to the south is San Jose hillsides. There Ppr branched lower and more prolifically on hillsides compared to the flats, and data from the southern site are not significantly different from those for Ppr at Rcho. Arenoso, in either of B, H and L (see Appendix A). I have no measurements of Ls specifically from hill-sides, but at Catavina, the next site to the south, Ls is no different in B, H or L from those at Rcho. Arenoso.

4.9. Effect of aspect on branching pattern

Besides the data from south-facing slopes summa-rized in Fig. 13a, I collected similar data on Ppr, Ls and Sg from similar slopes facing the other cardinal directions, within $\pm 5^{\circ}$ of north, east and west. On slopes with this particular range of angles ($18^{\circ} \pm 2^{\circ}$) my sample sizes are 51, 81 and 79 individuals for Ppr respectively, 13, 21 and 34 individuals respectively for Ls, and 3, 33 and 37 individuals respectively for Sg. Evaluating those statistics which are permitted by such sample sizes, I attempted to test the potential effect of aspect, the direction faced by a slope, on branching patterns. There were no discernible effects in three of the species, Mc (as stated above), Ls and Sg. In Ppr, however, there is a marked effect of aspect on branching pattern, in that individuals on south-facing slopes are sig-nificantly more branched over height and signifi-cantly taller than individuals on other slopes (see Fig. 13b). Further, Ppr on west-facing slopes are more branched than those on either north- or east-facing slopes, but there are no differences in branching between the Ppr on these last-men-tioned aspects. Included in Fig. 13b are standard error bars for the south and east slope data. The heights of peak branching are also higher on south and west slopes, and individuals on south-facing slopes are significantly larger than those grow-ing on slopes of other aspects. The slopes on

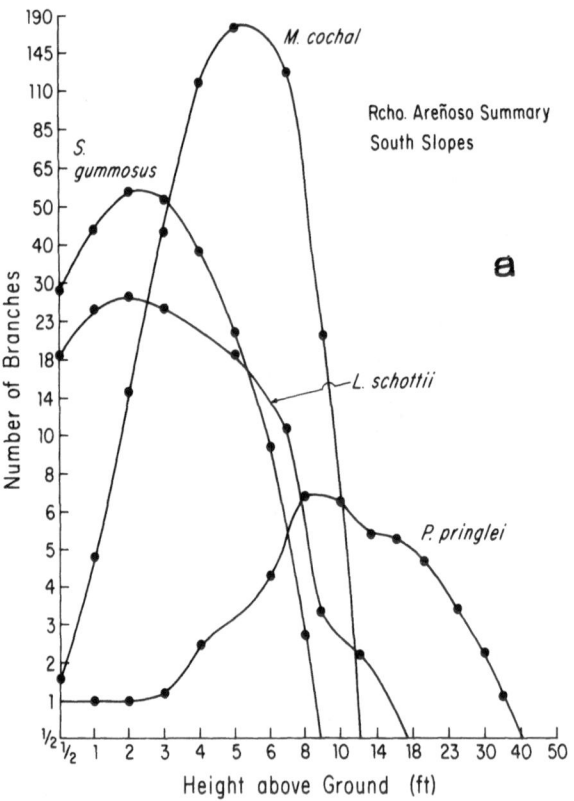

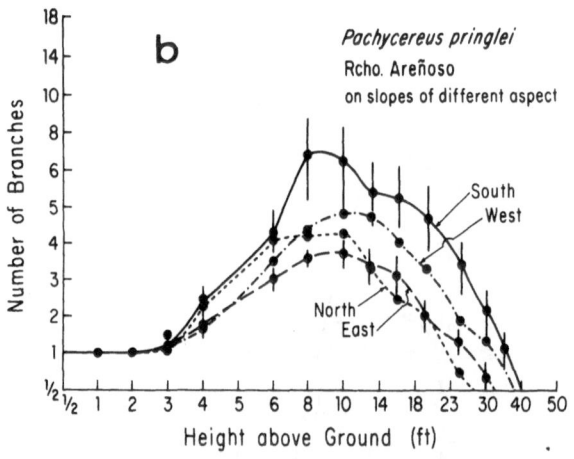

FIGURE 13. (a) Four species occur at Rancho Arenoso, with branching patterns summarized in this figure. (b) The effects of aspect on branching pattern of Ppr is shown here, on slopes of $18^{\circ} \pm 2^{\circ}$. Ppr is significantly more branched on south-facing slopes, less branched on west-facing slopes, and signifi-cantly less branched on north- and east-facing slopes.

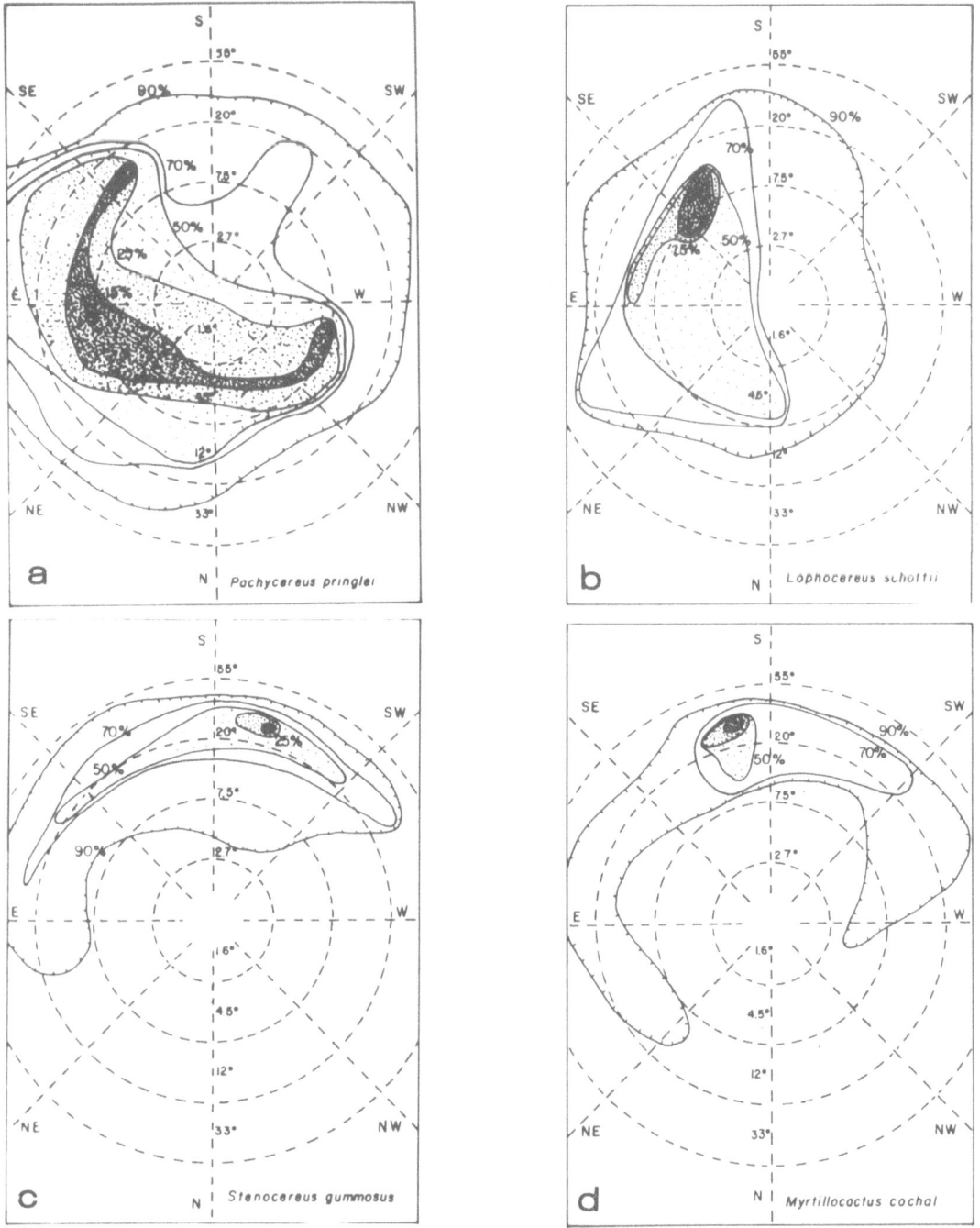

FIGURE 14. The distributions of the four columnar cactus species at Rancho Arenoso over slope and aspect. The graphs are viewed as if the observer were looking down into a bowl-shaped depression, with south-facing slopes to the top, north-facing slopes to the bottom, east-facing slopes to the left and west-facing slopes to the right. The concentric circles are of increasing slope angle from the center, with a log scale as indicated. Densely stippled areas are quadrats where a cactus occurs at densities 85% or above of the highest count recorded for the species (top 15%) in any quadrat. Successively lighter stippling shows 75% and 50% (or the top 25% and 50%) of maximum density, and further contours give the limits of quadrats with 30% and 10% of maximal density. Note how Mc and Sg are restricted to 20° south-facing slopes, which are those with maximal solar radiation, whereas Ppr is very widely distributed except on the south- and southwest-facing slopes, and Ls is commoner on lower angled slopes.

which Ppr are larger, taller and more branched are those which receive higher incident solar radiation and are warmer: south, then west, then north and east in descending order.

4.10. Cactus densities in relation to slope and aspect

A further data set that is relevant to the interpretation of branching pattern differences between various slopes and aspects is how cactus density itself changes with slope and aspect. With this information, it is possible to label some microsites (slopes, aspects) as preferred by a species and others as less preferred or avoided. At Rcho. Arenoso data were collected on the density of the four columnar cactus species over all available slopes and aspects within a square 1 1/2 km on a side. Within this area, quadrats of uniform substrate, slope and aspect were selected, with circular form of radius 12.6m and area 500m^2. In each quadrat, the number of individuals of Ppr, Mc, Ls and Sg were counted (without regard to size), and slope and aspect measurements at the quadrat recorded also. I collected these data for 216 quadrats that spanned the range of habitats available in the region. Species totals in the census are Ppr:595, Mc:142, Ls:267 and Sg:681; the overall total is 1685 individuals. Mean densities are Ppr:2.76 indiv/500m^2, Mc:0.66, Ls:1.24 and Sg:3.15 indiv/500m^2. More importantly the distributions of the four species are shown in Fig. 14a-d. These figures are viewed as if looking down into a bowl-shaped depression, with the south-facing slopes at the top of the figures, north-facing slopes at the bottom, and west- and east-facing slopes at the right and left respectively. The figures are contoured for relative abundances, that is cactus density relative to the highest density counts recorded. The highest counts per quadrat are Ppr:11, Mc:16, Ls:13 and Sg:38. Density contours enclose sites with 85%, 75%, 50% 30% and 10% the density of these maximum counts.

It is clear from Fig. 14 that the different cactus species have difference preferences for slope and aspect. Ppr reaches maximum densities on slopes from the south-east to northwest between 3° and 15°, but is broadly distributed over low and high angle slopes centered in the north-east facing part of the bowl. Moving sequentially around clockwise, one encounters the centers of distribution of Ls, Mc and Sg successively, with the former of comparably wide distribution to Ppr and the latter two of more restricted distributions. In particular, Ls does well in the flatter areas, whereas Mc and Sg are both restricted to slopes of higher angle; Mc does best on slopes between 20° and 35° and Sg on slopes of about the same magnitude. Clearly Mc and especially Sg are positioned in the bowl to receive maximal incident solar radiation. At this latitude, the slopes which receive maximum solar radiation are south-facing with slope angle 21°, with a yearly total of just under 160 kilolangleys (Holland and Steyn, 1975). The centers of distribution of Mc and especially Sg receive very little less than this maximum. South-facing slopes of 0-45° all receive 145 kilolangleys and above, while north-facing slopes where Ppr is common, on 5-10° slopes, the solar radiation totals 130-140 kilolangleys yearly, falling to values as low as 65 kilolangleys on steeper north-facing slopes still occupied by this species.

Referring back to the branching patterns of Ppr on different slopes at Rcho. Arenoso, an anomaly is immediately apparent. That is, Ppr individuals are largest and more branched, and by all reasonable estimates appear more vigorous, on south-facing slopes, are not quite so large and branched on west facing slopes, and are smallest and least branched on east- and north-facing slopes. Yet, as Fig. 14a shows, Ppr are least common on south facing slopes, somewhat commoner on west-facing slopes, but most common on north- and east-facing slopes. My inference is that Ppr are excluded from the south- and south-west facing slopes by other cacti, especially by Sg. These slopes are prefered by Sg, perhaps also by Ls and certainly by Mc, since they are warmer, more light-saturated and are likely to permit faster growth rates at this high latitude (see Nobel, 1980a). On the other hand, cacti with smaller radii are less apt to withstand the cooler temperatures of the north- and north-east facing slopes, and it seems probable that the smaller radii of Sg especially and Mc to a lesser extent would preclude their occupation of these cooler slopes. Ppr, with the largest radius of the four, does well on the cooler slopes, in terms of the densities it reaches there, but in terms of its morphology, it too would prefer to live on south-facing slopes. These points are further developed below.

5. DISCUSSION

5.1. The variables

In this final section I attempt to bring together certain diverse lines of argument, which all seem to be necessary components of an explanation for the various cactus branching patterns and their changes over habitat and amongst sites. These arguments help to illuminate different aspects of the selective basis of the branching patterns are with many general morphological characteristics, such as body size or proportions, branching patterns are likely to be under selection for different and perhaps conflicting ends; thus a general explanation is expected to be nonlinear with several facets, and is still beyond our under-standing.

In the plant itself, branching patterns are de-scribed by families of curves plotting branch number against height above the ground. Subsuming ontogenetic effects, branching pattern $B(h)$, a function of height h, appears to be no more than a two-parameter variable; species (size classes, populations) differ in an index of the position of peak branching, and in an index of how much they are branched. Thus branching pattern can be written $B(h) = {}^h e/h!$; this simple model, with the appropriate changes in the parameters and, fits most of the patterns described in this paper. An additional plant variable of considerable impor-

tance is stem radius Rd; Rd plus $B(h)$ comprise the most important variables in cactus morphology. If branching pattern B is written as some function $f(h)$, the area under this curve, L, is fdh, biomass is $(Rd)^2 fdh$, and surface area is $2(Rd)^2 fdh$. To a lesser but largely undetermined extent, rib number also may prove to be an interesting morphological variable.

The variables of the physical environment begin, surely, with moisture availability; a priori, it is difficult to see how one might interpret eco-morpho-logical aspects of cacti, with their most extreme forms amongst plants adapted to arid environments, without reference to water. The major variables of the physical environment of cacti appear to be precipitation and substrate; together they deter-mine the availability of moisture to the plants. Precipitation throughout the region under consider-ation varies from 100mm to around 300mm, with high-er rainfall producing the denser and shrubbier thorn scrub vegetation. The usefulness of this precipitation varies with the substrate; steep rocky hillsides have rapid runoff, and without fine sediments to soak up and retain water, little of the rainfall is usable. On the other hand, low-angle sandy flats have reduced surface runoff, and moisture percolates to depths where it is protected from evaporation and where some of it can be tapped by the deeper-rooted plants. Besides particle size and slope angle of the substrate, its composition in terms of propensity to repel water and promote sheet flow varies. In this respect, soils of de-composed granite seem to retain more and shed less water than soils of decomposed basaltic rock.

An additional physical variable of some importance is the temperature regime. Frosts of duration >24h have been shown to be important in determining the northern limits of Cg and possibly the upper eleva-tional limits of the columnar *Trichocereus* in the Chilean Andes (Nobel, 1980b,c). It is possible also that high summer temperatures adversely affect the heat loading of cacti, and increase the moisture lost during nocturnal gas exchange and thereby slow growth rates. A branching morphology with large sur-

face area and small stem radius would better dissi-
pate heat and reduce the detrimental effects of
high temperature.

There are several factors in the biotic environment
that may constitute selective effects on branching
patterns. Firstly water uptake through root systems may be reduced if the root systems of neighboring individuals overlap, either horizontally
or vertically, and there is competition for water.
Such competition seems likely throughout the precipitation range of the Sonoran Desert and thorn
scrub, and is presumably the major mechanism
regulating plant density, both intra- and interspecifically. Secondly, while light per se is not
usually regarded as limiting in desert habitats
(low latitude, low cloud cover, low vegetation
density), in thicker vegetation such as thorn
scrub shading may be significant and select for
changes in branching pattern over those of more
open habitats. Even in more open habitats, there
may be self-shading in well-branched cacti and an
orientation of branching to minimize this; and on
north-facing slopes of 40-50°, incident radiation
is 20%-50% lower than on 20° south-facing slopes,
and this difference might well be critical for
some species.

5.2. The basic adaptive syndrome

Different species of columnar cacti differ in
obvious ways in the basic morphological variables
just listed, and these morphological differences
presumably enable different species to live in
somewhat different habitats, and to use the resources of a given habitat in somewhat different
ways such that several can coexist in one place.
At one extreme is the morphotype of large stem
radius, large rib number, minimal branching, and
with strictly a surface root system. The best
paradigm for this extreme might be barrel cacti
(*Ferocactus* spp.), which are short (mostly around
1m), unbranched, with ca. 20cm radii and >30 ribs.
Amongst the columnar cacti, however, Cg and Ppr
are good examples. With their broad but shallow
root systems (see Yeaton et al., 1977) they can

make good use of water where runoff is rapid, and
with their large radii and rib numbers this water
can be stored for future use such that brief but
violent storms provide water for many weeks or
months. They are able to live on rocky hillsides,
and can grow tall and branch high off the ground
with the structural support of the large radius
stem or trunk. But with large radius, photosynthetic area (PSA) per unit volume (PSA/V) is
reduced.

At the other extreme are deep-rooted species which
tap a more constant water supply in alluvial fans,
deep sandy bajadas and flats. In this morphotype
large radius and rib number are not required for
water storage to the same extent, but these plants
(Ls and Sg are examples) may branch extensively to
multiply surface area, and with small radius PSA/V
will be higher and growth faster. With small radius,
they cannot structurally support such tall plants,
but individuals can spread horizontally and with
rapid growth pre-empt space and its inclusive water
and light catchment rights.

I wish to amplify this view of the adaptive nature
of branching patterns, summarize the evidence for
it, as well as speculate about its various aspects,
by centering the remaining discussion around
certain "postulates". These are statements which
seem reasonable given our existing knowledge, and
may serve to focus current arguments and specula-
tion, as well as perhaps future research efforts.
A summary figure, Fig. 15, is designed as a form of
algorithm for cactus branching patterns, and is
intended to accompany this final discussion section.

5.2.1. Postulate 1: Ecological segregation occurs via different branching and rooting strategies amongst columnar cacti, but there is ongoing competition for space (water, light).

Four species of cacti, from the large-radius Ppr
to the small-radius Sg, coexist in uniform sandy
flats at La Paz, and at most locations in the Sonoran Desert 3-4 species communities of columnar
cacti are the rule. Thus empirically coexistence
is possible within homogeneous habitat, evidently
by dint of different strategies of water and light
capture. Each species has access to resources not

229

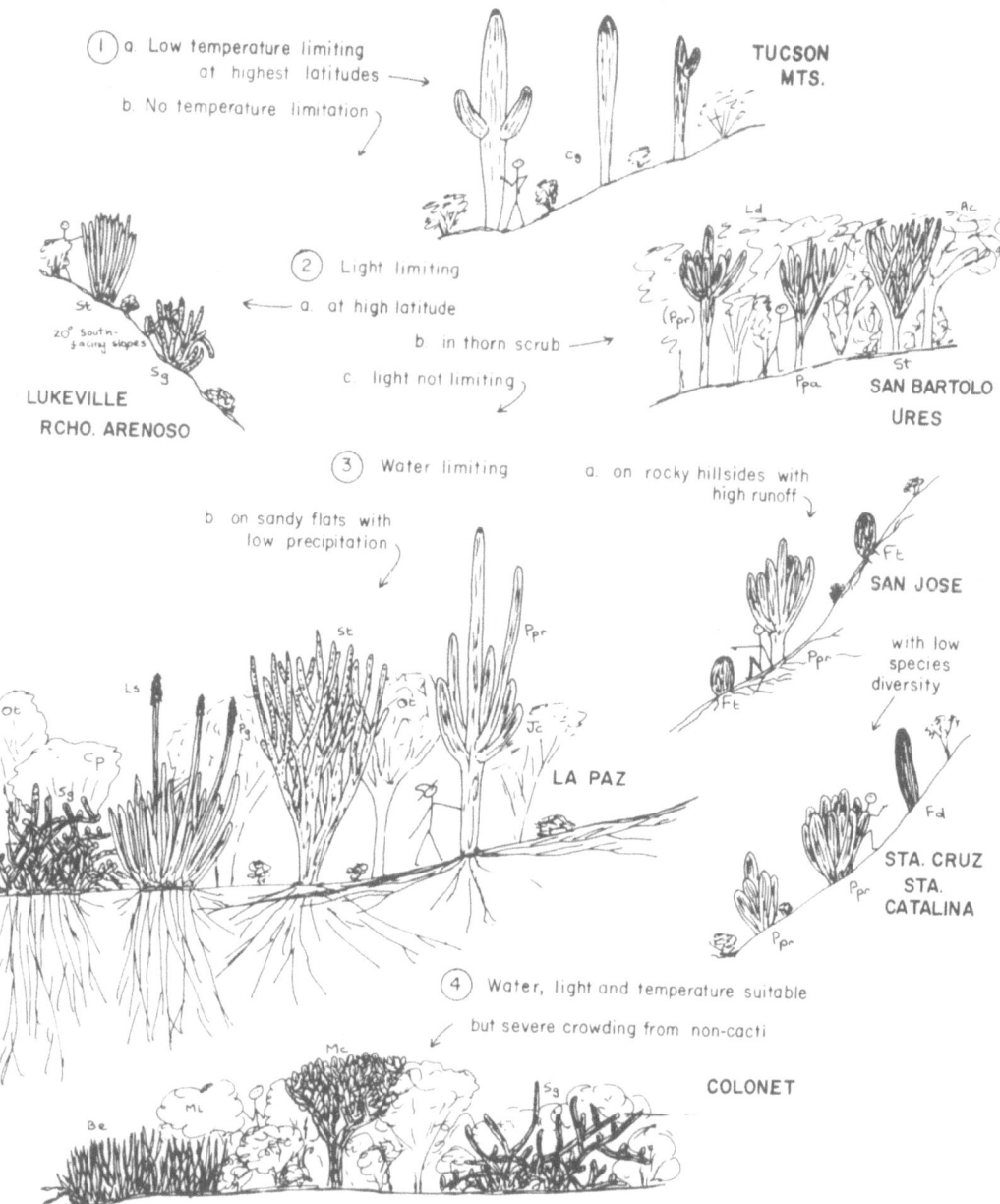

FIGURE 15. This figure illustrates how branching patterns change between species and within species between habitats, and suggests the selective factors involved in these changes. The figure may be read as an algorithm or a "recipe" for determining what branching patterns should be observed where. See text for further discussion. Species abbreviations are: Cg:*Carnegiea gigantea*; Ppr:*Pachycereus pringlei*; Ppa:*Pachycereus pecten-aboriginum*; St:*Stenocereus thurberi*; Sg: *Stenocereus gummosus*; Ld:*Lysiloma divaricata*; Ac:*Acacia cymbispina*; Ft:*Ferocactus townsendianus*; Fd:*Ferocactus diquetii*; Jc:*Jatropha cinerea*; Ot:*Olneya tesota*; Cp:*Cercidium peninsulare*; Pg:*Prosopis glandulosa*; Ml:*Malosma laurina*; Be:*Bergerocactus emoryi*; Mc:*Myrtillocactus cochal*; Ef:*Eriogonum fasciculatum*.

available to the others. At the one extreme, Ppr is presumed shallow-rooted, by overall comparison to Cg which is known to be and because of its relatively strong interaction with Cg at Pto. Libertad, and by its success on hillsides which necessitate shallow roots. Ppr could take advantage of light precipitation which would mostly evaporate and not percolate far into the soil, and of pre-cipitation which would otherwise be lost to runoff. It can efficiently store water because of large radius, which in turn enables it to support a tall structure, which might take advantage of more favorable light regimes above most of the surround-ing vegetation, over a longer growing season. In contrast and at the other extreme, Sg and Ls are likely to have deeper roots; they are the only two species that persist on deepsandy flats where pre-cipitation is lowest and best obtained seasonally from deeper in the ground. With reduced radius, these plants have higher PSA/V ratios, which would give them an edge over species with lower PSA/V if all other things were equal. But reduced radius means reduced stature, and an emphasis on spread-ing horizontally rather than growing tall. This might reduce growing season since there might be shading effects when the sum is low in the sky, although the fact that the taller non-cacti in the habitat, such as *Bursera* and *Jatropha* spp., and some leguminous trees, are opportunistically deciduous reduces this disadvantage, especially since the lower cacti can store some water and thus photosynthesize after the non-cacti have lost their leaves in drier seasons. Figures on PSA are available only from Lukeville (Yeaton and Cody, 1979), from which a PSA/V ratio in St is calculated at 8x that in Cg.

It is probable that the root systems of the cacti overlap both vertically and horizontally, such that there is direct competition for moisture. Species with wide-spreading and shallow root sys-tems, such as Cg and Ppr, should maintain lower densities than do other deeper-rooted species, and this is usually the case; intraspecific nearest neighbor distances should increase from Sg to Ls,

St and Ppr, but there are insufficient data on this. However, on some small Gulf islands such as San Pedro Martir and Partida Norte, their relatively flat tops are crowded with Ppr which, in the absense of other species, maintain near-neighbor distances no more than 1/3 those observed in diverse mainland communities. The evidence for direct competitive interactions amongst species in-cludes shifts in branching patterns as described above, changes in relative abundance with species changes, evident exclusion of Ppr from preferred south-facing slopes at Rcho. Arenoso and to some extent of Cg by St from south-facing slopes at Lukeville (Yeaton and Cody, 1979), and changes in cactus biomass. To illustrate the last point, con-sider that both Ppr and Cg maintain larger individ-ual biomass immediately outside their overlap zone at Pto. Libertad (261 and 203 respectively - see Appendix A) than either does inside the overlap zone (244 and 144 respectively).

5.2.2. Postulate 2: Larger radius is advantageous on hillsides.

At Rcho. Arenoso Ppr has a far larger distribution over slope and aspect than the other three columnar cacti present (Fig. 14), and likewise Cg has a wider distribution than St at Lukeville (Yeaton and Cody, 1979). Similarly, in the Mojave Desert large-radius *Ferocactus* ranges onto steeper slopes than *Echinocereus* and *Opuntia* spp., which have stem radii of decreasing size. In fact, of the two larger cylindropuntias of the Mojave Desert, *O. acanthocarpa* and *O. ramosissima*, the latter has on average a 60% larger PSA, smaller stem radius (0.6 cm vs. 1.5cm), is distributed on slopes of lower angle (Yeaton and Cody, 1979), and has a deeper and less spreading root system (Cody, 1983). Although comparable data have not yet been collected for all of the columnar cacti, the basic adaptive syndrome is just that hypothesized above.

All four of the cacti in the La Paz flats are present on Isla San Jose in the flats, but only two, Ppr and St, are common enough on the hillsides for a flats-hillsides comparison in branching pattern. There the Ppr are larger, more branched, but shorter, and on the rocky Isla Sta. Cruz Ppr are of comparable biomass but even more branched and shorter still. Steep hillsides are generally much more open in vegetational cover, and thus tall stature to avoid shading is less advantageous. Un Sta. Cruz Ppr is in fact the only common columnar cactus, where it practically monopolizes the habitat, but on Monserrate it coexists with 3 other species, and is less branched and much smaller in stature. Yeaton et al. (1980) found that where Cg occurred at the limit of its distribution on the steepest hillsides, it was unbranched with smaller biomass compared to larger and branched individuals on the flatter slopes. There are other plants capable of water storage with large stem radius and a preference for living on hillsides over flats. Examples are *Pachycormus discolor* (Anacardiaceae) and *Fouquieria columnaris* (Fouquieriaceae), both common at Catavina and both reaching maximum density on the rocky granitic hillsides. A further example is provided by *Bursera microphylla* and *B. hindsiana*(Burseraceae), which can be found together over much of the Baja California peninsula. The former has the most tapered growth form and softer tissues indicative of water storage ability, smaller leaves and reduced PSA/V, and while *B. microphylla* is usually found on hillsides, *B. hindsiana* is usually found on the lower bajadas.

5.2.3. Postulate 3: Temperature may be limiting for columnar cacti, but only at higher latitudes.

Besides being advantageous for rapid water uptake and storage, large radius may confer an ability to withstand low temperatures. Thus Cg, with the largest radius in columnar cacti of the southwestern U.S. and northwestern Mexico, reaches the most northerly latitudes, to where it is apparently frost-limited. But Ppr does not reach as far north as St or Ls in Sonora and Arizona, and both Sg and Ls extend further north than the larger-radius St in Baja California. Thus other considerations besides low temperatures restrict northern distributional ranges. Both Cg and Ppr are tall, and both have an apical fuzz which apparently serves to protect the meristem frol low temperature. But it seems that Cg, rather than physical factors, limits the northern range of Ppr (see above). The role of high temperatures is reducing stem radius and thus heat load has beed inferred for Ls (Felger and Lowe, 1967) but has not been thoroughly investigated. My data on St show a correlation (0.410, n = 9, p = 0.xx) between stem radius and latitude, but a better correlation between radius and precipitation (-0.44, n = 9, p = 0.xx); this indicates that water storage ability might take precedence over thermal regulation in stem radius.

5.2.4. Postulate 4: Light may be limiting in columnar cacti, but only at higher latitudes or in dense vegetation.

At the northern edge of its range at Lukeville, St shows a marked preference for 16°-32° south facing slopes, and similarly at Rcho. Arenoso both Mc and Sg are centered on 20° south-facing slopes with little tolerance of other aspects or lower slopes. These preferences coincide with those sites which receive the highest incident solar radiation. These are species that are much more restricted in habitat utilization at these sites than Cg or Ppr respectively; they are species that, because of their lower radii, should be less well equipped to deal with either low moisture availability or low temperatures. They appear to survive by selecting precisely those sites that give them the highest available solar radiation, and apparently only there can they withstand any competition from the species with larger radius.

The photosynthetically-active part of solar radiation appears to be limiting for the cacti in thorn scrub habitats, at Ures, San Bartolo and Rcho. Los Divisaderos. The first-mentioned two represent dense thorn scrub, and the effect is to limit the cacti to two species, Ppa and St, both with very similar stem radii and similar peak branching heights. At Ures the biomass of the two species is similar, but at San Bartolo St is significantly the larger. Where moisture availability is less limiting, the competitive edge should go to the species with the larger PSA/V, or smaller radius: St. Note that the biomass of Ppr, Ppa and St increases from La Paz to Rcho. Los Divisaderos to San Bertolo, with increasing precipitation, but species fare best with higher PSA/V, in inverse proportion to their stem radius. Ppr drops out first, and St appears to do better than Ppa in the dense vegetation where competition from *Acacia* and *Lysiloma* species must be intense.

5.2.5. Postulate 5: When water, light and temperature are undemanding, cacti must compete with non stem-succulents for space.

The diversity of columar cacti, and of cacti and stem succulents in general, falls rapidly on the edges of the deserts where the physical conditions are less harsh and especially water is more readily available. This happens at the southern end of the Baja California peninsula, where the desert merges into thorn scrub, and at the northwestern end where the Mediterranean-climate chaparral and coastal sage scrub dominate. The cacti that persist in the thorn scrub, as just described, are those with stem radii large enough to support plants of comparable stature to the *Acacia-Lysiloma* scrub, but small enough radii to give a high PSA/V ratio. The species that persist furthest in the Mediterranean-climate region are Sg, Be and Mc. The first two have the lowest stem radii of the columnar cacti considered, and are low and sprawling in growth form. The grow horizontally rather than vertically, and seem best designed to compete for an pre-empt space in the low but dense vegetation in which they occur at Colonet. Mc, on the other hand, has the larger stem radius of the three, and grows taller and candelabra-like in form. It persists by growing through and above the surrounding vegetation. As precipitation increases further north from Colonet, both Sg and Mc drop out. Be extends north into the U.S. but is found only in the low, coastal sage vegetation and fails to enter chaparral. The only common cactus in chaparral in the region is *Opuntia parryi* which morphologically is a taller (to 2 1/2m) and thinner (2cm stems) version of Be, and is pre-

sumably better adapted, with higher PSA/V and reduced stem radius, to compete with the non-succulent species of this vegetation.

These postulates require a good deal of further data before they can be either verified or disproved. Data on spacing patterns within communities such as at La Paz would be particularly helpful, and could illuminate the extent to which species interact for space within sites. Such data exist for some sites in the Mojave Desert (Yeaton and Cody, 1976; Cody, 1983; Yeaton, 1983a,b), and additional information from the nopaleros (arborescent *Opuntia streptacantha*) of San Luis Potosi (Yeaton, 1983c) suggest that microhabitat selection in the larger cacti might be an important factor in their coexistence. Further information on the distribution of cactus species from the southern Sonoran Desert is also badly needed, and such studies are currently underway.

ACKNOWLEDGEMENTS

Field work in Mexico has been supported by the National Science Foundation (through 1980) and by Resources for the Future (1981-1983), to which agencies I am very grateful. Collection of field data was made far more entertaining through the assistance of my family (Daryl Ann, Tom and Erin); Carl Biehl helped to collect the Colonet data, and Phyllin Nicholson the data from El Desengano. I am indebted to my colleagues Arthur Gibson, Park Nobel, Richard Yeaton, and especially Harry Thompson, for enlightening discussions about cactus.

REFERENCES

Alexander RMcN (1971) Size and Shape, London, Edward Arnold.

Biehl C and Cody ML (1983) Plant species-area curves in the Sonoran and Mojave Deserts (Ms in prep.).

Blom PE and Bratz RD (1976) Cacti of the central desert of Baja California, Mexico J. Idaho Acad. Sci. 12, 73-81.

Carlquist S (1965) Island Life, Garden City, New York, Natural History Press.

Case TJ and Cody ML (1983) Biogeography of the Islands in the Sea of Cortez, Univ. Calif. Press, (in press).

Cody ML, 1978. Distributional ecology of *Haplo-*

pappus and *Chrysothamnus* in the Mojave Desert. I. Niche positions and niche shifts on north-facing granitic slopes, Amer. J. Bot. 65(10), 1107-1116.

Cody ML (1983) Spacing patterns in Mojave Desert shrub communities (MS in review).

Cody ML, Moran R and Thompson HJ (1983) The Plants, Ch. 4 in Case and Cody, 1983, q.v. in References above.

Felger RS and Lowe CH (1967) Clinal variation in the surface-volume relationships of the columnar cactus *Lophocereus schottii* in Northwestern Mexico, Ecology 48, 530-536.

Fisher JB (1977) A quantitative analysis of *Terminalia* branching. In Tomlinson and Zimmerman, eds. q.v. in References below.

Gentry HS (1943) Rio Mayo Vegetation, Washington D.C., Carnegie Publ. Botany.

Gibson AC (1982) Phylogenetic relationships of Pachycereeae. In Barber JSF and Starmer WT, eds., Ecological Genetics and Evolution, Sydney and New York, Academic Press.

Gibson AC and Horak KE (1978) Systematic anatomy and phylogeny of Mexican columnar cacti, Ann. Missouri Bot. Gard. 65, 999-1057.

Halle F and Oldeman RAA (1970) Essai sur l'architecture et la dynamique de croissance des arbres tropicaux, Paris, Masson et Cie.

Halle F, Oldeman RAA and Tomlinson PB (1978) Tropical Trees and Forests: An Architectural Analysis, Berlin, New York, Springer-Verlag.

Holland PG and Steyn DG (1975) Vegetational responses to latitudinal variations in slope angle and aspect, J. Biogeogr. 2, 179-183.

Honda H and Fisher JB (1977) Tree branch angle: maximizing effective leaf area, Science 199, 888-889.

Horn HS (1971) Adaptive Geometry of Trees, Monogr. Pop. Biol. 3, Princeton, Princeton University Press.

Miller PC and Stoner WA (1979) Canopy structure and environmental interactions. In Solbrig O, Jain SA, Johnson GB and Raven PH, eds., Plant Population Biology, pp. 163-173. New York, Columbia University Press.

Nobel PS (1980a) Interception of photosynthetically active radiation by cacti of different morphology, Oecologia 45, 160-166.

Nobel PS (1980b) Morphology, surface temperatures, and northern limits of columnar cacti in the Sonoran Desert, Ecology 61, 1-7.

Nobel, PS (1980c) Influences of minimum stem temperatures on ranges of cacti in southwestern United States and central Chile, Oecologia (Berl.) 47, 10-15.

Nobel PS (1981) Influences of photosynthetically active radiation on cladode orientation, stem tilting, and height of cacti, Ecology 63, 982-990.

Shreve F and Wiggins I (1964) Vegetation and Flora of the Sonoran Desert, v. I,II, Stanford, Stanford Univ. Press.

Stevens SS (1957) On the psychophysical law, Psychol. Rev. 64, 153-181.

Tomlinson BB and Zimmerman MH (1978) Tropical Trees as Living Systems, New York, Columbia University Press.

Werner PK (1979) Competition and coexistence of similar species. In Solbrig OT, Jain SA, Johnson GB and Raven PH, New York, Columbia Univ. Press.

Wiggins I (1980) Flora of Baja California, Stanford, California, Stanford Univ. Press.

Weinstock R (1974) The Calculus of Variations, New York, Dover Publications

Woodhouse RM, Williams JG and Nobel PS (1980) Leaf orientation, radiation interception, and nocturnal acidity increases by the CAM plant *Agave deserti* (Agavaceae), Amer. J. Bot. 67, 1179-1185.

Yeaton RI (1983a) Ecomorphology and microhabitat utilization of *Echinocereus triglochidiatus* and *E. engelmannii* in southeastern California, Gt. Basin Naturalist 42(3), 353-359.

Yeaton RI (1983b) Microhabitat selection, shading and heat loading in *Opuntia litoralis* and *O. parryi* in chaparral vegetation in southern California, Southwestern Naturalist (in press).

Yeaton RI and Cody ML (1976) Competition and spacing in plant communities: the northern Mojave Desert, J. Ecol. 64, 689-696.

Yeaton RI and Cody ML (1979) Distribution of cacti along environmental gradients in the Sonoran and Mojave Deserts, J. Ecol. 67, 529-541.

Yeaton RI, Karban R and Wagner HB (1980) Morphological growth pattern of saguaro (*Carnegiea gigantea*: Cactaceae) on flats and slopes in Organ Pipe Cactus National Monument, Arizona, Southwestern Naturalist 25(3), 339-349.

Yeaton RI, Travis J and Gilinsky E (1977) Competition and spacing in plant communities: the Arizona upland association, J. Ecol. 65, 587-595.

APPENDIX A. CACTUS STATISTICS

In this appendix various statistics are given for species of columnar cactus at a series of study sites listed in order of decreasing latitude (see Fig. 2). For each study latitude is given to the nearest tenth degree, and precipitation in mm to the nearest cm by extrapolation within the iso-hyets of Fig. 2. The statistics written in columns from left to right are as follows.

Cactus species are listed in the first column, using abbreviations as follows: Cg=*Carnegiea gigantea*; Ppr=*Pachycereus pringlei*; Ppa=*Pachycereus pecten-aboriginum*; St=*Stenocereus thurberi*; Sg=*Stenocereus gummosus*; Ls=*Lophocereus schottii*; Mc=*Myrtillocactus cochal*; Be=*Bergerocactus emoryi*.

Mean maximum branch number B is given next, evaluated as the mean maximum of the individuals of the largest size class of the population sample. Mean + the standard error of the mean is included.

Mean height above ground H, in feet, + standard error, at which the individuals in the largest size class reach maximum branch number.

Mean total stem length L, in feet, of the individuals in the largest population size class. This statistic is proportional to mean biomass in the largest size class; to convert to biomass, multiply mean total stem length by (pi)(radius squared).

Total number of individuals in population sample N.

Number of individuals in largest size class n. This number is usually taken as 10 in the larger samples as about 15% of the total population in the somewhat smaller samples, and often rather more than 15% in the smallest samples.

Larger size class C gives the minimum total stem length of individuals that are included in the largest size class.

Mean number of ribs Rb, + standard error in larger samples, measured on cactus at various heights but never close to the distal end of stems.

Mean stem radius Rd measured at various heights in stems, where rib count was made, but never close to distal end of stem. Stem radius can vary considerably within individuals from one season or year to the next, as well as between individuals within populations and between populations. Stem radius is of most use in comparing species within sites, and of some use in comparing populations within species.

Correlation coefficient between rib number and stem radius for the larger samples of rib and radius.

Statistical comparison of figures within columns can be made by computing Student's t. This is done by calculating the difference between two means, and dividing by the geometric mean of the standard errors of the two means. This gives t with $n_1 + n_2 - 2$ degrees of freedom.

SPP	B	H	L	N	n	C	Rb	Rd	Corr
1. TUCSON MOUNTAINS, 32.2° north, 380mm									
Cg	3.0 +.35	12.0 +.74	61 +5.4	112	16	>40			
2. LUKEVILL, 31.9° north, 230mm									
Cg	4.7 +.38	15.1 +.48	81 +11.1	38	6	>57	17.6 +.44	22.3 +.83	+.316
St	34.5 +1.64	2.4 +.65	269 +17.2	25	4	>232	17.7 +1.2	7.2 +.65	+.218
3. COLONET, 31.0° north, 160mm									
Mc	113.8 +4.37	5.9 +.26	564 +63.8	50	10	>325	7.6 +.13	4.1 +.10	+.441
Sg	105.6 +21.0	1.6 +.20	285 +52.9	50	10	>80	9.5 +.18	3.7 +.16	+.442
Be	121.7 +17.6	0.9 +.08	268 +41.6	50	10	>135	16.2 +.18	1.2 +.10	+.612
4. RANCHO ARENOSO, 29.9° north, 90mm									
South Slopes:									
Ppr	6.9 +1.85	9.8 +.80	127 +24.9	71	10	>84			
Ls	27.5 +8.47	1.7 +.14	152 +61.8	40	6	>55			
Sg	56.2 +8.58	2.6 +.20	219 +21.4	107	16	>138			
All slopes:									
Mc	175.8 +27.2	5.6 +.30	639 +141.7	51	10	>340			
North slopes:									
Ppr	4.3 +.40	9.3 +.96	61 +8.4	51	7	>35			
Ls	11.5 at 2ft			13	2	>37			
East slopes:									
Ppr	3.8 +.43	8.8 +.53	64 +8.8	81	12	>44			
Ls	16 at 3ft			21	3	>66			
Sg	49.2 +14.9	2.6 +.40	159 +34.0	33	5	>89			
West slopes:									
Ppr	4.9 +.44	10.2 +.57	89 +14.3	79	12	>47			

Ls	29.2 +6.5	2.3 +.09	126 +21.2	34	5	>74			
Sg	62.6 +19.7	3.4 +.37	246 +57.2	37	5	>142			

5. PUERTO LIBERTAD, 29.8° north, 130mm

Overlap zone:

Cg	3.0 12ft - 17ft		55 +4.2	15	2	>48	17.7 +1.4	16.6 +.43	+.382
Ppr	10.7 +.99	5.8 +.83	103 +4.9	40	6	>88	13.6 +.21	15.1 +.52	+.140
St	22.8 +3.43	1.9 +.31	232 +40.4	33	5	>129	15.4 +.50	6.5 +.26	+.365
Ls	90.0 +10.4	1.3 +.28	467 +64.0	29	4	>271	5.5 +.12	5.6 +.13	-.168

Non-overlap zone:

Cg	3.3 +.16	15.7 +1.65	68 +7.9	34	6	>55	20.8 +.65	17.7 +.40	+.386
Ppr	10.9 +.37	8.4 +.85	127 +6.3	46	7	>97	13.8 +.25	14.7 +.45	+.339

6. CATAVINA, 29.7° north, 90mm

Ppr	11.9 +1.62	10.8 +.49	251 +47.8	86	10	>124	13.6 +.22	13.4 +1.2	+.621
Ls	32.4 +.71	1.6 +.26	145 +28.5	62	10	>68	5.4 +.17	7.0 +.22	+.509

7. URES, 29.4° north, 360mm

Ppa	15.6 +3.93	13.9 +.62	196 +35.2	20	5	>114	9.3 +.14	4.6 +.33	
St	22.8 +9.0	9.4 +.95	231 +72.3	21	5	>77	13.7 +.15	4.5 +.36	

8. EL DESENGANO, 28.9° north, 90mm

Ppr	5.4 +.74	13.1 +1.75	155 +22.3	30	5	>100			
Ls	22.0 +3.98	2.7 +.92	115 +14.5	24	4	>84			
Sg	37.5 +2.8	3.2 +.47	199 +29.6	17	4	>147			

9. POZO ALEMAN, 27.9° north, 160mm

Ppr	9.9 +1.86	15.5 +1.24	198 +18.8	65	10	>79			
Ls	9.2 +.97	2.2 +.25	40 +1.1	36	10	>29			
Sg	44.6 +5.0	3.0 +.28	204 +21.1	39	10	>117			

10. SAN IGNACIO, 27.3° north, 90mm

Ppr	6.3 +.89	15.5 +.78	109 +14.4	50	10	>57	13.3	8.6	
St	22 at 3 1/2ft		133	7	2	>113	13.8	7.9	
Sg	35.1 +4.14	1.3 +.16	82 +9.5	50	10	>53	8.6 +.25	2.9 +.14	+.560

11. ISLA MONSERRATE, 25.6° north, 120mm

Ppr	4.6 +.72	10.4 +1.33	67 +10.2	39	9	>20			
St	22.4 +1.80	6.2 +.59	202 +16.3	26	5	>150			
Ls	17.5 +3.33	1.8 +.34	58 +9.0	15	4	>32			
Sg	19.5 +1.15	1.8 +.12	49 +1.3	12	4	>44			

12. ISLA SANTA CRUZ, 25.2° north, 120mm

Ppr	10.8 +1.15	4.8 +.50	126 +30.2	54	10	>74	13.6 +.17	11.1 +.92	-.247

13. ISLA SAN JOSE, 25.0° north, 120mm

Flats:

Ppr	6.4 +1.51	11.9 +.18	73 +7.0	30	5	>61	13.8 +.28	9.2 +.75	+.262
St	31.0 +2.03	5.3 +.48	266 +67.3	16	4	>148	14.7	4.3	

Hillsides:

Ppr	8.5 +.87	10.1 +.81	89 +9.1 9	52	8	>64	13.6 +.23	9.2 +.43	-.014
St	27.0 +1.84	5.3 +.45	206 +17.6	36	7	>138	14.5 +.31	5.6 +.19	0.00

14. SAN JUAN CANYON, 24.6° north, 160mm

Ppr	7.4 +.83	9.4 +.94	71 +6.2	33	5	>55			

15. LA PAZ, 24.2° north, 180mm

Ppr	12.5 +1.57	12.3 +.44	157 +14.7	62	10	>50	14.5	10.0	+.554
St	24.5 +2.19	5.5 +.45	255 +46.6	56	10	>140	15.9	6.0	+.476
Ls	108.0	3.7 +.30	377 +58.6	12	3	>450	6.5	4.6	-.030
Sg	43.3 +8.26	2.9 +.42	204 +23.0	16	4	>200	8.8	3.4	

16. RANCHO LOS DIVISADEROS,   23.9° north,   250mm

Ppr	13.7 +1.27	16.3 +.64	205 +13.8	52	10	>134	12.5 +.25	9.3 +.37	+.297
Ppa	11.1 +1.05	10.8 +.60	123 +10.4	49	10	>89	10.8 +.24	4.8 +.17	+.160
St	34.1 +3.1	8.4 +.53	344 +36.5	50	10	>200	17.6 +1.02	3.5 +.28	+.729
Ls	40 at 6ft		89 +26.5	9	2	>200	6.9 +.30	3.4 +.14	-.025
Sg	110 at 5ft		352 +82.3	17	4	>400	8.8 +.19	2.3 +.10	-.116

17. SAN BARTOLO,   23.7° north,   300mm

Ppa	10.6 +.89	10.1 +.39	93 +5.0	64	10	>74	10.5	6.6	+.064
St	31.7 +4.4	9.8 +.59	317 +34.7	33	10	>200	14.9	6.1	+.345

Biomass Figures for largest size sample in population

Units are m^3 x 100.

2.Lukeville;	Cg: 383, St: 133.
3.Colonet;	Mc: 90, Sg: 37, Be: 4.
5.Pto. Libertad;	Cg: 144, Ppr:244, St: 93, Ls: 139 (over-lap zone).
Pto. Libertad;	Cg: 203, Ppr:261 (non-over-lap zone).
6.Catavina;	Ppr:429, Ls: 68.
7.Ures;	Ppa: 39, St: 45.
10.San Ignacio;	Ppr: 77, St: 79, Sg: 7.
12.Sta. Cruz;	Ppr:148,
13.San Jose;	Ppr: 59, St: 47 (flats); Ppr:72, St: 62 (Hillsides).
15.La Paz;	Ppr:149, St: 87, Ls: 76, Sg: 22.
16.Rcho. Los Divisaderos;	Ppr:169, Ppa: 27, St: 40, Ls: 10, Sg: 18
17.San Bartolo;	Ppa: 39, St: 112.

EDAPHIC RESTRICTION OF CUPRESSUS FORBESII (TECATE CYPRESS) IN SOUTHERN CALIFORNIA, U.S.A.--A HYPOTHESIS

PAUL H. ZEDLER, CLAYTON R. GAUTIER, PAULA JACKS (BIOLOGY DEPARTMENT, SAN DIEGO STATE UNIVERSITY, SAN DIEGO, CALIF. U.S.A.)

1. THE PROBLEM OF EDAPHIC RESTRICTION

The restriction of some terrestrial plant species to unusual or extreme substrates such as soils derived from ultramafic rocks has long fascinated ecologists (Proctor, Woodell, 1975). Sharply restricted plant distributions exemplify one of the most general ecological problems: What limits the local and regional abundance of species? There is reason to hope that deepening our knowledge of the causes of the dramatic distributions of rare plants will help us to understand the more subtle patterns of common species.

The problem of restricted species can be divided into two aspects: 1) What prevents the limited species from spreading? and 2) What prevents common species from appropriating the habitat of the rare species? On the most extreme soils, reasonably satisfactory answers to these questions have been found in some cases, as in the well-studied metal tolerant plants in North Wales, U.K. reviewed by Antonovics et al. (1971). Cupressus forbesii (Tecate cypress) is an example of a species confined to distinctive, but not strikingly inhospitable substrates (Armstrong, 1978). In these circumstances, answers may be more elusive and complex. We will review some of the literature which we feel is especially relevant to the question of edaphic restriction of cypress in California.

2. FACTORS CONTROLLING EDAPHIC RESTRICTION

The problem of edaphic restriction is a special case of gradient ecology, which is concerned in part with establishing the range of conditions over which species can grow. Since the most striking cases of endemism are associated with soils that are perceivably difficult environ-ments, it is natural, following Grime (1979), to take stress as the gradient of interest. To avoid circularity, it is necessary that stress be a habitat property independent of the species. Granting that such an objective stress gradient can be constructed, a general ecological principle can be applied. This is the idea implicit in many studies, but stated directly by Connell (1975) that along a gradient of stress, species will be limited by physical factors in the direction of increasing stress and by competition in the direction of decreasing stress.

The explanations for edaphic restriction have to a large extent conformed to this model. Because edaphic endemics can often be cultivated in normal soils and because of the sparsity of cover on some extreme substrates, it has been repeatedly suggested that competition from more vigorous species must limit the spread of endemics into less stressful habitats (e.g. Gankin, Major, 1964; Kruckeberg, 1969; Vogl, 1973; Raven and Axelrod, 1978; Goldberg, 1982). This could be called "the refuge from competition" hypothesis. Conversely, the spread of common species onto extreme soils is attributable to an inability to tolerate the physical stress.

Grime's (1979) views are in accord with the physical factor-competition model. He argues that common species are able, even on the stressful substrate, to be competitively dominant in the narrow sense. That is, when individuals of a common species and an endemic are in proximity, the common species manages to obtain energy and materials at the expense of the co-occurring edaphic endemic. The survival of the endemic against this competition must then be attributed to the inability of the common species to withstand the physical stress. Because mortality rates of the common species are high, they fail over the long-term to appropriate enough space to exclude the endemic.

The idea that edaphic endemics are doomed to be less competitive is supported by the concept of trade-off in the allocation of resources. To survive stress or to modify physiology to over-come a special problem of deficiency or toxicity may only be possible by accepting reduced performance in growth and resource acquisition. Evidence that this is the case has been obtained in studies of metal tolerance (Antonovics et al., 1971; Cook et al., 1972).

An alternative view which conflicts with the refuge from competition hypothesis holds that endemics are competitively superior by reason of their greater efficiency in obtaining resources (Moore, 1959; Hart, 1980). Theoretical argu-ments of Stewart and Levin (1973) show that differences in efficiency of resource use across a gradient of resource availability can account for coexistence of species limited by the same resource. If this concept is applicable, then species could be limited at both ends of their distributions along the stress gradient by direct short-term resource competition, and stable equilibrium is possible.

When toxicity is an aspect of the stress associated with unusual substrates, common species may perform poorly and appear to be at a competative disadvantage against endemics. Grime, however, (1979) argues that it would be wrong to conclude in such cases that relative competitive abilities reverse since competition cannot properly be considered the immediate cause of the poor showing of the common species. While this is perhaps only a technical distinc-tion, it underlines an important problem in dealing with competition across a stress gradient.

Biotic factors other than competition can also play a role in the outcome of species interaction. Seedlings of restricted species have been shown to be susceptible to soil pathogens encountered on less extreme substrates, (Hart, 1980, Zobel, 1969). Common species may be excluded or perform poorly on extreme soils because their microbial symbionts are adversely affected. Mycorrhizae and nitrogen fixing bacteria can be harmed by deficiency of toxicity to the detriment of their host plants (White 1971).

Ironically, the faster growth of stress-tolerant species when they are on better soils may render them more, not less susceptible to herbivores and parasites. Frankie (1968) has noted that Cupressus macrocarpa, when growing rapidly in cultivation often has broken branches which allow the invasion of a canker fungus and insects. The probability of death of such fast growing trees is much higher than that of trees on the native soil even though they also sustain some damage from the same organisms. McMillan (1956) noted the same tendency for increased attack on larger individuals in more favored microhabitats in native stands of other Cupressus species. A similar phenomenon is reported in halophytes which are heavily attacked by insects when growing rapidly in artificially improved conditions (Teal, Valiela, 1982; J. Zedler, pers. comm.).

Species interactions take place in a variable environment, and this variability is of critical importance. In attempting to explain why endemism on metalliferous soils of southern Africa seems lower than expected, Wild and Bradshaw (1977) are forced to conclude that evolutionary trends in favor of endemics are swamped by climatic vari-ations which cause shifts in species distributions allowing the invasion of non-endemics. Fitting with this interpretation is the fact that smaller enclaves of extreme soil often have few or no endemics present. Smaller populations would of course have higher probabilities of extinction in a variable environment.

The Wild and Bradshaw concepts are related to Whittaker's (1960) "shift toward the mesic", the tendency for species to occupy increasingly moist habitats on increasingly extreme substrates. For example, a species which occurred across the moisture gradient on dirorite might be found only in moist pockets on serpentine. If the same shifts occur through time on a single substrate, endemics secure from competition during dry periods could be inundated by more rapidly growing competitors when moisture conditions improved.

3. EDAPHIC RESTRICTION IN TECATE CYPRESS

Tecate cypress, like many other conifers, shows substrate restriction. In San Diego County it occurs only on gabbro and metavolcanic substrates, even though both of these rock types are much less abundant than the granitic rocks characteristic of most of southern California. Following a long tradition, we have asked the obvious question: Why is Tecate cypress absent from habitats where it would seem possible for it to grow? A reasonable place to begin to answer this question is to test the widely held hypothesis that patterns of restriction are primarily the result of competitive pressures of species occupying the common substrates. We have carried out a preliminary pot experiment with this aim.

4. NATURAL HISTORY OF TECATE CYPRESS

A brief sketch of the natural history of Tecate cypress is necessary so that our views on its edaphic restriction may be understood. Further information may be found in Zedler (1977,1981). In brief, Tecate cypress is a closed-cone conifer in which seed dispersal is limited and mostly ineffectual except after crown fires in which the mortality of trees is virtually complete. Seedling establishment is primarily in the first growing season after fire, and stands are even aged. Cone production begins about six years after the fire, and the trees

seem capable of living for more than a century. Tecate cypress generally forms dense populations, from small clumps to large stands covering hundreds of hectares. The dominant associated species are shrubs, most of which are evergreen. A key feature of Tecate cypress is its ability to grow past and overtop the shrub species with which it is found. In older stands with a large admixture of shrubs there is a distinct double canopy, with the surviving shrubs at 1-3 m and the cypress at 4-6 m, with only moderate overlap.

The fire dependent life history of Tecate cypress makes it very sensitive to extinction when fires burn at short intervals, and like seed regenerators elsewhere severe decline of some stands has been recorded as a result of fires of human origin (Zedler, 1977,1981). Since cone opening is inefficient without fire, extremely long intervals between fires might also pose a threat, but there are no known stands old enough to be in this peril. If populations do disappear locally, the relatively poor dispersal capacity of the large seeds limit the possibility that local losses can be offset by dispersal from adjacent areas.

The soils on which Tecate cypress grows in San Diego County are derived from gabbro and metavolcanic rock. These soils differ in a number of features but have in common relatively high clay contents (c. 20-30% U.S.D.A. clay) and macronutrient deficiency, especially of phosphorous. Romaine lettuce grown on cypress soils was severely deficient in phosphorous (Zedler, unpublished data). Heavy metal toxicity does not seem to be a factor (C. Gautier, unpublished data). The vegetation on the soils is not conspicuously more sparse than that of granitic areas, though the rate of growth may be lower at equivalent moisture levels.

4. EXPERIMENTAL TEST OF THE COMPETITION HYPOTHESIS FOR TECATE CYPRESS

The details of our pot experiment will be reported elsewhere (Gautier et al., MS) and we give here a summary of the major results. In the experiment Tecate cypresses from two populations were grown from seed, as were plants of Helianthemum scoparium, a common associate of Tecate cypress also found in abundance on the more common granitic substrates. The Helianthemum was collected at three sites: the two cypress populations and an area of granitic soils adjacent to one of the cypress stands. A standard diallel experiment at a single density was run in which the seedlings of Helianthemum and Tecate cypress were grown in mixed and pure culture on three soils: the gabbro and metavolcanic soils from the cypress stands, and granitic soil on which cypress did not naturally occur. The competition treatments pitted cypress from each population against the granitic Helianthemum on granitic soils and the Helianthemum and cypress from gabbro and metavolcanic soils against each other on their native soils.

A simple way to present the results is to compare the outcome of competition for each population of cypress on its own soil with the outcome on the granitic soil. Three situations were observed: 1) no change (0)-the outcome of competition on the granitic and cypress soils was the same; 2) impairment of cypress on granitic soil (C-)--cypress with a competitive edge on both soils, but with a decreased advantage on the granitic soil; 3) gain by Helianthemum (H+)-no difference between the cypress and Helianthemum on cypress substrates, but Helianthemum with a competitive edge on granitic substrates. We present three measures of performance: mean per plant yield (oven-dry) at the end of the experiment, height of tallest stem, and total length of all branches.

Population of cypress	Yield	Tallest Stem	Branch Length
Metavolcanic	0	H+	H+
Gabbro	C-	H+	H+

Table 2. Outcome of competition on granitic soil relative to the outcome on native cypress soils as measured by three plant features. Symbols explained in text.

From these results (Table 1) we have concluded that competition probably is a factor in the restriction of Tecate cypress. However, we suspect that competitive pressure would not be so strong as to make the survival of Tecate cypress to reproductive maturity on granitic substrates unthinkable. As with other cypresses (McMillan, 1956), the yield of Tecate cypress was higher on the more fertile non-native substrate. Considering this, Tecate cypress might be expected to do well unless competition was especially intense and inevitable. But in the chaparral, fire acts to create gaps in which conditions for seedling establishment are unusually favorable. While seedling densities can be very high, distributions are patchy. It is hard to accept that chaparral on granitic soils can be closed to invasion of any species physiologically able to grow on them if viable seeds are available. Thus direct and immediate elimination by competition seems unlikely. There is also evidence which makes us doubt that Tecate cypress would be unduly susceptible to competition after the seedling stage. When planted into chaparral on granitic soils Tecate cypress has overtopped the associated shrub species, just as it does on its native soils. We therefore feel that Tecate cypress, if introduced into many natural habitats in which it does not now occur, would have a good chance of establishing in many of them. A satisfactory explanation of edaphic restriction of Tecate cypress, and probably many other species, must consider more than short-run competitive advantage.

5. A REVISED HYPOTHESIS OF EDAPHIC RESTRICTION
IN TECATE CYPRESS

The results of our experiment, combined with our knowledge of the natural history of the species and information in the literature allow a more sophisticated hypothesis of edaphic restriction to be formulated. As a first step we factor stress into two components, moisture and nutrients. Were toxicity a factor, as it can be in other situations, this would probably deserve to be a third axis. The factoring is important because moisture and nutrient deficiency are to a large degree independent. That is, there are moist microhabitats on infertile substrates and dry ones on fertile substrates.

The distribution of Tecate cypress in the landscape may be represented in this two-axis system and contrasted with that of the other major woody growth forms (Fig. 1).

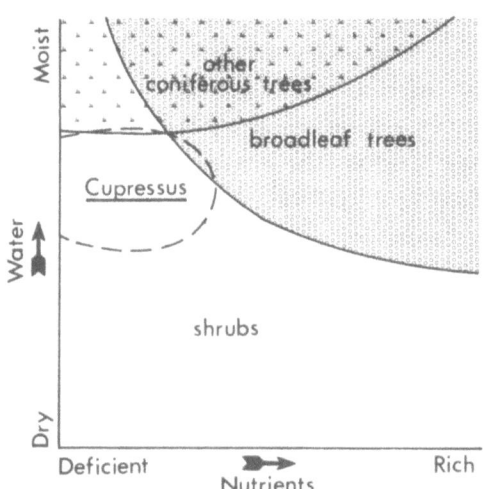

FIGURE 1. Diagrammatic representation of the hypothesized distribution of Tecate Cypress (Cupressus forbesii) in the landscape with respect to soil moisture and nutrient levels and other woody growth forms. Shrubs can occur over the entire range of moisture and nutrients conditions depicted in the figure but are generally restricted to dry habitats by competition from tree forms. Tecate cypress is restricted as the richer ends of the moisture and nutrient gradients by competition from trees and shrubs. Physiological intolerance probably

prevents Tecate cypress from occurring on the driest sites.

Cypress occurs on the low nutrient to moderate moisture sites in the region (300-1400 m elevation, mostly north-facing slopes). The vegetation of sites less deficient in nutrients of equivalent moisture tend to be occupied by large shrubs (e.g. Quercus dumosa, Heteromeles arbutifolia) and trees (mainly Quercus agrifolia). This is the case even though, as the graph suggests, there is almost no overlap in the distribution of cypress with other tree species.

On drier, but equally deficient sites, cypress encounters mostly shrubs. The overlap with shrubs (e.g. Adenostoma fasciculatum) is extensive. On equally deficient but moister habitat higher in the mountains cypress would encounter other coniferous species, chiefly pines (Pinus coulteri, P. jeffryi). It is probably significant that Tecate cypress in southern California is on sites well-removed from contact with other conifers.

Keeping in mind that Tecate cypress has a very limited capacity for dispersal, it would follow that the heaviest seed rain would be from cypress stands into areas dominated by shrubs, but mostly having rather poor moisture conditions. In agreement with the model presented earlier, we imagine that Tecate cypress is restricted in this direction by physical factors related to drought stress. At present, since cypress does not abut extensive areas dominated by trees, dispersal into these habitats must be rare. When it does occur, the evidence suggests that survival of the seedling to reproductive size might be quite probable. However, when pitted against plants whose maximum size equals or exceeds its own, Tecate cypress would have little chance of moving to a dominant position in the canopy as it does in shrub dominated vegetation. Despite improved growth, Tecate cypress could not be expected to be an integral part of forests of either broadleaf or coniferous trees,

242

and if the population persisted, it would have to be on the drier, more nutrient deficient fringes of the forests or groves. Here the argument of Wild and Bradshaw (1972) regarding the effect of climatic variations becomes relevant. Such marginal populations of Tecate cypress would be subject to rapid local extinction whenever an improvement in climatic conditions allowed the trees to expand into formerly unfavorable habitat. Because of its restricted dispersal, Tecate cypress may be limited to isolated peaks well removed from large populations of trees, since on these sites they have not been overgrown during periods of more favorable moisture conditions.

If it is true that cypress has increased susceptibility to pathogens when grown in more favorable conditions, then the possibilities for expansion in the direction of increasing nutrient or water supply would be even more severely limited. A phenomenon related to this, and possibly of equal or greater importance is the tendency noted in southern California (Zedler, unpublished data) and South Africa (E. Moll, pers. comm.) for many shrub species to have decreased life expectancy when they grow faster, a botanical expression of the "rate of living" concept which has long been held by zoologists (Lamb, 1977). Because Tecate cypress depends on fire for seedling establishment, there is the possibility that rapid aging on good sites may cause the populations to disappear before the next fire if there should be a long waiting time. The other, possibly closely related possibility is that rapidly growing cypresses might be susceptible to disease or herbivory, as discussed earlier.

There is little doubt that fire pattern is a factor affecting cypress distributions. The local population depletions which we have seen in San Diego County (Zedler, 1981) are the result of man-made fire, but similar reductions

doubtless occurred naturally. If a change in weather conditions increased the number of fires (it is difficult to specify precisely what would do this, most probably increased variation in rainfall) then the reduction in cypress populations in drier portions of the habitat discussed in Zedler (1981) may be counterbalanced by expansion into more nutrient rich habitats where both the competition of trees and the possibility of death before the next fire would be decreased with increased fire frequency.

Since this is a symposium on adaptation, we conclude by asking if the available evidence suggests that Tecate cypress has evolved as a specialist for nutrient deficient, fire-prone, sites. While there has undoubtedly been some genetic accommodation to its present ecological circumstances, the overall impression is that Tecate cypress is a paleoendemic which has been pushed to the fringes of forest habitat by the shifting process of long-term competition (in the broad sense). One could argue that the compression into unusual habitat forced by competition could lead to genetic adjustment to sterile habitats to a degree that preclude its ability to exploit other ecological opportunities should they arise. Then, the argument might go, there would be a downward spiral which would end in oblivion when the last patch of extreme soil slides down the hill. But counteracting these tendencies is the demonstrated ability of Tecate cypress to grow in a wide variety of conditions, and the likely fact that its accommodation to deficient sites probably involves very minor adjustment of ancestral genetic programs. The threat to continued existence is undoubtedly not competition, but evolution of a new pathogen or pest that can bypass the chemical and mechanical defenses that make Tecate cypress a long lived tree in nutrient deficient and semi-arid conditions.

6. CONCLUSION

The edaphic restriction of many woody plants will be explainable only if all aspects of its natural history are considered, and only if the consequences of climatic variation, including fire, are taken into account.

7. ACKNOWLEDGEMENTS

We thank T. Ebert, E. Moll, D. Goldberg, and especially J. Zedler for insights, and suggestions. This work was supported by NSF grant DEB-79-13424.

8. LITERATURE CITED

Antonovics J, Bradshaw AD and Turner RG (1971) Heavy metal tolerance in plants. Adv. Ecol. Res. 7:1-85.

Armstrong WP (1978) Southern California's vanishing cypresses. Fremontia 6:24-29.

Connell JH (1975) Some mechanisms producing structure in natural communities: A model and evidence from field experiments. In Coty ML and Diamond JM eds. Ecology and evolution of communities, pp. 460-490. Cambridge, Harvard Univ., Press.

Cook SCA, Lefevre C and McNeilly T (1972) Competition between metal tolerant and normal plant populations on normal soil. Evolution 26: 366-372.

Frankie GW (1968) Investigations on the ecology of the cypress bark moth Laspeyresia cupressana (Kearfott)(Lepidoptera: Olethreutidae). Ph.d. Thesis, Univ., of California, Berkeley. 126 pp.

Gankin R and Major J (1964) Arctostaphylos myrtifolia, its biology and relationship to the problems of endemism. Ecol. 45:792-808.

Goldberg DE (1982) The distribution of evergreen and deciduous trees relative to soil type: An example from the Sierra Madre, Mexico, and a general model. Ecol. 63:942-951.

Grime JP (1979) Plant strategies and vegetation processes. J. Wiley & Sons, Chichester. pp 222.

Hart R (1980) The coexistence of weeds and restricted native plants on serpentine barrens in southeastern Pennsylvania. Ecol. 61:688-701.

Kruckeberg AR (1969) Soil diversity and the distribution of plants with examples from western North America. Madroño 20:129-154.

Lamb MJ (1977) Biology of ageing. Halsted Press, John Wiley & Sons, New York. 184 pp.

McMillan C (1956) The edaphic restriction of Cupressus and Pinus in the coast ranges of central California. Ecol. Mono. 26:177-212.

Moore CWE (1959) Interaction of species and soil in relation to the distribution of eucalypts. Ecol. 40(4):734-735.

Raven PH and Axelrod DI (1978) Origin and relationships of the California flora. Univ. of California Pub. in Botany, Vol. 72. Univ. of California Press, Berkeley. pp. 134.

Stewart FM and Levin BR (1973) Partitioning of resources and the outcome of interspecific competition: A model and some general considerations. Amer. Nat. 107:171-198.

Teal J and Valiela I (1982) Long-term effects of wastewater application in salt marshes. Paper presented at the workshop on "Ecological considerations in wetlands treatment of municipal wastewater," June 23-25, 1982, sponsored by USF&WS and EPA. Held at Univ. of Mass., Amherst. Proceedings in press.

Vogl RJ (1973) Ecology of knobcone pine in the Santa Ana Mountains, California. Ecol. Mono. 43: 125-143.

White CD (1971) Vegetation-soil chemistry correlations in serpentine ecosystems. Ph.D. Thesis. Univ. of Oregon. 274 pp.

Whittaker RH (1960) Vegetation of the Siskiyou Mountains, Oregon and California. Ecol. Mono. 30: 279-338.

Wild H and Bradshaw AD (1977) The evolutionary effects of metalliferous and other anomalous soils in south central Africa. Evol. 31:282-293.

Zedler PH (1977) Life history attributes of plants and the fire cycle: A case study in chaparral dominated by Cupressus forbesii. p. 451-458. In Mooney HA and Conrad CE eds. Proceedings of the symposium on the environmental consequences of fire and fuel management in Mediterranean ecosystems. U.S.D.A. Forest Service. General Tech. Rept. WO-3.

Zedler PH (1981) Vegetation change in chaparral and desert communities in San Diego County, California. pp. 406-430 in West DC, Shugart HH and Botkin DB eds. Forest Succession. Springer Verlag, New York. pp. 517.

Zobel DB (1969) Factors affecting the distribution of Pinus pungens, an Appalachian endemic. Ecol. Mono. 39:303-331.

ULTRASTRUCTURE AND CYTOCHEMISTRY OF RETINACLES AND POLLINIA FROM MEDITERRANEAN ORCHIDS ·THEIR RELATION WITH POLLINATION

Maria Salomé S. PAIS - Laboratório de Botânica - Faculdade de Ciências de Lisboa
1294 Lisboa Codex Portugal (Granted by I N J C)

1. INTRODUCTION

Submicroscopic studies from Orchid retinacles and pollinia are quite scarce. In a previous paper (Pais, 1976) we described come inframicroscopic aspects from *Ophrys lutea* retinacles.

In this note some features from *Oph. lutea* and *Serapias lingua* retinacles and pollinia are presented and their relation with pollination is discussed.

2. PROCEDURE

2.1. Material and methods

Retinacles and pollinia from two Orchids, *Ophrys lutea* and *Serapias lingua*, have been submitted to the histochemical characterisation from mucopolysaccharides according to Stempien (1962). Induced fluorescence by treatment with aqueous solution 1:1000 of acridin orange has been observed by means of an epifluorescence microscope with a BP 450-490 nm exciter filter and ultra-violet.

For submicroscopic observations, the material has been fixed by a glutaraldehyde solution 4% in phosphate buffer 0.1M pH 7.2 during 20 hours at 4ºC, and post-fixed by an osmium tetroxyde solution 2% in veronal buffer during 2 hours. After dehydration material was embedded in a mixture Epon-Araldite.

Ultra-thin sections have been made in an ultra microtome Porter-Blum MT_2 and observed in an electron microscope Hitachi HU-12 under 75 KV.

The sections obtained were either coloured by potassium permanganate / lead citrate, or submitted to the Thiery test for detection of polysaccharides, according to Thiery (1967).

2.2 Results

Retinacles

Ophrys lutea flowers have two retinacles, each one for each pollinium (fig. 1a).

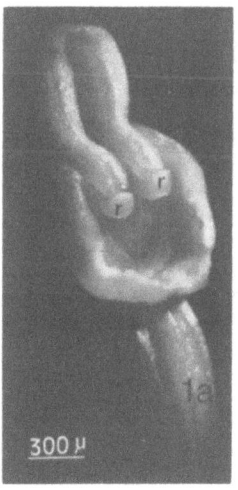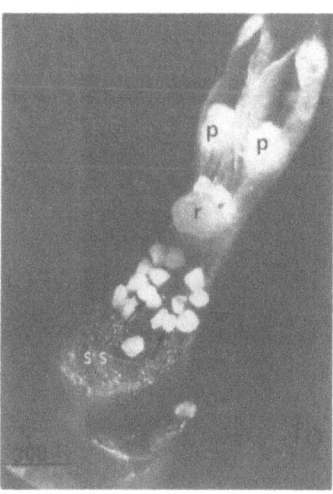

FIGURE 1.a) *Oph. lutea* flower devoided from perianth. Two retinacles (r) and two pollinia (p) are seen.
Figure 1.b) *S. lingua* flower without perianth showing one retinacle for two pollinia and pollen masses break down on the stigmatic surface (arrow).

Those from *Serapias lingua* show only one retinacle for two pollinia (fig. 1b). The abundant secretion from *Oph. lutea* retinacles have a mucilaginous character (fig.2). However, the secretion of *Serapias* retinacles is less abundant (fig.3,4) even if its composition is the same. This is related to the smaller adhesivity of these retinacles.

246

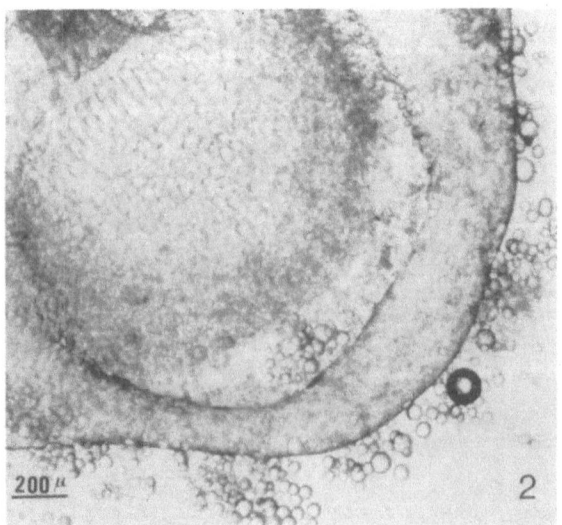

FIGURE 2. Microscopic aspect from *Oph. lutea* retinacle. Secretion is very abundant.

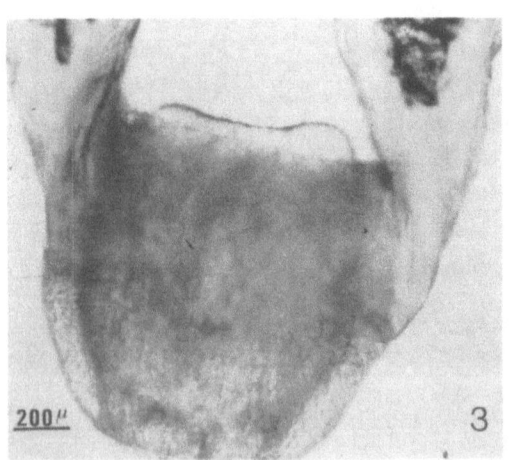

FIGURE 3. Microscopic aspect from *S.lingua* retinacle.

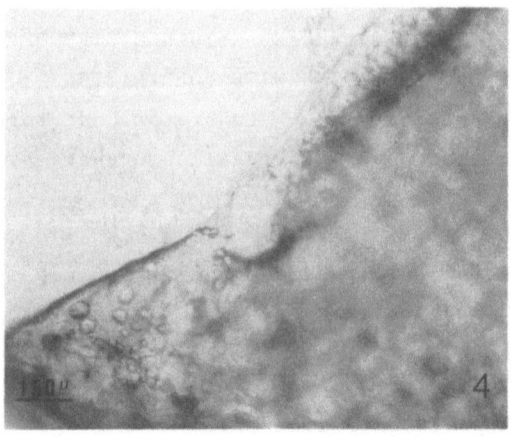

FIGURE 4. Magnification from fig.3 showing few secretion.

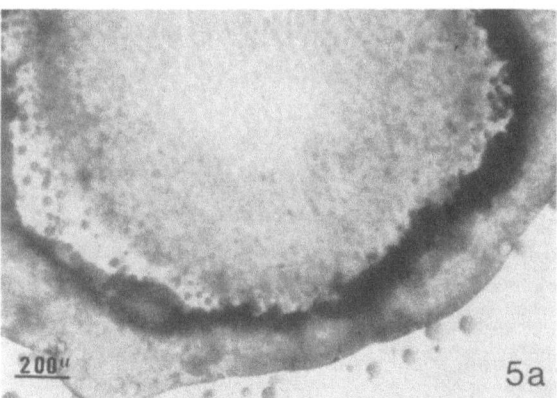

FIGURE 5.a) Histochemical coloration revealing the presence of polysaccharides.

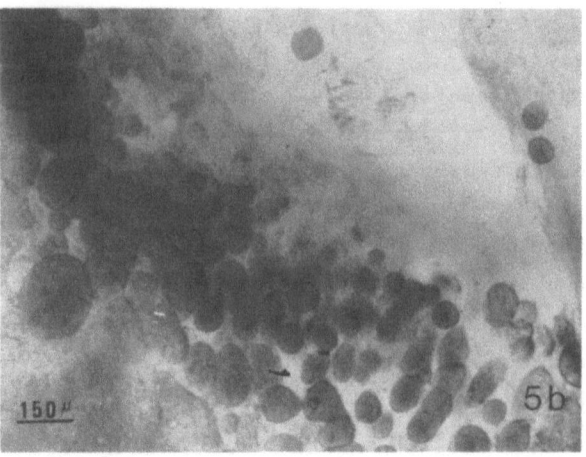

FIGURE 5.b) Magnification from fig.5a show-some droplets in the secretate well coloured.

Histochemical tests and fluorescence observations show the presence of polysaccharides (fig. 5a, 5b, 6). The Thiery test confirms the presence of a polysaccharide fraction in the retinacle heterogenous secretion (fig. 7, 8).

Inframicroscopic observations of retinacles show the typical features from secreting cells namely the dictyosomes relation with polysaccharidic secretion. In young cells, vesicles anastomose with plasmalemma. Older cells undergo a lytic process: degrading materials form part of the secretion.

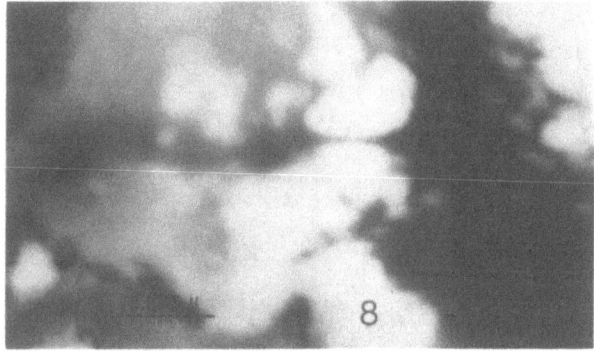

FIGURE 8. Inframicroscopic aspect from secretion after Thiery test. Black portions correspond to polysaccharidic fraction.

Pollinia

Pollinia from *Oph. lutea* flowers are consistent. Pollen masses are firmly attached by threads.

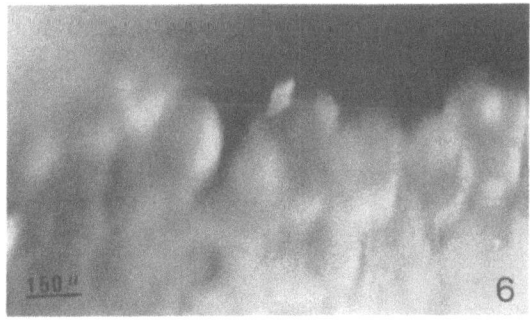

FIGURE 6. Red yellowish fluorescence after treatment with acridin orange.

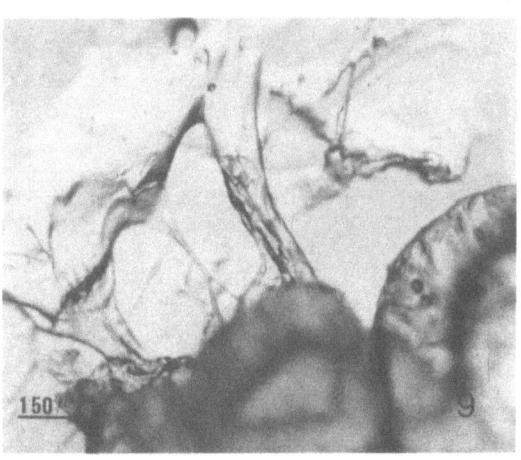

FIGURE 9. Microscopic aspect from threads in *Oph. lutea* pollinia.

Threads from *S. lingua* pollinia are less abundant and pollen masses tend to break down on the stigmatic surface (fig. 1b, 10). Induced fluorescence and inframicroscopic Thiery test reveal the polysaccharidic character from threads (fig. 11).

During pollinia ontogenesis the cells wich possess a large nucleus produce abundant secretion that accumulates between adjacent cell walls (fig.12). La-

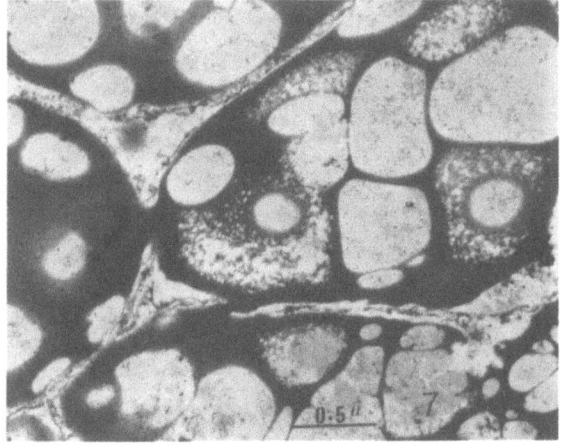

FIGURE 7. Inframicroscopic aspect from heterogenous retinacles secretion.

248

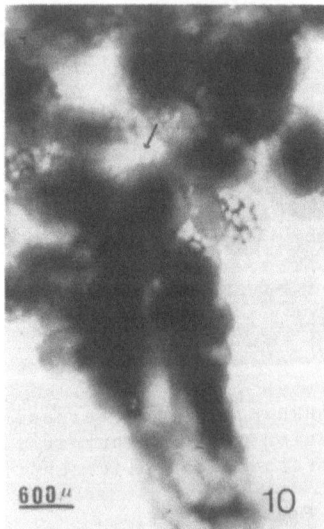

FIGURE 10. Pollen masses from *S. lingua*
with threads almost invisible (arrow).
FIGURE 11. Acridin orange induced fluorescence from threads show the polysaccharidic character from them.

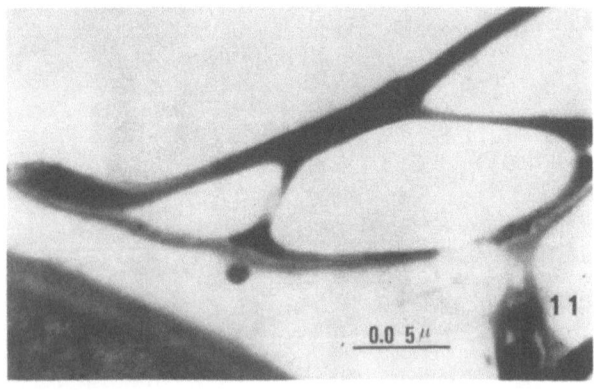

FIGURE 11. Inframicroscopic aspect from
a thread after Thiery test.

ter on, mucilaginous secretion form the
threads (fig. 13). The plasma membrane
sinuous aspect suggests that secreted
material is transported via vesicles
(fig. 13, arrow).

2.3 Discussion

The features reported to retinacles
and pollinia show that there is a close
relation between them and pollination.

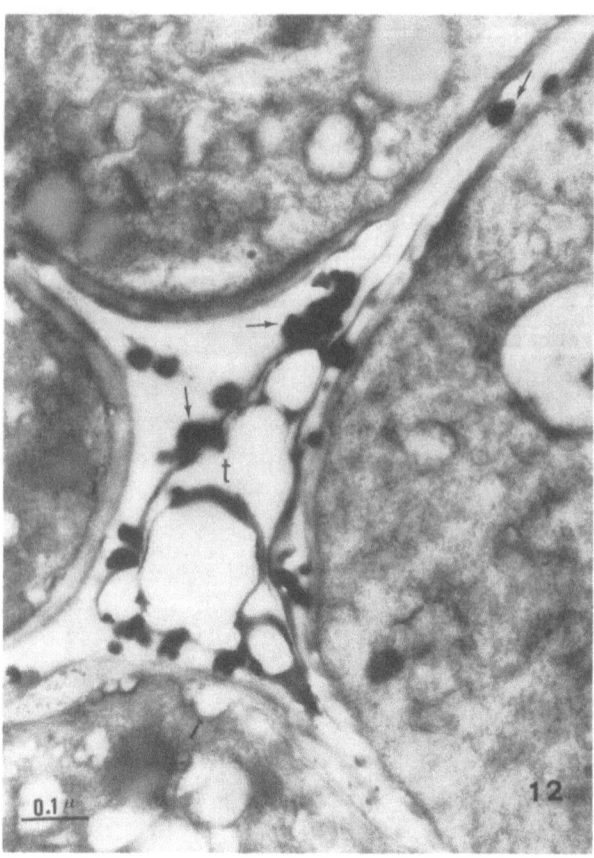

FIGURE 12. Inframicroscopic aspect from
pollinium cells showing the formation of
threads. Secretion accumulates in the in-
tercellular space (arrow).

The granulo-holocrine secretion from retinacles is responsible for the retinacle adhesivity. So, the cross pollina-
ted Ophryds possess one pollinium for
each retinacle , its secretion being very abundant. In consequence, the retinacles can firmly adhere to the pollinator. Another important feature is the
maintenance of pollinia integrity during
transport. This is due to the attachment of pollen masses to one another and to
the pollinium spindle by numerous threads.

In *S. lingua* flowers, where two pollinia are attached to the same retinacle

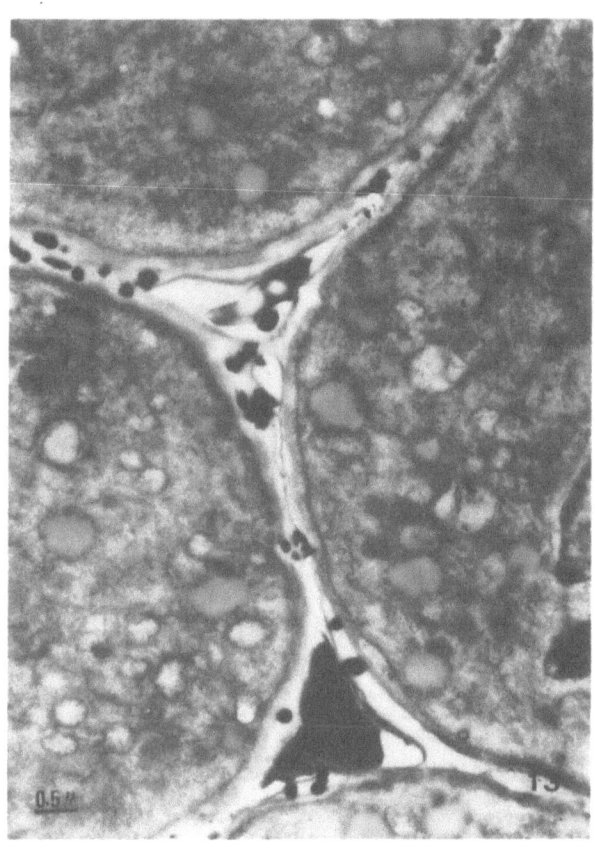

FIGURE 13 .   Inframicroscopic aspect
from a later ontogenic stage from threads
formation. Note the sinuous aspect of
plasmalemma (arrow).

the little amount of retinacles secretion determines the smaller adhesivity
to the pollinator (or, at least, the retinacle remains adherent during shorter
periods). Pollen masses are slightly attached to one another and to the central
spindle of pollinium, and break down
(even before anthesis) on the stigmatic
surface, and self pollination occurs. It
does not mean, however, that cross pollination cannot occur. When it happens
only the basal part from pollinium is
transported.

In the sequence, we can consider
that the following features are related:

A retinacle to one pollinium / abun-

dant retinacle secretion / pollen masses
firmly attached to one another and to the
central spindle of pollinium by threads
/ cross pollination.

Two pollinia joined to a single retinacle / small retinacle secretion / pollen masses slightly attached to one another and to the spindle of pollinium by
threads / self pollination.

In terms of community continuity it
is obvious that orchids with cross pollination only are more dependent upon the
co-existence of pollinators than those
with both cross and self pollination.

3. REFERENCE

Pais M S (1976) Quelques données sur
la sécrétion chez les Orchidées, Soc.
Bot. Fr. Coll. Sécrét. végét. ,149-159
Stempien M F (1962) in Ganter P, Jollès
G Histochimie Normale et Pathologique
vol. 1 pp. 480 Gauthier - Villars 1969
Thiery JP (1967) Mise en évidence des polysaccharides sur coupes fines en micros-
pie électronique , J. Microscopie 6, 987-
- 1018

250

Harper JL (1977) Population biology of plants, London, Academic Press.

Ingrain P (1978) Observations sur la glandée, la germination, la croissance et la survie du semis de chêne (Quercus sessiliflora Ehrl.) dans deux parcelles de la Forêt de Fontainebleau, differant par l'humidité édaphique, These Doctorat de Spécialité, Orsay, Université Paris-Sud.

Jacquiot C (1973) Un probleme d'écologie appliquée: la gestion et la sauvegarde des forêts françaises, Bull. Ecol. 4. 203-207.

Larson MM and Whitmore FW (1970) Moisture stress affects root regeneration and early growth of red oak seedlings, For. Sci., 16, 495-498.

Lavarenne S, Champaquat P and Barnole P (1971) Croissance rythmique de quelques végétaux ligneux de régions témperées cultivés en chambres climatiques a temperature élevée et constante et sous diverses photopériodes, Bull. Soc. Bot. For., 118, 131-162.

Lemée G (1978) La hêtraie naturelle de Fontaine-bleau. In Lamotte M and Bourliere F, eds., Problemes d'écologie:structure et fonctionnement des écosystèmes terrestres,pp 75-128,Paris, Masson.

Musselman RC and Gatherum GE (1969) Effects of light and moisture on red oak seedlings, Iowa Sta. J. Sci., 43, 273-284.

Oswald H (1978) Régénération des chênaias-hêtraias de la France moyenne, Paris, D.G.R.S.T.

Seidel KW (1970) The drought resistance and in-ternal water balance of oak seedlings, Diss. Abstr., 30B, 3450.

Solbrig OT, ed. (1980) Demography and evolution in plant populations, Berkeley, University of California Press.

APPENDIX. PLANT COMPOSITION OF EACH STAND

NS: Natural sowing

Arborescent stratum:

Covering: 90%

Fagus silvatica	4	5
Carpinus betulus	3	1

Herbaceous stratum:

Covering: <25%

Quercus sessiliflora(seedlings)	2	5
Ruscus aculeatus	1	3
Rubus fruticosus	+	2
Ilex aquifolium	+	2

ANS: Assisted natural sowing

Herbaceous stratum:

Covering: 90%

Quercus sessiliflora(seedlings)	3	5
Rubus fruticosus	3	5
Carpinus betulus(seedlings)	3	4
Ruscus aculeatus	1	3
Ilex aquifolium	+	3
Deschampsia flexuosa	+	2
Fagus silvatica(seedlings)	+	1

SR: Sowing in rows

Arborescent stratum:

Covering: 25% (canopy)

Fagus silvatica	+	1

Herbaceous stratum:

Covering: 60%

Quercus sessiliflora(ssedlings)	3	5
Deschampsia flexuosa	+	2
Rubus fruticosus	+	1
Fagus silvatica(seedlings)	+	1

Explanation to Appendix:

a. First column: Abundance-dominance: measure of the occupied area:
   + present
   1 abundance: low covering
   2   "   : covering <5% of the surface
   3   "   :   "   25-50% "   "   "
   4   "   :   "   50-75% "   "   "
b. Second column: clustering: measure the way of the repartition of the individuals:
   1 isolated
   2 small groups
   3 aggregates
   4 small colonies
   5 communities

OBSERVATIONS ON THE DYNAMICS OF THREE POPULATIONS OF OAK SEEDLINGS (*QUERCUS SESSILIFLORA* EHRL.) IN THE FONTAINEBLEAU FOREST

H. CHEVILLOTTE and J.P. CANCELA DA FONSECA (Laboratoire de Biologie Végétale et d'Ecologie Forestière*, Université Paris 7, Route de la Tour Dénecourt, F-77300 Fontainebleau, France)

1. INTRODUCTION

The observations on the dynamics of three populations of oak seedlings (*Quercus sessiliflora* Ehrl.) were carried out during 1981 on the plot no. 291 of the Fontainebleau Forest. They are both of practical and theoretical interest.

On the practical side, the National Forestry Council (ONF) has tried different regeneration techniques concerning the beech and the oak. Their aim is to replant large areas with oak. The first trials by clearing the land out of all trees were catastrophic, either from the point of view of the regeneration or the ecological point of view (Jacquiot, 1973). Thus, this technique was abandoned. Three main techniques are used nowadays:

a. Natural sowing (NS): after some tress are cut out, sowing comes only from natural local pirducus;

b. Assisted natural sowing (ANS): after the cut out of a greater number of trees, sowing of seeds from some other origins as a complement to the natural sowing; during several years pulling up by hand of the blackberry bush, a forest weed, twice a year;

c. Sowing in rows (SR): once the thinning is over (a small number of trees is left) the soil is plowed with, sometimes, the incorporation of a green mulch into the soil, the sowing being made in rows; the plot is treated at regular intervals with pesticides (weed-killers) or mechanically (pulling up blackberries bushes and Graminae).

All these techniques are more or less empirical and some experiments have been and are conducted by the Forestry Services. This study is then a

*Research Group "Analyse des Systémes Ecologiques, Ecologie du Sol".

contribution to the solution of such a problem. Our aim is to compare the dynamics of these three types of seedlings.

On the theoretical side, the studies of plant population dynamics are fewer than those of animal populations (Barbault, 1981; Begon and Mortimer, 1981). Among the first ones, the studies made are in a greater number on herbaceous species (annual vegetation, quick reproduction cycles,...) than on the trees, as their evolution is slow and consequently their study needs a long term scientific investment (Causton and Venus, 1981; Harper, 1977; Solbrig, 1980). In general, more research work was done on the beech than on the oak. It seems that the regeneration problems are more complex and difficult with the latter. In France, several studies were done on broad-leaf trees, mainly the beech but they take into account chiefly the influence of the biological, ecological and climatic factors on the seedling development (D.C. R.S.T., 1974; Lemée, 1978; Oswald, 1978), not the population dynamics of the seedlings, and put aside the biomathematical point of view.

2. MATERIAL AND METHODS

All experiments and observations were made in the field. The forestry environment of the plot no. 291 is quite heterogeneous: remnant woods, big trees, clearings. So we have chosen stands as homogeneous as possible(soil, light, wind, rainfall) and a sampling method based on continuous transects. In all the three stands (NS, ANS, SR) the soil is a sandy leached soil with a mull-moder type of humus on the non-exploited stand (NS). Each of the experimental fields in the exploited stands (SR and ANS) was situated in the middle of clearings. They were constituted of two transects of five contiguous quadrats of one square meter each. The spatial distribution of the oak seedlings remain an uncontrolled variable. The number of seedlings sampled is the one shown on Table 1. The classes of ages present in the three stands were somewhat different.

TABLE 1. Number of seedlings sampled by stand and by class of age.

Stand	Classes of age (years)					
	1	2	3	4	5	6
SR	10	10	10	-	-	-
ANS	-	-	10	10	10	10
NS	-	10	8*	6*	-	-

*The number of seedlings sampled is equal to the total number of seedlings recorded.

The measurements carried out were: a) initial length of the stem (early April); b) rate of increase of the stem once a week (May to July); c) number of leaves by seedling sampled (end of June and end of August), with determination of the number of leaves partially eaten (predation) and the number of leaves attacked by *Oidium* sp. (parasitism); d) number of plant species in competition with the oak seedlings in each quadrat (July-August); e) moisture content of the soil (July-August).

Two kinds of statistical analysis were employed for the data analysis: a) analysis of variance over the stem population, and b) x^L analysis over the leaf population.

3. RESULTS

3.1. Seedlings growth

At first, we have compared the increase rate of the seedlings of the same class of age. The only class of age present in the three stands is the seedlings of three years age (Table 2). The mean rate of increase (m) was significantly different, being higher in the assisted natural sowing and lower in the natural sowing.

$$m_{ANS} > m_{SR} > m_{NS}$$

where

$\Delta(m_{ANS} - m_{SR})$ is significant at the 5% level, and $\Delta(m_{SR} - m_{NS})$ is significant at the 1% level. Then, the comparison of the different classes of age inside each stand gave the following results (of Table 2):

TABLE 2. Seedlings growth: mean rate of increase for each stand and class of age.

Stand	Classes of age (years)					
	1	2	3	4	5	6
SR	1.34	3.50	2.83	-	-	-
ANS	-	-	3.63	3.80	3.84	3.51
NS	-	1.25	1.71	2.28	-	-

a. Sowing in rows: the rates of increase were independent of the age, but significantly different:

$$m_2 > m_3 > m_1$$

where

$\Delta(m_2 - m_3)$ is significant at the 5% level, and $\Delta(m_3 - m_1)$ is significant at the 0.1% level;
b. Assisted natural sowing: the rates of increase do not differ significantly;
c. Natural sowing: the rates of increase differ significantly, the higher rates corresponding to the older seedlings:

$$m_4 > m_3 > m_2$$

where

$\Delta(m_4 - m_3)$ and $\Delta(m_3 - m_2)$ are both significant at the 5% level

3.2. "Healthy state" of the seedlings

3.2.1. Demographic aspects. The "natality" and mortality rates were estimated from the leaf population census at the time t (June) and the time t+1 (August), according to the following equations:

$$\text{"Natality" rate}: \nu = \frac{\Sigma B}{0.5(\Sigma N_t + \Sigma N_{t+1})}$$

$$\text{Mortality rate}: \mu = \frac{\Sigma \Delta}{0.5(\Sigma N_t + \Sigma N_{t+1})}$$

where: b is the number of "births"; B, the number of deaths; N_t, the number of leaves at the time t; and, N_{t+1}, the number of leaves at the time t+1.

For the natural sowing, the number of leaves between t and t+1 was stable, but for the other two types of sowing a certain number of leaves by seedling appear between t and t+1. In both stands the "natality" rate increased with the age, except for the five years old at the assisted natural sowing (Tables 3 and 4). On the contrary, the mortality rate was quite variable from one class of age to the other (Tables 3 and 4). The highest rate of mortality was observed at the sowing in rows for the seedlings of one year age. Their stem mortality was also higher.

TABLE 3. Sowing in rows: "natality"and mortality rates of the leaf population.

Rates	Classes of age (years)		
	1	2	3
ν	0.054	0.316	0.323
μ	0.108	0.000	0.056

TABLE 4. Assisted natural sowing: "natality" and mortality rates of the leaf population.

Rates	Classes of age (years)			
	3	4	5	6
ν	0.155	0.173	0.103	0.189
μ	0.000	0.025	0.088	0.004

3.2.2. Predation and parasitism. Though the predators were not identified, the predation was measured by the number of leaves partially eaten by the invertebrates. The parasites were a species of *Oidium*.

Predation took place in all the stands, but the parasitism was only observed at the sowing in rows and the assisted natural sowing; none at the natural sowing (Table 5).

As before, the main comparison study was done for the seedlings aged of three years. The differences between the three stands were significative:

a. Predation: It was higher at the natural sowing (70%), followed by the assisted natural sowing

TABLE 5. Rates of predation and parasitism of the leaf population (%).

Stands	Classes of age (years)						Overall average
	1	2	3	4	5	6	
Leaves eaten (%)							
SR	61.0	59.0	45.0	–	–	–	55.0
ANS	–	–	52.0	52.0	46.5	42.0	48.0
NS	–	80.0	70.0	76.0	–	–	75.0
Leaves parasitized (%)							
SR	19.0	30.0	49.0	–	–	–	33.0
ANS	–	–	16.0	14.0	9.5	16.0	14.0
NS	–	0	0	0	–	–	0
Leaves non-attacked (%)							
SR	20.0	11.0	6.0	–	–	–	12.0
ANS	–	–	32.0	34.0	44.0	42.0	38.0
NS	–	20.0	30.0	24.0	–	–	25.0

(52%) and the sowing in rows (45%).

b. Parasitism: The attack was inversed, the higher percentage being observed at the sowing in rows (49%) followed by the assisted natural sowing (16%). Taking into account the overall average (all classes of age present at each stand), the assisted natural seedlings seem the "healthy" ones, the number of leaves non-attacked being higher (38%) than in the other two types of sowing (25% and 12%) (Table 5).

4. DISCUSSION AND CONCLUSIONS

From the results, the assisted natural sowing seems the better technique of either natural or artificial regeneration of the exploited oak forests.

There, the seedlings rate of growth is higher and the number of leaves attacked relatively lower. The vegetation cover seems to be a very important factor (see plant composition of each stand in Appendix). Thus, for the three years seedlings, the intensity of the predation and of the parasitism was significantly different at the three types of sowing according to their degree of vegetation cover (plants other than oak seedlings):

a. Sowing in rows: vegetation cover practically absent, lowest percentages of predation and highest of parasitism;

b. Assisted natural sowing: plant cover abundant, mainly herbaceous, higher percentage of predation and lower of parasitism;

c. Natural sowing: mainly arborescent plant cover; highest percentage of predation, and no parasitism. The causes of lower parasitism with increasing vegetation cover are of two kinds: The *Oidium* can not develop at dim light and attack chiefly the young leaves. At the natural sowing conditions the environment is darker than at the other two conditions, and the growth of seedlings stopped earlier. At light conditions the seedlings in rows and the assisted natural seedlings have had at least two periods of growth: one in July-August when the *Oidium* started the attack. These extra periods of growth are very often observed in the oak seedlings (Lavarenne et al., 1971). The lower rate of parasitism of the assisted natural seedlings can be explained then by a higher degree of cover.

As the oak is a species which needs light conditions for its development (Aussenac and Durcey, 1978; Musselman and Gatherum, 1969), though only 30 to 50% on the first years (Jacquiot, 1973), it is natural that the rate of growth of the seedlings at natural sowing conditions be lower. However, the difference between the other two types of seedlings cannot be so easily explained. As a matter of fact, the light and the soil moisture conditions (Table 6) were the same in both stands, the only differences being the lateral darkness of the assisted natural seedlings due to plant cover, which normally speeds growth (Jacquiot, 1973), and the biological activity and the presence of litter in soil, higher at the assisted natural sowing (Massot, personal communication). Better growth conditions were also observed this year for one year age seedlings sowed in rows in a plot were some kind of organic matter (ground wood) was added before the sowing (Chevillotte,

TABLE 6. Average soil moisture content, in per cent, of the three stands (July-August 1981).

Stands	Dates		
	30.07.81	12.08.81	27.08.81
SR	17.24±2.76	19.43±2.46	13.61±2.77
ANS	18.33±5.12	19.71±4.82	16.41±4.82
NS	15.09±6.39	19.10±3.69	8.84±1.22*

OBS.-Arithmetic mean ± standard deviation.
*Significantly different at the 0.1% level.

personal communication).

Finally, drought is a limit condition to the implantation and the growth of the oak seedlings (Aussenac and Durcey, 1978; Ingrain, 1978; Larson and Whitmore, 1970). But the root system of the oak, a deep tap root, is a form of resistance to the drought (Seidel, 1970). Then, the artificial sowings can have some advantage over the natural seedlings where the young root system, quite sensible to the moisture conditions (Aussenac and Durcey, 1978), is in competition with the lateral root system of the beech. In reality, for the same climatic conditions, the average soil moisture content of the natural sowing stand was in general lower and at the end of the summer significantly lower (Table 6).

In conclusion, it seems that the oak seedlings are well adapted to the environmental conditions created by the assisted natural sowing technique, conditions similar to the natural forest clearings.

REFERENCES

Aussenac G and Durcey M (1978) Etude de la croissance de quelques espèces forestières cultivées à différents niveaux d' éclairement et d'alimentation hydrique , C.R. 103e Congr. Nat. Soc. Savantes, 1,105-117.

Barbault R (1981) Ecologie des populations et des peuplements, Paris, Masson.

Begon M and Mortimer M (1981) Population ecology: a unified study of animals and plants, Oxford, Blackwell.

Causton DR and Venus JC (1981) The biometry of plant growth, London, Arnold.

D.G.R.S.T. (1974) Projet "Régéneration des chènaias et hêtraias", Nancy, CNRF.

Harper TL (1977) Population biology of plants, London, Academic Press.

Ingrain P (1978) Observations sur la glandée, la germination, la croissance et la survie des semis de chêne (*Quercus sessiliflora*, Ehrl.) dans deux parcelles de la Forêt de Fontainebleau, different par l'humidité édaphique, Thèse Doctorat de spécialité, Orsay, Université Paris-Sud.

Jacquiot C (1973) Un problème d'écologie appliquée: la gestion et la sauvegarde des forêts francaises, Bull. Ecol., 4, 203-207.

Larson MM and Whitmore FW (1970) Moisture stress affects root regeneration and early growth of red oak seedlings, For. Sci, 16, 495-498.

Lavarenne S, Champagnat P et Barnola P (1971) Croissance rythmique de quelques végétaux ligneux de régions tempérées cultivés en chambres climatiques à temperature élevée et constante et sous diverse photopériodes, Bull, Soc. Bot. Fr., 118, 131-162.

Lemée G (1978) La hêtraie naturelle de Fontainebleau. In Lamotte M and Bourlière F, eds., Problèmes d'écologie: structure et fonctionnement des écosystèmes terrestres, pp. 75-128. Paris, Masson.

Musselman RC and Gatherum GE (1969) Effects of light and moisture on red oak seedlings, Iowa Sta. J.Sci., 43, 273-284.

Oswald H (1978) Régénération des chênaies-hêtraies de la France moyenne, Paris, D.G.R.S.T.

Seidel KW (1970) The drought resistance and internal water balance of oak seedlings, Diss. Abstr., 303, 3450.

Solbrig OT, ed. (1980) Demography and evolution in plant populations, Berkeley, University of California Press.

APPENDIX. PLANT COMPOSITION OF EACH STAND

NS : Natural sowing	column 1	2
Arborescent stratum:		
Cover: 90%		
Fagus silvatica	4	5
Carpinus betulus	3	1
Herbaceous stratum:		
Cover: <25%		
Quercus sessiliflora (seedlings)	2	5
Ruscus aculeatus	1	3
Rubus fruticosus	+	2
Ilex aquifolium	+	2

ANS : Assisted natural sowing		
Herbaceous stratum:		
Cover: 90%		
Quercus sessiliflora (seedlings)	3	5
Rubus fruticosus	3	5
Carpinus betulus (seedlings)	3	4
Ruscus aculeatus	1	3
Ilex aquifolium	+	3
Deschampsia flexuosa	+	2
Fagus silvatica (seedlings)	+	1

SR : Sowing in rows		
Arborescent stratum:		
Covering: 25% (canopy)		
Fagus silvatica	+	1
Herbaceous stratum:		
Cover: 60%		
Quercus sessiliflora (seedlings)	3	5
Deschampsia flexuosa	+	2
Rubus fruticosus	+	1
Fagus silvatica (seedlings)	+	1

LEGEND:

a. First column: Abundance - dominance: measure of the occupied area:

+   present
1   abundance: low covering
2   "   : covering   < 5% of the surface
3   "   :   "   25-50% "   "   "
4   "   :   "   50-75% "   "   "

b. Second column: Clustering: Measure the way of the repartition of the individuals:

1   isolated
2   small groups
3   aggregates
4   small colonies
5   community - coherent cover

EFFECT OF PLANT VOLATILES ON RHIZOSPHERIC AND PHYLLOSPHERIC MYCOFLORA

A. F. Afifi (Botany Department, Faculty of Education, Roxy, Heliopolis, Cairo, Egypt)

1. EFFECT OF VOLATILES PRODUCED BY GERMINATING SEEDS AND ROOTS ON FUNGAL SPORE GERMINATION

During studies on interaction of plant and soil fungi, little attention has so far been devoted to volatile and gaseous metabolites of plants. Germinating seeds release, in addition to carbon dioxide, a series of other volatile and gaseous metabolites into the environment.

The effect of carbon dioxide on fungi depends on its concentration (Gundersen, 1961), and on other factors such as concentration of oxygen in the environment (Hollies, 1948; Burges, Fenton, 1953; Newcombe, 1960; Gundersen, 1961; Papavizas, Bavey, 1962; Julian, Phillips, 1963; Leuvet, Bulit, 1964; Teler et al., 1966; Griffin, 1966; Macauley, Griffin, 1969; Mitchell, Zontmyer, 1971), so that carbon dioxide may act as an inhibitor. When oxygen is not present in the environment, carbon dioxide may be required for development of some fungi (Gunner, Alexander, 1964; Gundersen, 1961; Andersen, Ordal, 1961). For example, carbon dioxide was required for germination of perithecia in *Chaetomium glubosum* (Buston et al., 1966). Germination of spores and growth of certain fungi were usually stimulated under the influence of carbon dioxide (Platz et al., 1927; Stover, Freiberg, 1958; Barinova, 1960, 1961; Newcombe, 1960; Eaves, Lockhart, 1961; Stanek, 1963; Michell, Zontmyer, 1971; Hodgkiss, Harvey, 1972).

Of the other metabolites of germinating seeds, alcohols, aldehydes, olefins and volatile organic acids appear as most important (Grinová, 1964; Vančurá, Statzky, 1971, 1974; Doireau, 1972; Holm, 1972; Afifi, 1977). The stimulatory effect of organic volatile acids on fungi was studied by Glasare (1970). Certain fungi are very sensitive to ethylene, which suppresses their development (Lockhart et al., 1968; Lockhart, 1970; Afifi, El-Gindy, 1978). Spores of other fungi were stimulated to germinate in presence of ethylene (Afifi, El-Gindy, 1978).

Volatile and gaseous exudates of swelling seeds of *Pisum sativum*, *Phaseolus vulgaris*, *Triticum vulgare*, *Zea mays*, *Cucumis sativus*, *Lens esculenta*, *Solanum lycopersicum*, *Daucus carota*, *Capsicum annuum*, and *Lactuca sativa* differ in their effect on germination of spores of the following fungi: *Mucor racemosus*, *Rhizopus arrhizus*, *Fusarium oxysporum* var. *orthoceras*, *Trichoderma viride*, *Verticillium dahliae* and *Botrytis cinerea* (Čatská et al., 1975), but many more considerable differences were found in sensitivity of spores of different species of fungi to volatile and gaseous metabolites released during the first hours of swelling of the seeds. Considerable inhibition of germination of spores of the fungus *Mucor racemosus* and particularly of *Botrytis cinerea* occurred in the atmosphere of swelling seeds of all studied plants. The inhibitory effect of seed exudates of some plants on germination of spores of *Trichoderma viride* and *Verticillium dahliae* was lower. On the other hand, germination of spores of these two fungi was stimulated by exudates of lettuce seeds. The least sensitive were spores of *Fusarium oxysporum*, the germination of which was significantly inhibited only by exudates of lentil and carrot, and stimulated by exudates of peas and bean.

The inhibitory effect of volatile and gaseous exudates was decreased or removed completely in an atmosphere enriched by exudates of swelling seeds and in the presence of the absorption agent potassium permanganate ($KMnO_4$), particularly in *Botrytis cinerea*. A partial decrease of the inhibitory effect of the seed exudates of most plants was observed also in *Mucor racemosus*. When potassium hydroxide (KOH) was used as absorption agent, spores of *Trichoderma viride* and *Verticillium dahliae* did not germinate at all, and the inhibitory effect increased in *Mucor racemosus* and *Fusarium oxysporum*. However, in *Botrytis cinerea* the inhibitory effect of exudates was not increased in the presence of KOH. It seems that the germination of the fungal spores was inhibited and even stimulated by some compounds absorbed in a solution of $KMnO_4$. With the exception of *Botrytis cinerea*, the germination of spores of the studied fungi was more strongly inhibited in the absence of the compounds absorbed in KOH. The results obtained indicate that the volatile and gaseous exudates of swelling seeds may influence, under certain conditions and at certain concentrations, the development of suppression of some fungi, thus playing a role even during the initial phase of the rhizosphere effect, during colonization of the germinating seeds and outgrowing roots by fungi.

Results obtained by Afifi and Abdulla (1978) indicated that the gaseous and volatile metabolites of the germinating seeds and roots of two *Lupinus termis* varieties stimulated the spore germination of all rhizospheric fungi in the root zone of the plant. Stimulation was increased when CO_2 from the gaseous metabolites was avoided by using 5 percent KOH solution. Mitchell and Mitchell (1973) reported that the variability of pathogens probably enables many of them to develop in habitats differing in CO_2 levels. Nair and Fahy (1972) showed that the growth rate of the mycelia of *Sclerotium rolfsii* was not inhibited by low oxygen concentration down to 3 percent, but was as the CO_2 concentration increased above 0.03 percent. High CO_2 levels may favor the pathogen over less tolerant microorganisms (Dubrin, 1959).

Similarly, when some organic volatiles (especially these with unsaturated double bonds), which might emanate from either the germinating seeds or roots were excluded by use of 0.15 percent $KMnO_4$ solution, growth was stimulated. Smith (1973) found that sclerotia of *Sclerotium rolfsii* remained dormant on soil over which passed air containing 1 ppm ethylene. Spores of *Helminthosporium sativum* on field soil failed to germinate until ethylene was removed by passing air over the surface. Coley-Smith (1971) showed that volatiles from onion and garlic stimulated germination of sclerotia of *Sclerotium cepivorum*.

2. ETHYLENE PRODUCTION DURING SEED GERMINATION AND ITS EFFECT ON FUNGAL SPORE GERMINATION

Detection of ethylene was carried out by Afifi et al. (1977), using gas chromatography technique, in the volatile and gaseous exudates of the germinating seeds of cotton(*Gossypium barbadense*), "Menoufi" and "Giza 68" varieties, maize (*Zea mays*) "Hybrid 69" and "Hybrid 80", and broad bean (*Vicia faba*) "Romi" and "Giza 1".

The results obtained indicate that during the first 72 hours of germination, the germinating seeds of the experimental plants emanate ethylene among the other volatile and gaseous exudates. The amount of ethylene decreases with the age of the germinating seeds. Ethylene production stopped completely after 72 hours. It was also found that the amount of ethylene varied according to the seed variety. "Menoufi" cotton variety exuded more ethylene than "Giza 68" during the first 18 hours of germination, after which the reverse occurred. "Hybrid 80" maize variety exuded more ethylene than "Hybrid 69" maize variety. Similarly, "Giza 1" bean variety exuded more ethylene than "Romi" bean variety.

Burg and Clagett (1967) and Lieberman and Kunishi (1966) found that methionine increased ethylene production in several fruits. Yang (1969) stated that methionine is converted to ethylene. The C-3 and C-4 of the labelled methionine gives rise to ethylene (Yang, 1969). Up to 80% of the labelled methionine could be converted by apple tissues to ethylene within an hour (Burg, Burg, 1965). Analysis with methionine, bearing ^{14}C in different positions, shows that C-1 goes off as CO_2, the $S-CH_3$ is transferred to acceptors and not lost as volatiles, and the C-3 and C-4 take part in ethylene formation (Burg, Burg, 1969; Burg, Clagett, 1967; Yang, 1969). Ethionine, a powerful antimetabolite of methionine, inhibits ethylene formation in fruit tissues, which further confirms methionine as natural precursor. Experimental studies carried out by Afifi and El-Gindy (1978) showed that ethylene stimulated spore germination of *Fusarium solani* and *Humicola fusco-atra* and inhibited germination of the spores of *Alternaria alterana* and *Trichoderma aureoviride* (Table 1).

3. EFFECT OF PLANT EMANATIONS ON COMPOSITION AND DISTRIBUTION OF RHIZOSPHERIC AND PHYLLOSPHERIC MICROFLORA

The qualitative and quantitative composition of the fungal differs between the rhizosphere of *Phaseolus vulgaris* and control soils, as well as between its phyllosphere and the surrounding atmosphere (Afifi, 1975). The results obtained by the author show that these differences are also pronounced in individual species of the isolated fungi in response to the volatile substances emanated from *Origanum majorana* and *Ocimum basilicum*. The fungal species detected belong to the families Choanephoraceae (*Cunninghamella elegans*), Moniliaceae (*Aspergillus* spp., *Penicillium* spp., and *Verticillium glaucum*), Dematiaceae (*Alternaria tenuis* and *Cochliobolus spicifer*), Mucoraceae (*Rhizopus nigricans*) and Tuberculariaceae (*Fusarium* spp.). The most representative fungus in *Phaseolus vulgaris* rhizosphere was *Fusarium solani* whereas *Aspergillus luchuensis* was the most abundant fungus in control soil. The most representative fungus in *Phaseolus vulgaris* phyllosphere was *Aspergillus flavus*.

Counts of the rhizospheric fungi were made from the soil around the plant root from the soil surface to 10cm depth, where the volatile compounds were expected to have the greatest effect. The inhibitory effect of the volatile substances of *Origanum majorana* on the germination of fungal spores was shown by their frequency and distribution. This effect overcomes the action of root exudates of *Phaseolus vulgaris* except in the case of *Rhizopus nigricans*, where the action of root exudates is clear. The inhibitory effect of volatile substances of *Ocimum basilicum* was clear on most fungi. The decrease of *Aspergillus clavatus*, *Aspergillus terreus* and *Fusarium solani* cannot be explained by the action of emanations from *Ocimum basilicum* nor root exudates of *Phaseolus vulgaris*, but other edaphic factors may play a role in this respect.

The composition and distribution of rhizospheric mycoflora of *Origanum majorana* and *Ocimum basilicum* were affected mostly by the combined action of exudates of their roots and volatile substances emanating from their leaves. The effect of actively growing roots on the formation of the rhizosphere population is particularly marked.

The inhibitory effect of emanations from *Origanum majorana* on the fungal population in the rhizosphere of *Phaseolus vulgaris* was manifested in most cases except for *Penicillium citrinum*, where the emanations seem to have a stimulatory effect on spore germination, accompanied by an increase in fungal population. Emanations of *Ocimum basilicum* inhibited the germination potentialities of spores of *Aspergillus niger*, *Cochliobolus spicifer*, *Cunninghamella elegans*, *Fusarium moniliforme*, *Fusarium semitectum* and *Penicillium citrinum*, leading to their disappearance or decrease in *Phaseolus vulgaris* phyllosphere. On the other hand, spore germination potential and the number of the other phyllospheric fungi, *Alter-*

TABLE 1. Effect of ethylene on spore germination of some soil fungi.

Fungus	Percent spore germination		Percent of the control
	Control	Treated	
F. solani	63.35	95.60	158.40
A. alternata	55.65	53.36	90.49
T. aureoviride	50.53	40.80	80.74
H. fuscoatra	58.70	85.00	144.80

naria tenuis and *Penicillium chermesinum* increased under the influence of volatiles from *Ocimum basilicum*.

The data obtained by Muller (1965) indicated that the bacterial population was larger and fluctuated more in the grass zone than in both the bare zone and zone of *Salvia leucophylla* shrubs. Under the shrubs and grass, approximately 80 percent of the organisms isolated were gram-positive rods, while the rest were largely gram-negative rods and actinomycetes. In the soil of the bare zone, however, over half of the organisms were actinomycetes, and the rest consisted almost exclusively of gram-positive rods and small gram-negative rods. The relative number of gram-negative rods did not fluctuate to any great degree among the three zones. Although a few gram-variable rods were found occasionally, no coccoid forms were discovered.

Of the 44 bacterial isolates that were examined, *Salvia* volatiles inhibited growth in 32, stimulated 5, and had no effect on 7. All of the 11 gram-negative rod types were inhibited. There was no detectable general difference in response to isolates from the three zones. Thirty-six of the isolates were inoculated into nutrients broth containing cineole. With 5 γ of cineole present, 18 of the isolates were inhibited, 8 were stimulated, and 10 were not affected. When the amount of cineole were raised to 10 γ, 8 of the previously unaffected isolates were inhibited, and one was no longer stimulated. At 25 γ, growth of all isolates were inhibited, with 11 not growing at all. Six of eight isolates which responded with increased growth to the presence of minute concentrations of cineole were obtained from soil of the bare zone or that beneath *Salvia* shrubs. The effect of minute quantities of cineole stimulating the growth of some of the isolates from either the bare or shrub zone also suggests an influence of *Salvia* shrubs on the distribution of bacteria. Muller and Muller (1964) tested the presence of six terpenes in the volatiles of leaves of *Salvia leucophylla, S. apiana* and *S. mellifera* shrubs.

They indicated the following as most prominant: α-pinene, camphene, β-pinene, dipentene, cineole and camphor.

The therapeutic and bacteriostatic properties of the substances isolated from *Eucalyptus* plant show phenolic and polyphenolic nature. These compounds exert a definite antibacterial action against many gram-positive bacteria (Gyselinck, 1952). Guenther (1955) noticed that the most favorable time for collecting the leaves of *Eucalyptus glabulus*, for the preparation of the oil, was found from April to September, when the yield of the oil reached about 0.8 percent. The work of Abdulla (1979) confirmed this result. He also showed that the number of phyllospheric fungi of *Eucalyptus* leaves was less than the number of fungi in the surrounding air. The decrease of fungal flora on the leaf surface from December to August, may be attributed to the inhibitory effect of the high concentration of the essential oil content during this period of the year.

4. EFFECT OF PLANT VOLATILES ON SOME PHYSIOLOGICAL PROCESSES DURING SPORE GERMINATION

Afifi and Dowidar (1978) studied the effect of volatile substances extracted from leaves of *Origanum majorana* and *Ocimum basilicum* on the germination and respiration of the spores of some soil fungi i.e. *Fusarium moniliforme, Fusarium oxysporum* var. *orthoceras, Fusarium semitectum, Fusarium solani, Mucor racemosus* and *Trichoderma viride*. *Origanum majorana* volatiles reduced spore germination of *Fusarium moniliforme, F. oxysporum, F. solani* and *Mucor racemosus*, whereas the spores of both *F. semitectum* and *Trichoderma viride* were stimulated to germinate. Under the influence of *Ocimum basilicum* volatiles, the germination potentialities of spores of *F. moniliforme, F. oxysporum* and *F. semitectum* were reduced, whereas spores of *F. solani* and *Mucor racemosus* were stimulated to germinate. *Trichoderma viride* spores were not affected.

In all cases, oxygen uptake of the spores of all fungi increased under the influence of these volatiles in comparison with the control. This increase was greater in case of volatiles from *Ocimum basi-*

licum than in case of volatiles from *Origanum majorana* except in case of *F. semitectum* where the reverse occurred. The drop in respiratory quotient (R.Q.) of all the germinating spores of the fungi used in this study under the influence of the volatiles in comparison with the control suggests the utilization of these volatiles as a source of energy by the spore during germination.

In an attempt to study the role played by the volatile and gaseous exudates of the germinating seeds of some plants i.e. *Pisum sativum*, *Zea mays*, *Prunus armeniaca*, *Cucumis sativus* and *Corchorus olitorius* on spore germination of some fungi in relation to their absorption of some sugars (glucose and ribose) and nitrogen sources (ammonium sulphate and Dl-alanine), Afifi (1977) found that volatiles from germinating seeds of *Z. mays*, *P. armeniaca*, *C. sativus* and *C. olitorius* stimulated the spore germination of *Botryotrichum piluliferum*, whereas volatiles from *P. sativum* seeds inhibited the fungus spore germination. The increase or decrease in spore germination was found to be concomitant with the increases and decreases in the sugar and nitrogen source absorbed.

In *Botrytis cinerea*, it seems that the interaction between the volatile and gaseous exudates of the germinating seeds, on one hand, and the sugar and nitrogen source absorbed by the fungus spores, on the other hand, reflects its effect on fungus spore germination.

By comparing these findings with the results of other authors, Hawker (1950) and Cochrane (1958, 1960) showed that germination of certain fungal spores can be accelerated by external supplies of sugars, amino acids, growth factors, and other substances. The fact that such substances are commonly excreted by plants (Flentje, 1959; Rovira, 1962; Spencer, 1962) may explain the activation of spore germination caused by a number of plants. Cheo and Leach (1950) suggested that spore exposure to dung infusion is believed to increase the permeability of the spore wall, allowing more rapid absorption of water. The work of Hooker et al. (1945) indicated that mustard oils, which are present as glucosides in the roots of certain brassiceae, stimulate spore germination if supplied at low concentrations. At high concentrations, however, they have an inhibitory effect (Mac Farlane, 1952). Afifi (1976) demonstrated that germinating seeds of some plants emanate some volatile and gaseous exudates which act as growth regulators like ethylene. While it is possible that increased spore germination is primarily due to some stimulatory substances, it seems likely that certain nutrients are also involved.

5. THE ROLE PLAYED BY PLANT VOLATILES IN HOST-PARASITE RELATIONSHIPS

The volatile and gaseous exudates of either the germinating seeds or roots of some plants seem to play an important role in host-parasite relationships. A correlation could be established between the susceptibility of *Prunus armeniaca* to *Fusarium* wilt on one hand, and the influence of these volatile and gaseous exudates on *Fusarium solani* conidial germination on the other hand (Afifi, 1977). The volatile and gaseous exudates of the germinating seeds or roots of *Prunus* plant were found to have a stimulatory effect on fungus conidial germination. This stimulatory effect is higher for germinating seeds than for roots, and decreases when the age of the seeds or roots exceeds 10 days.

Qualitative chemical analysis of these volatile and gaseous exudates indicates the presence of benzaldehyde, ethanol, acetaldehyde, ethylene in addition to carbon dioxide.

The role played by volatile and gaseous exudates of either the germinating seeds or roots of okra plant (*Hibiscus esculentus*) is deciding the extent of either susceptibility or resistance of the two okra varieties; "Kafr El-Dawar" and "Gold Coast" to four *Fusarium* spp.: *F. oxysporum*, *F. moniliforme*, *F. solani* and *F. semitectum* was studied (Afifi, 1976). This physiological factor may operate during the early pre-penetration phase of host-parasite relationship by production of certain stimulatory or inhibitory volatile and gaseous exudates for the fungal conidial germination. The relative pathogenicities of the *Fusarium* spp. to the two okra varieties were tested. "Kafr El-Dawar" variety was found to be susceptible to the four experimental *Fusarium* spp. *Fusarium oxysporum* was found to be highly pathogenic to this variety, followed by *F. solani* and *F. semitectum*, and the least pathogenic *Fusarium* sp. was *F. moniliforme*. On the other hand, the resistance of "Gold coast" okra variety was broken by *F. moniliforme* only.

A correlation was observed between the pathogenic potentialities of the experimental *Fusarium* spp. and the volatile and gaseous exudates of either the germinating seeds or roots of the two okra varieties. It was found that the volatile and gaseous exudates of either the germinating seeds or roots of the susceptible "Kafr El-Dawar" okra variety stimulated the conidial germination of the four *Fusarium* spp., but the degree of stimulation did not run parallel to the relative susceptibility of this variety to these *Fusarium* spp. The volatile and gaseous exudates of either the germinating seeds or roots of resistant "Gold coast" okra variety inhibited the conidial germination of *F. oxysporum*, *F. solani* and *F. semitectum*. These species cannot attack this okra variety, whereas the conidial germination of *F. moniliforme*, which attack this variety, was stimulated. The maximum stimulatory effect of these exudates was found to be approximately at the end of the incubation period of the pathogenic fungus. It has been noticed that cotton-*Fusarium* wilt incidence increased in fields where *Artemisia herba-alba* plants were grown. The pathogenic potentialities of three *Fusarium* species for inducing *Fusarium* wilt of cotton were tested in presence of *Artemisia herba-alba* association (A (Afifi, 1976). The experimental *Fusarium* spp. were isolated from soil cultivated with cotton plants and identified as *F. moniliforme*, *F. oxysporum* and *F. solani*. The possible role played by the associative effect of *Artemisia* on cotton plant, in influencing the host-parasite relationship, has been elucidated by carrying out soil inoculation experiments, in presence of *Artemisia* plants

or its fragments. Table 2 shows that *F. moniliforme* caused the highest infection percentage to cotton plant, followed by *F. oxysporum*. The percentage infection increased when cotton plants were cultivated in presence of *Artemisia* plants. Under these conditions *F. solani* seems to cause infection also. When *Artemisia* fragments were used as a source of volatiles, the percentage infection increased in *F. moniliforme* with respect to the control, but less than in presence of whole *Artemisia* plants. A comparison was made between treated cotton plants and control plants to evaluate the host growth vigour as influenced by the association of *Artemisia* and as manifested by host growth criteria. Table 3 shows that the association of either the whole *Artemisia* plant or its fragments decreased in general the host vigour in comparison with the control.

The stimulatory effect of both volatiles and extract of *Artemisia* plant on fungal spore germination seems to have a strong influence on the extent of pathogenicity. Table 4 shows that both *Artemisia* volatiles and extract increased the germination potentialities of *Fusarium* spores. The germination potentialities followed the following descending order: *F. moniliforne*, *F. oxysporum*, then *F. solani*. At the same time, when the root exudates of either cotton or *Artemisia* were used singly, the process of spore germination was stimulated, but a combination of root exudates of both plants increased this stimulatory effect (Table 5). Pathogenicity of the experimental *Fusarium* spp. in presence of *Artemisia* plants was found to be more than that induced by them in control soil. This may be attributed to the effects of volatiles and root exudates of *Artemisia* plants, during the pre-penetration phase, in establishing conditions that reduce host vigor and stimulate pathogen spore germination and parasitic potentiality.

El-Gindy (1981) analysed the volatile exudates emanated from mint leaves by gas liquid chromatography technique. His results show that infection of mint leaves by *Pestalotiopsis* caused about 25 percent reduction in the quantity of the essential oils extracted (Table 6). The analysis was carried

TABLE 2. Percentage of infection of cotton plants with the experimental *Fusarium* spp.

Treatment	S p e c i e s		
	F. moniliforme	*F. oxysporum*	*F. solani*
Control	35	5	-
Artemisia fragments	45	10	-
In *Artemisia* field	55	10	5

TABLE 3. Effect of *Artemisia* plants and fragments on cotton growth after inoculation with *Fusarium* spp.

Host growth criteria	Plant	Cotton growth under different inoculation treatments			
		Uninoculated	*F. moniliforme*	*F. oxysporum*	*F. solani*
Length of root (cm)	Control	11.7±0.3	13.5±0.4	15.9±0.8	13.4±0.7
	Artemisia fragments	9.4±0.4	12.0±0.5	13.0±0.6	11.8±0.6
	Artemisia field	9.2±0.4	11.6±0.5	12.7±0.4	12.0±0.4
Length of shoot (cm)	Control	14.2±0.5	15.0±0.7	13.9±0.8	15.3±0.5
	Artemisia fragments	13.5±0.5	12.5±0.6	13.8±0.5	14.5±0.9
	Artemisia field	12.8±0.5	12.4±0.5	13.4±0.6	31.6±0.8
Fresh weight (mg)	Control	2905±210	2167±200	2912±290	2884±180
	Artemisia fragments	2215±150	2245±120	2785±150	2375±120
	Artemisia field	2060±120	2130±120	2540±160	2130±110
Dry weight (mg)	Control	338±30	356±25	385±25	454±35
	Artemisia fragments	294±30	336±20	334±20	330±30
	Artemisia field	220±20	335±20	315±20	325±25

TABLE 4. Effect of *Artemisia* volatiles and extract on spore germination of *Fusarium* spp.

Fungus	Item		*Artemisia* volatiles	*Artemisia* extract
F. moniliforme	Percent germination	Control	44	51
		Treated	69	65
	Percent difference		+57	+27
F. oxysporum	Percent germination	Control	49	46
		Treated	55	53
	Percent difference		+12	+15
F. solani	Percent germination	Control	52	48
		Treated	54	50
	Percent difference		+ 4	+ 4

TABLE 5. Effect of root exudates of cotton and *Artemisia* plant, either singly or in combination, on spore germination of *Fusarium* spp.

Fungus	Item		Single exudates		Exudates in combination
			Cotton	*Artemisia*	
F. moniliforme	Percent germination	Control	44	52	50
		Treated	59	65	73
	Percent difference		+34	+25	+46
F. oxysporum	Percent germination	Control	52	53	48
		Treated	58	62	64
	Percent difference		+ 8	+17	+33
F. solani	Percent germination	Control	55	47	51
		Treated	60	61	67
	Percent difference		+ 5	+30	+31

TABLE 6. Volatile oil content of healthy and *Pestalotiopsis*-infected mint leaves.

Mint species	Volatile oil content (ml per 100 gm fresh wt. of leaves)
Mentha piperita	
Healthy	0.75
Infected	0.56
Mentha viridis	
Healthy	0.70
Infected	0.53

out for two species of mint: *Mentha piperita* and *Mentha viridis* (Table 7). The volatile oils of both species of mint stimulated spore germination of the fungus (Table 8). Stimulation was more pronounced for volatile oils of *M. piperita*. In addition, the effect of each constituent of the volatile oils on the fungus spore germination was tested singly, with results shown in Table 9.

6. INTERACTION BETWEEN PLANT VOLATILES AND ORGANO_ PHOSPHORUS INSECTICIDES

Another study was carried out by Afifi and Abdulla (1977) on the interaction between the insecticide "thiolane" and volatiles of *Artemisia vulgaris* on its phyllospheric fungi. The results obtained showed that spraying *Artemisia* plants with the field dose of thiolane resulted in a highly significant increase in the fungal number in comparison with the unsprayed plants. The fungal population in the phyllosphere of the plants, either sprayed or unsprayed, as well as in the surrounding atmosphere, differs quantitatively and qualitatively. *Alternaria alternata*, *Aspergillus niger* and *Helminthosporium* sp. were represented in equal amounts in the surrounding atmosphere, whereas the representation of fungi on the surface of the leaves of unsprayed plants was as follows: *Aspergillus flavus* (25%), *Aspergillus niger* (12.5%) and *Synsporium* sp. (62.5%). This disturbance in fungal mycoflora may be attributed to the volatile substances present in *Artemisia* tissues. Fungal representation on the leaf surface sprayed with thiolane was changed to the following form: *Aspergillus flavus* (12.5%), *Aspergillus niger* (50%), *Fusarium solani* (12.5%) and *Penicillium frequantans* (25%).

Concerning the effect of thiolane in the concentration of the field dose (1.88 ml per 250 ml dist. water) on fungal spore germination, it was found that the insecticide increased the percentage of spore germination of *Aspergillus niger*, *Fusarium solani* and *Penicillium frequantans*. These results agreed with those obtained by Verona and Picci (1952) and Naumann (1958). They stated that

organophosphorus insecticides such as pestox, schardane and parathion in concentrations as high as 1000 ppm stimulated soil microorganisms. On the other hand, results obtained from the author's work (1977) showed that the spores of *Alternaria alternata*, *Aspergillus flavus*, *Helminthosporium* sp. and *Synsporium* sp. were inhibited by this insecticide. It was found also that the effect of *Artemisia* volatiles differs according to the fungal species. These volatiles caused stimulation of *Aspergillus niger*, *Fusarium solani* and *Penicillium frequantans*, and inhibition of germination of the spores of *Alternaria alternata*, *Aspergillus flavus* and *Helminthosporium* sp. Spores of *Synsporium* sp. were not affected. With application of a combination of thiolane and *Artemisia* volatiles, the percentage of spore germination of *Alternaria alternata*, *Aspergillus niger* and *Penicillium frequantans* increased more than from each treatment singly. The contrary occurred in *Aspergillus flavus* and *Fusarium solani*. Germination of *Helminthosporium* spores was inhibited under the influence of all experimental treatments. Germination of *Synsporium* spores showed a highly significant decrease with application of the combination of the insecticide and *Artemisia* volatiles and also on applying volatiles singly in comparison with the control.

REFERENCES

Abdulla M El-S (1979) Seasonal variation of the volatile oil of *Eucalyptus* plant and its role on phyllosphere fungi. The Journal of Faculty of Education, Ain Shams University, No 2, 43-52.

Anderson RL and Ordale EJ (1961) CO_2 dependant fermentation of glucose by *Cytophaga succinicans*, J. Bacteriol. 31, 139.

Afifi AF (1975) Effect of volatile substances from species of Labiatae on rhizospheric and phyllospheric fungi of *Phaseolus vulgaris*, Phytopath. Z. 83, 296-302.

Afifi AF (1976) Volatile and gaseous exudates of either germinating seeds or roots of okra plant in relation to okra wilt desease, caused by some *Fusarium* species, Zbl. Bakt. Abt. II, Bd. 131, S. 555-564.

Afifi AF (1976) Cotton-*Fusarium* wilt-*Artemisia* relationship, Egypt. J. Bot. 19(2-3), 187-193.

Afifi AF (1977) Effect of volatile and gaseous exudates of germinating seeds of some plants on the germinative potentialities of some fungal

TABLE 7. Relative percentage of volatile oils detected in both healthy and *Pestalotiopsis*-infected mint leaves.

Volatile oils	Relative percentage of volatile oils			
	Mentha piperita		*Mentha viridis*	
	Healthy	Infected	Healthy	Infected
Camphene	-	-	1.69	1.98
α-pinene	1.46	2.80	-	-
β-pinene	2.93	4.45	2.57	2.47
Myrcene	61.03	18.00	-	-
Limonene	0.07	0.16	19.56	22.94
Terpinolene	-	-	0.05	0.85
Caryophyllene	-	-	0.02	0.04
Cineole	15.60	21.10	0.93	0.42
p-cymene	0.83	0.03	-	-
Linalol	25.60	23.76	-	-
Menthone	11.40	14.20	1.65	13.17
Farnesol	-	-	26.38	7.79
Menthol	13.60	13.10	-	-
Carvone	-	-	42.20	49.57
Menthofuran	0.02	0.82	-	-
Menthylacetate	1.46	3.90	-	-
Dihydrocarvone	-	-	3.29	0.21
Terpineol	0.66	0.19	0.69	0.04
Citronellol	0.04	0.13	0.15	0.04
Pulegone	0.39	0.13	0.72	0.04
Geraniol	0.88	0.49	-	-
Piperitone	5.81	0.82	-	-
Eugenol	0.98	0.13	-	-
Unidentified oil	-	-	0.05	0.42

TABLE 8. Effect of volatile oils extracted from mint plant on *Pestalotiopsis* spore germination.

Volatile oil	Volatile oil concentration (ml)			
	0.02	0.05	0.10	0.50
	Spore germination as percentages of the control			
Mentha piperita	125.60	155.30	150.60	125.60
Mentha viridis	105.00	116.50	110.50	105.50

TABLE 9. Effect of each volatile oil detected in both mint species on *Pestalotiopsis* spore germination.

Volatile oil	Volatile oil concentration (ml or mg)				
	0.02	0.05	0.10	0.50	
	Spore germination as percentages of the control				
Camphene	150.6	157.6	170.8	190.3	
α-pinene	155.5	160.8	180.6	210.8	
β-pinene	145.0	160.9	165.9	176.6	
Myrcene	165.3	185.3	195.5	220.3	
Limonene	128.6	136.3	120.5	110.5	
Terpinolene	120.6	120.6	135.8	140.8	
Caryophyllene	165.8	170.8	185.3	205.3	
Cineole	130.6	120.6	115.5	92.8	
p-cymene	160.8	180.9	205.3	215.3	
Linalol	60.3	50.6	30.4	30.4	
Menthone	148.8	155.6	125.3	110.5	
Farnesol	130.6	121.2	110.5	110.5	
Menthol	40.6	30.3	30.5	20.6	
Carvone	110.8	135.5	140.5	156.7	
Menthofuran	145.3	160.8	175.8	180.6	
Menthyllacetate	110.8	106.5	100.6	98.6	
Dihydrocarvone	110.6	105.6	100.6	98.5	
Terpineol	50.5	40.6	30.7	20.5	
Citronellol	60.5	50.9	45.8	45.8	
Pulegone	30.8	20.3	20.3	20.3	
Geraniol	35.3	30.5	32.3	28.7	
Piperitone	45.3	40.6	28.6	25.3	
Eugenol	30.5	20.6	20.4	15.9	

spores in relation to their ability of absorption of some sugars and nitrogen sources, Zbl. Bakt. Abt. II, Bd. 132, S. 308-316.

Afifi AF (1977) *Fusarium* wilt of *Prunus armeniaca* seedlings, Zbl. Bakt. Abt. II, Bd. 132, S. 184-188.

Afifi AF and Abdulla ME (1977) Effect of the insecticide thiolane on the spore germinating potentialities of *Artemisia vulgaris* phyllospheric fungi, Egypt. J. Bot. 20 (2), 121-126.

Afifi AF and Abdulla ME (1978) The role of gaseous and volatile root metabolites of *Lupinus termis* on the spore germination of rhizospheric mycoflora, Indian Phytopath. 31(3), 334-338.

Afifi AF and Dowidar AE (1976) Effect of volatile materials produced by some members of Labiatae on spore germination and spore respiration of some soil fungi, Egypt. J. Physiological Sci. 3 (1,2), 81-92.

Afifi AF, Ebid MF and El-Gindy AA (1977) Ethylene production during seed germination of some plants, Zbl. Bakt. Abt. II, Bd. 132, S. 673-676.

Afifi AF and El-Gindy AA (1978) Effect of volatile and gaseous exudates of both germinating seeds and roots of some crops on spore germination of some soil fungi, J. Indian Bot. Soc. 57, 266-271.

Barinova SA (1960) Growth stimulation of moulds by carbon dioxide with reference to the effect of mesotartaric acid upon this process, Mikrobiologiya 29, 161.

Barinova SA (1961) The significance of carbon dioxide for growth of moulds (in Russian), Izv AN SSSR, Ser. biol. 4, 561.

Burg SP and Burg EA (1965) Ethylene action and the ripening of fruits, Science 148, 1190-1196.

Burg SP and Clagett CO (1967) Conversion of methionine to ethylene in vegetative tissue and fruits, Biochem. Biophys. Res. Comm. 27, 125-130.

Burges A and Fenton E (1953) The effect of carbon dioxide on the growth of certain fungi, Trans. Brit. Mycol. Soc. 36, 104.

Buston HW, Moss MO and Tyrrell D (1966) The influence of carbon dioxide on growth and sporulation of *Chaetomium globosum*, Trans. Brit. Mycol. Soc. 49, 387.

Čatská V, Afifi AF and Vančurá V (1975) The influence of volatile and gaseous exudates of swelling seeds on germination of fungal spores, Folia Microbiol. 20, 152-156.

Cochrane VW (1958) Physiology of fungi, New York, Wiley.

Cochrane VW (1960) Spore germination, Plant Pathol. 2, 167-202.

Cheo PC and Leach JG (1950) The stimulatory effect of dung infusion on the germination of spores of *Ustilago striiformis*, Phytopath. 40, 584-589.

Coley-Smith JR and Cooke RC (1971) Survival and germination of fungal sclerotia, Ann. Rev. Phytopath. 9, 65-92.

Doireau MP (1972) Utilization metabolique de l'ethanol forme au cours de phenomenes fermentaire, lors de la germination du Haricot, C. R. Acad. Sci. Paris 275D, 907.

Dubrin RD (1959) Factors affecting the vertical distribution of *Rhizoctonia solani* with reference to CO_2 concentration, Amer. J. Bot. 56, 22-25.

Eaves CA and Lockhart CL (1961) Storage of tomato in artificial atmospheres using the calcium hydroxide absorption method, J. Hort. Sci. 36, 85.

El-Gindy AA (1981) Studies on *Pestalotiopsis* leaf spot disease of mint, Ph.d.Thesis, Ain Shams Univ., Cairo, Egypt, p. 47-59.

Flentje NT (1959) Plant Pathol. Probl. Progr. 1908-1958.

Glasare P (1970) Volatile compounds from *Pinus silvestris* stimulating the growth of wood rotting fungi, Arch. Microbiol. 72, 333.

Griffin DN (1966) Soil physical factors and the ecology of fungi. IV. Influence of the soil atmosphere, Trans. Brit. Mycol. Soc. 49, 115.

Grinova GM (1964) Accumulation and secretion of alcohols by plant roots during an insufficient supply of oxygen (in Russian) Dokl. AN SSSR. 156, 1225.

Guenther E (1955) The essential oils, London, D. Nostrand Co. Inc.

Gundersen K (1961) Growth of *Phomes annosus* under reduced oxygen pressure and the effect of CO_2, Nature 190, 649.

Gunner HB and Alexander M (1964) Anaerobic growth of *Fusarium oxysporum*, J. Bacteriol. 37, 1309.

Gyselinch A (1952) Agricultural fungicides, Belg. Sto. 744.

Hawker LE (1950) Physiology of fungi. University of London Press.

Hodgkiss IJ and Harvey R (1972) Effect of CO_2 on the growth and sporulation of certain coprophilous Pyrenomycetes, Trans. Brit. Mycol. Soc. 59, 409.

Hollis JP (1948) Oxygen and carbon dioxide relations of *Fusarium oxysporum* Schlecht. and *Fusarium uemertii* Carp., Phytopath. 38, 761.

Holm RE (1972) Volatile metabolites controlling germination in buried weed seeds, Plant Physiol. 50, 293.

Hooker WJ, Walker JC and Link KP (1945) Effect of two mustard oils on *Plasmodiophora brassicae* and their relation to resistance of clubroot, J. Agr. Res. 70, 63-78.

Julian JB and Phillips WR (1963) Note on the effect of CO_2 and O_2 mixtures on the growth of apple scab cultures, Can. J. Plant Sci. 43, 227.

Lieberman M, Kunishi AT, Mapson BW and Wardale DA (1966) Stimulation of ethylene production in apple tissue slices by methionine, Plant Physiol. 41, 376-380.

Lockhart CL (1970) Suppression by ethylene of *Fusarium oxysporum* in culture and rots of tomato in controlled atmosphere storage, Can. J. Plant Sci. 50, 347.

Lockhart CL, Forsyth FR and Eaves CA (1968) Effect of ethylene on development of *Gloeosporium album* in apple and on growth of the fungus in culture, Can. J. Plant Sci. 48, 557.

Louvet J and Bulit J (1964) Researches sur l'ecologie des champignons parasites dans le sol. I. Action du gas carbonique sur la croissance et l'activité parasitaire de *Sclerotinia minor* et de

Fusarium oxysporum F. *Melonis*, Ann. Epiphyt. 15,21.

Macaulay BJ and Griffin DM (1969) Effect of CO_2 and O_2 on the activity of some soil fungi, Trans. Brit. Mycol. Soc. 53, 53.

Mitchell DJ and Mitchell JE (1973) Oxygen and carbon dioxide concentration effects on the growth and reproduction of *Aphanomyces eutoiches* and certain other soil-borne plant pathogens, Phytopathology 63, 1053-1059.

Mitchell DJ and Zentmyer GA (1971) Effects of oxygen and carbon dioxide tensions on growth of several species of *Phytophthora*, Phytopathology 61, 787.

Muller WH (1965) Volatile materials produced by *Salvia leucophylla*. Effect on seedling growth and soil bacteria, Bot. Gaz. 126, 195.

Muller WH and Muller CH (1964) Volatile growth inhibitors produced by *Salvia* spp., Bull. Torrey Bot. Club 91, 327.

Naire NG and Fahy PC (1972) Bacteria antagonistic to *Pseudomonas tolasii* and their control of brown blotch of the cultivated mushroom *Agaricus bisporus*. I. Appl. Bact. 35, 4396442.

Naumann K (1959) Effect of pesticides on the soil microflora. "Mitt. Biol. Bundesanstalt Land, Forst. Wirtsch." Berlin-Dahlem 97.

Newcombe M (1960) Some effects of water and anaerobic conditions on *Fusarium oxysporum* f. *cubense* in soil, Trans. Brit. Mycol. Soc. 43, 51.

Papavizas GC and Davey CR (1962) A mycostatic role for CO_2 in the suppression of *Rhizoctonia* in the soil, Phytopathology 52, 165.

Platz GA, Durell LW and Howe MF (1927) Effect of CO_2 upon the germination of chlamydospores of *Ustilago zeae* (Beckm.), Unger. J. Agr. 34, 137.

Rovira AD (1962) Plant root exudates in relation to the rhizosphere microflora, Soils and Fert. 25, 167-172.

Smith AM (1973) Ethylene: A cause of fungistasis, Nature 246, 311.

Spencer DM (1962) Antibiotics in seeds and seedling plants. In Spencer DM, ed. Antibiotics in Agriculture, pp. 125-145.

Stanek M (1963) Germination of chlamydospores of the fungus *Ustilago zeae* (Beckm) in rhizosphere of maize (in Chech.), Rostlinna Vyroba 7-8, 721.

Toler RW, Dukes PD and Jenkins SF (1966) Growth response of *Fusarium oxysporum* f. *tracheiphilum* in vitro to varying O_2 and CO_2 tensions, Phytopathology 56, 183.

Vančurá V and Stotzky G (1971) Excretion of germinating plant seed, Folia Microbiol. 16, 512.

Veroná D and Picci G (1952) The effect of systematic insecticides on microflora of soil, Soil and Fert. 16, 36.

Yang SF (1969) Biosynthesis of ethylene. In Yang SF, ed. Biochemistry and Physiology of plant growth substances, pp 1217-1218, Ottawa, Runge Press.

MOSS GROWTH AND DEVELOPMENT IS FACILITATED BY NATURAL BACTERIAL FLORA

L.D. SPIESS, B.B. LIPPINCOTT, J.A. LIPPINCOTT (Northwestern University, Evanston, IL 60201 U.S.A.)

Optimal conditions for growth, development or reproduction in many organisms often depend on association with a second unrelated organism in either a parasitic, mutualistic or protocooperative relationship. Several examples of such natural relationships between plants and a second organism which allows them to adapt and survive in a particular environment have been described. The lichen results from a symbiotic association in which the algal cells supply photosynthates to the fungus and the fungus absorbs moisture and shields the algae from intense light, allowing both to exist together in an otherwise unfavorable environment (Ahmadjian 1967). Rhizobia aid legume growth through symbiotic nitrogen fixation (Nester, Kosuge 1981). Spirillum lipoferum (Smith et al. 1976) and Azospirillum brasilene (Kapulnik et al. 1981) function in a similar manner in association with grasses. Pseudomonas interacts with plant roots, inducing a mechanism for choline sulfate uptake (Nissen, 1973). Beggiatoa oxidizes sulfides released by Desulfovibrio which are toxic to rice plants, while peroxides produced by Beggiatoa are decomposed by rice root catalase (Pitts et al. 1972). Disease suppression and increased growth and productivity of many plants is influenced by soil rhizobacteria (Burr et al. 1978; Schroth, Hancock 1982).

Although moss physiology, development and ecology have not been studied to the extent of that of higher plants, a few examples of positive cooperation between mosses and other organisms in their immediate environment have been demonstrated. The algae, Anabaena variabilis and Nostoc muscorum, growing in association with the moss, Funaria hygrometrica, cause rapid growth of the moss and an increase in the number of gametophores while Funaria in turn has a growth promoting effect on cyanobacteria (Rodgers, Henriksson 1976). Cyanobacteria as epiphytes or intracellular symbionts of mosses are the dominant nitrogen fixing organisms in the Arctic (Jordan et al. 1978). Fixation and uptake of nitrogen in Sphagnum is increased by a blue-green algal association (Brasiler 1980). Chlorochytrium, a green alga, has been found to be endophytic in moss (Reese 1981). Various fungal-moss associations have also been documented (Chopra et al. 1978; Parke, Lenderman 1980; Rabitin 1980; Grasso, Scheirer 1981; Redhead 1981), although details of their benefit to the moss remain to be ascertained. Scheirer and Dolan (personal communication) identified a unicellular green alga (Chlorella ellipsoidea), a fungus (Trichoderma viride), and a rod shaped bacterium as a consistent part of the epiflora of the moss Polytrichum commune. Again, the significance of this association with the moss is unknown.

We have found that several species of soil bacteria belonging to the genus Agrobacterium and Rhizobium promote growth and either gametophore or abnormal bud development in the moss Pylaisiella selwynii (Spiess et al. 1971, 1977a,b, 1981b). We have also shown that bacteria found associated in nature with two other species of moss will promote growth and either gametophores or abnormal bud development. Attempts to identify these promotive strains have not been successful to date, but they appear most similar to either avirulent agrobacteria or Pseudomonas. In this paper we review our investigations on the effects of Agrobacterium, Rhizobium and natural moss-associated bacteria (MAB) on moss development and consider some of the bacterial products that might account for these effects.

Response of mosses in axenic culture to bacteria --- Specificity

The moss Pylaisiella selwynii produces buds that

develop into callus, abnormal, or normal gametophores in response to strains of A. tumefaciens
which are tumorigenic on higher plants, while
strains which are non-infectious (e.g., A. radio-
bacter) are less effective (Spiess et al. 1977a,
1977b, 1981a). A. rubi, another tumorigenic
species, induces many abnormal gametophores and
callus. A. rhizogenes, the causal agent of the
hairy root disease, has little effect on gametophore development but promotes the elongation of
moss filament cells with loss of chloroplasts,
resulting in the conversion of branches and
filaments into rhizoid-like structures. A marked
increase in normal gametophores also occurs when
A. tumefaciens is added to the mosses Entodon
seductrix or Heterophyllum haldaneanum but it
has no effect on the development of Atrichum un-
dulatum, Thuidium delicatum, Climacium americanum,
Funaria hygrometrica or Polytrichum commune
(Spiess 1976). Protonemal growth of all of these
mosses, however, is promoted by agrobacteria
(Spiess et al. 1981b, unpublished). Several
species of Rhizobium are also effective in inducing gametophores on Pylaisiella. The witches
broom pathogen, Corynebacterium fasciens, causes
bud-callus formation (Spiess et al. 1977a) and
like Agrobacterium and Rhizobium is a common
component of the soil microflora.

Bacteria were isolated from Pylaisiella, Polytrichum and Funaria collected at several locations. The moss were thoroughly washed with
sterile distilled water, fragmented and plated
on either media reported to be selective for
agrobacteria or an enriched medium. All of these
MAB proved to be short, motile, gram negative
rods (Spiess et al. 1981b) producing yellow,
creamy or white colonies on nutrient broth media.
Similar bacteria have been found on Polytrichum
by Scheirer and associates (personal communication). The taxonomic identity of MAB isolates
is not clear although their size, motility,
growth on selective media and colony characteristics are generally similar to those of agro-

bacteria and Pseudomonas. Physiological and biochemical tests that are indisputably indicative of
Agrobacterium were negative for all MAB, although
many agrobacteria also give negative results in all
of these same tests (Spiess et al. 1981b). Several
MAB isolates compete with Agrobacterium for binding
sites essential for tumorigenesis on beans whereas
most other species of bacteria tested do not (Lippincott, Lippincott 1969). Further tests will be
required to establish the identity of the MAB.

When MAB isolates were added to sterile cultures of
Pylaisiella, Funaria or Polytrichum either on agar
or in liquid culture medium, increased protonemal
growth of at least one moss resulted. Many MAB,
however, also induce development of caulonemata,
formation of callus, or an increase in number of
either normal or abnormal gametophores or callus on
chloronemata. Only 3 of 32 strains tested were
found to induce normal gametophores on all three
species of moss. Eleven of twelve MAB isolates
from Pylaisiella, ten of eleven from Polytrichum
and eight of nine from Funaria induce normal
gametophores when tested on their homologous moss.
Less than half of these isolates show similar
effects on non-homologous mosses, suggesting a
specific relationship exists in nature between a
particular species of moss and its associated
bacteria (Fig. 1).

Physiological effects of bacteria on moss
The specificity of the moss-bacterium interaction
may be due to a particular substance(s) the bacterium produces. Since different mosses may vary
in the kinds of compounds they respond to, this
could account for the apparent specificity of
particular mosses for their own unique flora. The
three mosses studied in detail do exhibit different
growth and developmental patterns. Polytrichum
buds develop at the center of the plant whereas
those of Funaria and Pylaisiella form further out
on the filaments. Under our laboratory conditions
(Spiess et al. in press), Polytrichum and Funaria
grow optimally on Bold's medium, pH 6.5, (Bold 1967)

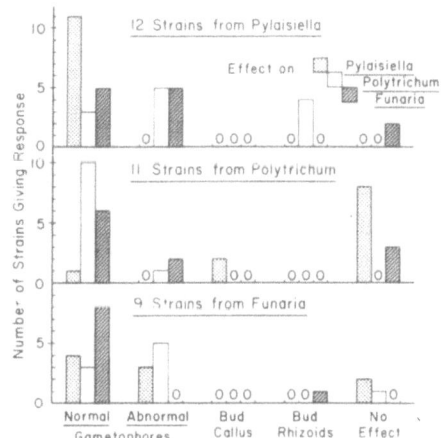

FIGURE 1. Developmental changes produced on the
mosses Pylaisiella, Polytrichum and Funaria by
bacteria isolated from each moss in nature. In
each set of bacterial isolates more were effective in inducing normal gametophores on the
original host moss than on the other two mosses,
indicating that a selective association between
moss and bacteria occurs in nature.

with most responding plants forming gametophores
within 17-19 days. Pylaisiella grows best on a
modified Nebel and Naylor medium at pH 5.0 (Spiess
et al. 1971) and only a very few plants form
gametophores under control conditions by 28 days
after spore germination.

Plant hormones

The role of hormones in moss development has been
reviewed by Bopp (1980) and Richardson (1981).
In general, cytokinins and auxins appear to have
effects on moss growth and development similar
to those observed in higher plants. These hormones affect protonema differentiation (chloronema, caulonema), bud initiation and organization and development of the leafy gametophore
and of rhizoids.

Spiess (1975, 1976) and Spiess et al. (1971, 1972,
1973) have described the normal stages in development of Pylaisiella and the growth of abnormal
structures resulting from treatment of plants
with auxin and cytokinins. Vitamin B_{12} increases
early production of normal gametophores but cyto-

kinin and auxin in various proportions, while affecting bud formation, do not permit normal gametophore development (Spiess et al. 1973). Pylaisiella
responds to all cytokinins tested including zeatin,
ribosyl-zeatin and N-6-γ γ-dimethylallylaminopurine
(i^6Ade) (Spiess et al. 1973; Spiess 1976). However,
Funaria and Polytrichum fail to respond to ribosyl-
zeatin treatment but do respond to the free base
(Whitaker, Kende 1974; Spiess 1975, 1976). Funaria
is more sensitive to auxin than the other mosses,
the growth of the plant being stunted at 1 uM
auxin (Table 1).

Free living bacteria (Pseudomonas, Azotobacter
and Bacillus) found in association with plants
produce plant growth hormones (Brown 1974). The
cytokinins, i^6Ade and ribosyl-zeatin, are synthesized by A. tumefaciens (Nester, Kosuge 1981).
Rhizobium synthesizes zeatin and ribosyl-zeatin
and Corynebacterium fasciens produces ribosyl-
zeatin (Nester, Kosuge 1981). Agrobacteria synthesize auxin (Nester, Kosuge 1981) as do the rhizobia
(Triplett et al. 1981). Many bacteria including
rhizobia require and/or produce vitamin B_{12} (Cowles
et al. 1969) as do most of the blue-green algae and
some green algae (Evans, Kliewer 1963).

Amino acids

Spiess et al. (1972, 1981a) showed that octopine
and lysopine (unusual N-substituted derivatives of
arginine and lysine that are found in crown gall
tumors induced by A. tumefaciens) greatly increase
the number of gametophores on Pylaisiella. These
develop normally, unlike the callus structures
induced by cytokinin. An increase in protonemal
growth can be demonstrated with octopine at concentrations as low as 0.001 uM. When added with
cytokinin at concentrations greater than 0.1 uM,
octopine increases the amount of bud-callus produced and decreases the time to callus appearance.
With 0.01 uM cytokinin and octopine (1 to 100 uM)
Pylaisiella forms clusters of normal gametophores
which are similar to those obtained by adding A.

TABLE 1. Comparative effect of hormones and certain amino acids on the development of three species of moss

Substances added[a]	Pylaisiella		Polytrichum		Funaria	
	First appear (days)	Mean no. structures per plant[b]	First appear (days)	Mean no. structures per plant[b]	First appear (days)	Mean no. structures per plant[b]
None		None	14	0.3 N[c]	17	0.5 N[c]
Auxin	28	0.03 N[c]	17	0.1 N	stunts plants	
Zeatin	19	2.2 C	10	1.0 C	10	5.0 C
Octopine	24	0.6 N	12	0.5 N	14	1.5 N
Glutamate		None	14	0.7 N		None
Zeatin + auxin	17	2.4 C	10	1.0 Ab, N	10	2.0 C
Zeatin + octopine	17	3.4 C	10	1.0 N	10	5.0 C
Zeatin + glutamate		None	10	1.0 N		None
Zeatin + glut. + auxin	21	1.3 C	14	0.5 Ab		None
Zeatin + glut. + oct.	26	1.5 C	10	1.0 C	17	0.5 C

[a] Substances added at the time spores were placed in culture. Concentrations: auxin (IAA), 1 uM; zeatin, 1 uM; other compounds, 100 uM. Abbreviations: oct. = octopine; glut. = glutamate.
[b] Mean no. of gametophores or bud callus structures per 30 plants scored. Pylaisiella scored 28 days after sowing; Polytrichum at 17 days; and Funaria at 17 days.
[c] Abbreviations: N = normal gametophores; Ab = abnormal gametophores; C = bud-callus.

tumefaciens to moss. Octopine has a slight effect in increasing and accelerating gametophore formation on Funaria and Polytrichum but no synergistic effect with cytokinin (Table 1). Octopine is produced in tumors induced by octopine-type strains of Agrobacterium and promotes conjugational plasmid transfer in the bacteria (Nester, Kosuge 1981).

Glutamate alters the growth and branching of Pylaisiella and Funaria and inhibits cytokinin induced bud-callus formation. This effect of glutamate appears to be specific as it is not duplicated by other amino acids or Krebs cycle intermediates except aspartate (Spiess unpublished). The glutamate inhibition is partially prevented by octopine or auxin in Pylaisiella but octopine is less effective in this respect and auxin ineffective on Funaria. Glutamate also prevents normal development of gametophores on Polytrichum in the presence of auxin and zeatin but has no effect with zeatin alone (Table 1).

Bacteria release amino acids and other organic molecules to the culture media and may elaborate compounds from substances released by the moss. The bacteria may also induce the moss to produce compounds that are atypical or in normally low supply, such as octopine and lysopine which are produced by plants infected with Agrobacterium. A chemical analysis of the medium of bacteria with moss remains to be done.

Cyclic nucleotides

Although the presence and role of nucleotides such as cAMP in higher plants is questionable, Handa and Johri (1979) identified a factor from Funaria which is indistinguishable from 3':5'-cyclic adenosine monophosphate (cAMP) and further showed that differentiation of chloronema cells (which have 4 to 7 times as much cAMP as caulonema) is affected by cAMP. Uptake and degradation of cAMP by Funaria cells has been demonstrated (Sharma, Johri 1982). Spiess (1979) demonstrated that cyclic nucleotides and cAMP phosphodiesterase inhibitors reduce callus formation on Pylaisiella when added with cytokinin to the moss.

cAMP is produced by many bacteria (DeCrombrugge et al. 1969), fungi such as Dictyostelium (Konijin et al. 1969) and yeasts (Takai et al. 1974), and algae such as Laminaria (Becker, Ziegler 1973) and Chlamydomonas (Amrhein, Filner 1973). cAMP causes cell differentiation, morphogenesis or changes in growth patterns in these organisms. Adenyl cyclase and cAMP phosphodiesterase activity have been detected in extracts of Agrobacterium (Rutherford et al. 1976)

Bacterial contact with plant cells

Contact between the moss and the bacterial cell is necessary in the case of Agrobacterium, Rhizobium and most MAB isolates before they can induce promotion of gametophores by Pylaisiella (Spiess et al. 1976; 1981b). Lipopolysaccharide (LPS) from cell envelopes of virulent strains of A. tumefaciens block the effect of intact virulent bacteria on Pylaisiella while LPS from strains which do not adhere to higher plants are not inhibitory (Whatley, Spiess 1977). Similarly, LPS from Rhizobium reduces the number of gametophores when added to moss along with viable Rhizobium.

Lippincott and Lippincott (1977) reported that a pectic component of middle lamellar origin in dicots may constitute the attachment site for agrobacteria in the initiation of tumors on higher plants. Spiess et al. (in press) found that protonemal cell walls of Pylaisiella have a thick surface layer which, based on cytological tests, is similar to pectin. Electron microscope pictures show agrobacteria or MAB associated with or imbedded in the moss cell wall (Spiess et al. 1977b; unpublished).

Polygalacturonic acid (1 mg/ml) added to virulent agrobacteria prior to inoculation on Pylaisiella reduces the effectiveness of bacteria to induce gametophores by 80%, presumably by preventing

contact of bacteria with the plant cell wall. Cell walls from mature pinto bean leaves or other dicots have a similar effect but walls from embryonic beans or monocots are ineffective in both moss and pinto bean tests with agrobacteria. Cell wall fragments from Pylaisiella protonema also prevent agrobacteria from inducing developmental changes on Pylaisiella but cell walls from mature Pylaisiella or protonemal or mature Polytrichum are less effective. If Polytrichum cell walls are treated with pectinesterase prior to inoculation with agrobacteria, they become inhibitory, presumably because polygalacturonic acid sites for bacterial adherence have been produced (Lippincott, Lippincott 1977, 1978, Fig. 2). Similar changes are seen when non-inhibitory higher plant cell walls are treated with pectinesterase and appear to reflect the fact that agrobacteria adhere

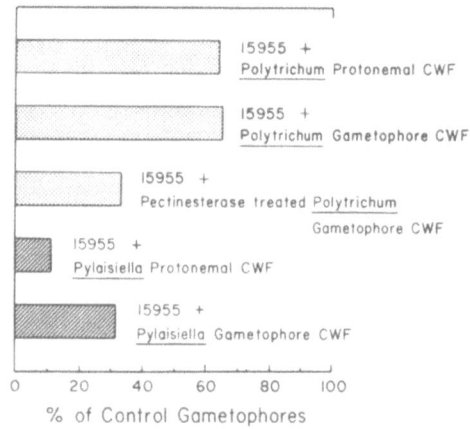

% of Control Gametophores

FIGURE 2. Inhibition of A. tumefaciens (strain 15955 induced gametophore formation on Pylaisiella by moss cell wall fragments (CWF) added to the bacteria immediately before inoculation. Results are expressed relative to the control number of gametophores (= 100%) induced on Pylaisiella by 15955 in the absence of CWF. Bacterial adherence to the CWF rather than to the moss appears responsible for this effect. CWF from Pylaisiella, which responds to this bacterium, are more inhibitory than those from Polytrichum which does not respond. Treatment of Polytrichum with pectinesterase presumably exposes attachment sites for bacteria and preliminary evidence indicates this treatment of the intact moss allows it to respond to the bacterium.

to polygalacturonic acid but not to methylated polygalacturonic acid (Rao et al. in press). Although cell wall components have not been extensively studied in moss, Inoue et al. (1981) found differences in staining of the walls of nine mosses and liverworts due to unknown differences in composition. Inoue et al. (1980) also found differences between juvenile and mature cell walls of mosses. Such differences may account for the fact that agrobacteria associate with the cell walls of protonema but not gametophores (Spiess et al. 1977b).

Physical attachment of bacteria is necessary for Rhizobium induction of nodules on legumes (Purchase, Nutman 1957) and for the bacterial induction of choline sulfate uptake in barley roots (Nissen 1973). The bacterium-plant surface to surface interaction may also initiate changes important for moss transformation much as fungal carbohydrate elicitors induce biochemical changes in higher plants (Albersheim, Anderson-Prouty 1975).

Our evidence indicates that some bacterial moss associations found in nature are specific. Such associations can promote growth of the juvenile stage of the moss and in many cases induce rapid initiation of the reproductive gametophore stage of development. These responses in turn may provide a competitive advantage for the moss in particular environments. Increased protonemal growth in some interactions seems to be the result of substances released or elaborated by the bacteria in the moss environemnt. In other cases, especially when a rapid increase in the number of gametophores occurs, bacteria must first adhere to the moss in a step that appears to involve specific recognition between cell wall polymers of both the bacterium and the plant. The exact nature of this attachment which can determine specificity and its possible role beyond mechanical anchoring of the bacterium to its host are

under investigation. The identification of compounds produced by the MAB and the physiological responsiveness of different mosses to various hormones and organic compounds can provide information about the mechanism of the moss response to active bacteria, specificity, and adaptive significance of the moss-bacterium interaction.

Summary

Interaction of moss with bacteria, algae or fungi may be an important factor in the colonization of an environmental niche by specific species of moss. Watson (1980, 1981) suggests that the juvenile or protonemal period of the life cycle of mosses is critical for the selective establishment of a species of moss and influences the patterns of Bryophyte community diversity. Chemical factors leaked by juveniles may be important in regulating the number of species of mosses that co-exist and the rate of protonemal growth may control the production of such factors (Watson 1980). Fungi and bacteria accelerate growth and development (Chopra et al. 1978; Spiess et al. 1981a). In the case of the bacteria-moss interactions, the natural bacterial flora of the moss seems to be specifically adapted to the promotion of moss development. To establish an adaptive significance for such a relationship, it will be necessary to demonstrate the specificity of these interactions and to identify the physiological processes involved. Continuing studies with the MAB isolates, with new MAB isolates and with products active on moss which these bacteria release into culture media will help to resolve these relationships and their significance.

Acknowledgements

This work was supported by United States National Science Foundation Grant #PCM-77-24792. The authors wish to thank Dr. Malcolm Sargent of the University of Illinois, Urbana, for Funaria spores.

References

Ahamdjian V (1967) The lichen symbiosis. Boston, Mass., Ginn/Blaisdell.

Albersheim P and Anderson-Prouty AJ (1975) Carbohydrates, proteins, cell surfaces, and the biochemistry of pathogenesis, Ann. Rev. Plant Physiol. 26, 31-52.

Amrhein N and Filner P (1973) Adenosine 3':5'-cyclic monophosphate in Chlamydomonas reinhardii: isolation and characterization, Proc. Nat. Acad. Sci. USA 70, 1099-1103.

Becker D and Zeigler H (1973) Cyclic adenosine-3':5'-monophosphate in translocation tissues of plants, Planta 110, 83-89.

Bold HC (1967) A laboratory manual for plant morphology, New York, Harper and Row.

Bopp M (1980) The hormonal regulation of morphogenesis in mosses. In Skoog F, ed. Plant growth substances, 1979, pp. 351-369. New York, Springer-Verlag.

Brasiler K (1980) Fixation and uptake of nitrogen in Sphagnum blue-green algal associations, Oikos 34, 239-242.

Brown ME (1974) Seed and root bacterization, Ann. Rev. Phytopathol. 12, 181-187.

Burr TJ, Schroth MN and Suslow T (1978) Increased potato yields by treatment of seedpieces with specific strains of Pseudomonas fluorescens and P. putida, Phytopathology 68, 1377-1383.

Chopra RN, Kumra PK and Rikhi A (1978) Influence of Rhodotorula and Aerobacter on protonemal growth and bud initiation in two mosses, Curr. Sci. 47, 735-737.

Cowles JR, Evans HJ and Russell SA (1969) B$_{12}$ coenzyme-dependent ribonucleotide reductase in Rhizobium species and the effects of cobalt deficiency on the activity of the enzyme, J. Bacteriol. 97, 1460-1465.

DeCrombrugge B, Perlman RL, Varmus HE and Pastan I (1969) Regulation of inducible enzyme synthesis in Escherichia coli by cyclic adenosine 3',5'-monophosphate, Jour. Biol. Chem. 244, 5828-5835.

Evans HJ and Kliewer M (1963) Vitamin B$_{12}$ compound in relation to the requirements of cobalt for higher plants and nitrogen-fixing organisms, Ann. N. Y. Acad. Sci. 112, 735-755.

Grasso SM and Scheirer DC (1981) Scanning electron microscopic observations of a moss-fungus association, The Bryologist 84, 348-350.

Handa AK and Johri MM (1979) Involvement of cyclic adenosine-3',5'-monophosphate in chloronema differentiation in protonema cultures of Funaria hygrometrica, Planta 144, 317-324.

Inoue S, Ishida A and Kodama M (1980) Polysaccharide composition of the bryophyte cell-wall, Proc. Bryol. Soc. Japan 2, 169-170.

Inoue S, Ishida A and Kodama M (1981) Cellulose and uronic acid contents of cell-wall isolated from gametophytes of some mosses and liverworts, Jour. Hattori Bot. Lab. 49, 141-145.

Jordan DC, McNicol PJ and Marshall MR (1978) Biological nitrogen fixation in the terrestrial environment of a high Arctic ecosystem (Truelove Lowland, Devon Island, N.W.T.), Can. J. Microbiol. 24, 643-649.

Kapulnik Y, Okon Y, Kigel J, Nur I and Henis Y (1981) Effects of temperature, nitrogen fertilization, and plant age on nitrogen fixation in Setaria italica inoculated with Azospirillum brasilense (strain cd), Plant Physiol. 68, 340-343.

Konijin TM, Chang YY and Bonner JT (1969) Synthesis of cyclic AMP in Dictyostelium discoidium and Polypondylium pallidum, Nature 224, 1211-1212.

Lippincott BB and Lippincott JA (1969) Bacterial attachment to a specific wound site as an essential stage in tumor initiation by Agrobacterium tumefaciens, J. Bacteriol. 97, 620-628.

Lippincott JA and Lippincott BB (1977) Nature and specificity of the bacterium-host attachment in Agrobacterium infection. In Solheim B and Raa J, ed. Cell wall biochemistry related to specificity in host-pathogen interactions, pp. 439-451. Oslo, Universitetsforlaget.

Lippincott JA and Lippincott BB (1978) Cell walls of crown-gall tumors and embryonic plant tissues lack Agrobacterium adherence sites, Science 199, 1075-1078.

Nester EW and Kosuge T (1981) Plasmids specifying plant hyperplasias, Ann. Rev. Microbiol. 35, 531-565.

Nissen P (1973) Bacteria-mediated uptake of choline sulfate by plants, Scientific Reports of the Agricultural University of Norway 52, 1-52.

Parke JL and Linderman RG (1980) Association of vesicular-arbuscular mycorrhizal fungi with the moss Funaria hygrometrica, Can. J. Bot. 58, 1898-1904.

Pitts GA, Allam I and Hollis JP (1972) Beggiatoa: occurrence in the rice rhizosphere, Science 178, 990-991.

Purchase HF and Nutman RS (1957) Studies on the physiology of nodule formation. VI. The influence of bacterial numbers in the rhizosphere on nodule initiation, Ann. Bot. N.S. 21, 439-454.

Rabatin SC (1980) The occurrence of the vesicular-arbuscular-mycorrhizal fungus Glomus tennuis with moss, Mycologia 72, 191-195.

Rao SS, Lippincott BB and Lippincott JA (1982) Agrobacterium adherence involves the pectic portion of the host cell wall and is sensitive to the degree of pectin methylation, Physiol. Plant. in press.

Redhead SA (1981) Parasitism of bryophytes by agarica, Can. J. Bot. 59, 63-67.

Reese WD (1981) "Chlorochytrium," a green alga endophytic in musci, The Bryologist 84, 75-78.

Richardson DHS (1981) The biology of mosses. New York, John Wiley and Sons, pp. 119-133.

Rodgers GA and Henriksson E (1976) Associations between the blue-green algae Anabaena variabilis and Nostoc muscorum and the moss Funaria hygrometrica with reference to the colonization of Surtsey, Acta Bot. Isl. 4, 10-15.

Rutherford B, Jenkins J, Zorich N and Galsky A (1976) The possible involvement of adenylcyclase and cyclic-AMP phosphodiesterase in the formation of crown-gall tumors on the primary leaves of pinto beans, Plant Cell Physiol. 17, 1111-1117.

Schroth MN and Hancock JG (1982) Disease-suppressive soil and root-colonizing bacteria, Science 216, 1376-1381.

Sharma S and Johri MM (1982) Uptake and degradation of cyclic AMP by chloronema cells, Plant Physiol. 69, 1401-1403.

Smith RL, Bouton JH, Schank SC, Qusenberry KH, Tyler ME, Milan JR, Gaskins MH and Littell RC. (1976) Nitrogen fixation in grasses inoculated with Spirillum lipoferum, Science 183, 1003-1005.

Spiess LD (1975) Comparative activity of isomers of zeatin and ribosyl-zeatin on Funaria hygrometrica, Plant Physiol. 55, 583-585.

Spiess LD (1976) Developmental effects of zeatin, ribosyl-zeatin and Agrobacterium tumefaciens B_6 on certain mosses, Plant Physiol. 58, 107-109.

Spiess LD (1979) Antagonism of cytokinin induced callus in Pylaisiella selwynii by nucleosides and cyclic nucleotides, The Bryologist 82, 47-53.

Spiess LD, Lippincott BB and Lippincott JA (1971) Development and gametophore induction in the moss Pylaisiella selwynii as influenced by Agrobacterium tumefaciens, Amer. J. Bot. 58, 726-731.

Spiess LD, Lippincott BB and Lippincott JA (1972) Influence of certain plant growth regulators and crown-gall related substances on bud formation and gametophore development of the moss Pylaisiella selwynii, Amer. J. Bot. 59, 233-241.

Spiess LD, Lippincott BB and Lippincott JA (1973) Effect of hormones and vitamin B_{12} on gametophore development in the moss Pylaisiella selwynii, Amer. J. Bot. 60, 708-716.

Spiess LD, Lippincott BB and Lippincott JA (1976) The requirement of physical contact for moss gametophore induction by Agrobacterium tumefaciens, Amer. J. Bot. 63, 324-328.

Spiess LD, Lippincott BB and Lippincott JA (1977a) Comparative response of Pylaisiella selwynii to Agrobacterium and Rhizobium species, Bot. Gaz. 138, 35-40.

Spiess, LD, Turner JC, Mahlberg PG, Lippincott BB and Lippincott JA (1977b) Adherence of agrobacteria to moss protonema and gametophores viewed by scanning electron microscopy, Amer. J. Bot. 64, 1200-1208.

Spiess LD, Lippincott BB and Lippincott JA (1981a) Bacteria isolated from moss and their effect on moss development, Bot. Gaz. 142, 512-518.

Spiess LD, Lippincott BB and Lippincott JA (1981b) Promotion of Pylaisiella selwynii growth and gametophore formation by octopine and cytokinin. Physiol. Plant. 51, 99-105.

Spiess LD, Lippincott BB and Lippincott JA Bacteria-moss interactions in the regulation of protonemal growth and development, Jour. Hattori Bot. Lab. in press.

Takai Y, Yamanura H and Nishizuka Y (1974) Adenosine 3':5'-monophosphate dependent protein kinase from yeast, Jour. Biol. Chem. 249, 530-543.

Triplett EW, Heitholt JJ, Evensen KB and Blevins DG (1981) Increase in internode length of Phaseolus lunatus L. caused by inoculation with a nitrate reductase deficient strain of Rhizobium sp., Plant Physiol. 67, 1-4.

Watson M (1980) Patterns of habitat occupation in mosses - relevance to consideration of the niche, Bull. Torrey Bot. Club 107, 346-372.

Watson M (1981) Chemically mediated interactions among juvenile mosses as possible determinants of their community structure, J. Chem. Evol. 7, 367-376.

Whatley MH and Spiess LD (1977) Role of bacterial lipopolysaccharide in attachment of Agrobacterium to moss, Plant Physiol. 60, 765-766.

Whitaker B and Kende H (1974) Bud formation in Funaria hygrometrica. A comparison of the activity of three cytokinins with their ribosides, Planta 121, 93-96.

COMPETITIVE ABILITY OF *RHIZOBIUM PHASEOLI* STRAINS

D. CHANOVA (N. Poushkarov Institute of Soil Science and Yield Prediction, Sofia, Bulgaria)

1. INTRODUCTION

Recently the view has become increasingly common that the positive effect of nitrogen depends not only on the N-fixing activity and virulence of the introduced nodule bacteria but also on their ability to compete with nodule bacteria naturally occurring in soil, in the event of infection of leguminous plants. Hence many authors (Chanova, 1981; Johnson et al., 1965; Makarova, 1974; Novikova, Vassilieva, 1974, 1978; and others) have been interested in studying and selecting strains on the basis of their competitive ability. In this country such investigations have been carried out only for *Rhizobium leguminosarum pisum* (Chanova, 1981). The study reported here dealt with the competitive ability of seven active *R. phaseoli* strains of different origin.

2. MATERIALS AND METHODS

Strains of *R. phaseoli* to be used were selected by determining their antigenic properties, using the precipitation in agar reaction (PR) method of Ouchterlony as described by Vyazov (1967). The specific serums required for this precipitation reaction (PR) were obtained through immunization of rabbits; the immunization course consisted of six injections one applied every fourth day. The total number of introduced cells was 20 billion. Suspensions of 3-day-old cultures of investigated strains in a physiological solution were used as antigens in the immunization. Suspensions of 10-day-old cultures of nodule bacteria and homogenates prepared from nodules in a physiological solution were used as antigens in PR tests. Before use, to free them of non-specific H antigens, the suspensions and homogenates were heated in a boiling water-bath for 30 minutes, after which they were placed overnight in a refrigerator.

The competitive ability of *R. phaseoli* strains was determined using a pot experiment with sterile sand. The nutrient medium of Hellriegel modified by Fedorov (1952) was used; the trace elements manganese, zinc and cobalt were added to it following the method of Clark (1965). The experiment was carried out with beans of the Dobrudjanka 2 variety in pots with a volume of 1 kg in five replications. The seeds were inoculated with a mixture of two serologically different strains taken in equal amounts in different combinations. At flowering, 40 and 50 nodules on the average were analyzed from each treatment in the experiment. Nodule bacteria were isolated from every nodule and afterwards these bacteria were used as antigens in the serological identification of cultures in order to determine the dominant strain in the formation of nodules on bean roots.

According to the modified method of Means and Johnson (1964) homogenates from nodules can also be used as antigens in PR tests. To confirm this finding for bean roots, in four treatments in this experiment the two methods for preparing antigens needed for PR were compared: (1) use of pure culture of nodule bacteria isolated from nodules and (2) use of nodule homogenates (Means, Johnson, 1964). In method (2), a homogenate was prepared from each nodule using 0.5 to 1 ml sterile physiological solution. Both fresh and dried nodules were used. The latter were placed in a physiological solution and left overnight at room temperature. After that they were handled in the same way as fresh nodules.

3. RESULTS AND DISCUSSION

Part of the results obtained from the study of the serological properties of nodule bacteria of *R. phaseoli* species are given in Table 1. It is obvious from the table that in a cross somatic reaction with the studied strains the seven strains (305, 318, 328, 354, 361, 369 and 372) do not react, i.e., each unique with respect to their antigenic properties. The results of the analysis for competitive

Table 1. Serological properties of *R. phaseoli* strains

Antigens	Immune serums														
	305	324	371	318	361	309	317	328	384	369	354	378	349	372	368
305	+	-	-	-	-	-	-	-	-	-	-	-	-	-	-
324	-	+	+	-	-	-	-	-	-	-	-	-	-	-	-
371	-	+	+	-	-	+	-	-	-	-	-	-	-	-	-
318	-	-	-	+	-	-	-	-	-	-	-	-	-	-	-
361	-	-	-	-	+	-	-	-	-	-	-	-	-	-	-
309	-	-	+	-	-	+	-	-	-	-	-	-	-	-	-
317	-	-	-	-	-	-	+	-	'+	-	-	+	-	-	-
328	-	-	-	-	-	-	-	+	-	-	-	-	-	-	-
384	-	-	-	-	-	-	+	-	+	-	-	-	-	-	-
369	-	-	-	-	-	-	~	-	-	+	~	-	-	-	-
354	-	-	-	-	-	-	-	-	-	-	+	-	-	-	-
378	-	-	-	-	-	-	+	-	-	-	-	+	-	-	+
349	-	-	-	-	-	-	-	-	-	-	-	-	+	-	+
372	-	-	-	-	-	-	-	-	-	-	-	-	-	+	-
368	-	-	-	-	-	-	-	-	-	-	-	+	+	-	+

Note: +, positive PR (precipitation reaction); -, negative PR.

ability of these strains the following year, are given in Table 2.

It can be seen in the table that the competitive ability of the other six strains is lower than that of 305. This was expected from earlier studies which showed it to have high N-fixing activity and good virulence (Chanova, 1981). The obtained data indicate that the investigated strains have different compatitive abilities. Strain 305 has the highest competitive ability and it also has the highest N-fixing activity as already mentioned. The competitive ability of strain 328 is very close to that of strain 305: 54% of the nodules formed on bean roots are due to that of strain 305 and 46% are due to strain 328. These strains, as reported in another paper (Chanova, 1981), do not differ significantly in terms of their N-fixing activity and virulence. However, when beans are inoculated with strains 305 and 361, only 20% of the nodules formed are due to strain 361, i.e., this strain has a much lower competitive ability that strain 305. The

difference does not correspond to the difference in the N-fixing activities of these strains 354 and 372 are compared with strain 305. All this confirms the view that the selection of strains for fixing nitrogen requires a study of the competitive ability of nodule bacteria.

The presented data indicate that the two types of antigens used in PR tests to study the competitive ability of strains 305, 318, 328, 369 and 372 show identical results. Therefore, in our further investigations with this leguminous crop, nodule homogenates can be used as antigens instead of pure cultures isolated from nodules. This will considerably shorten the period of investigation.

The results of the study give us reason to propose strains 305 and 328 for use in future investigations and for eventual practical application because these strains have the highest competitive ability, N-fixing activity, and virulence. These rank among the most important properties of nodule bacteria in obtaining positive effect when inoculating leguminous plants.

Table 2. Competitive ability of R. *phaseoli* strains

Strains used for inoculation	Number of nodules analyzed	Antigen pre-paration method	Percent of nodules responding positively to immune serums							Comparative com-petitive ability
			305	318	361	328	369	354	372	
305+318	48	a	75	25	-	-	-	-	-	305 > 318
		b	75	25	-	-	-	-	-	
305+361	45	a	80	-	20	-	-	-	-	305 > 361
305+328	50	a	54	-	-	46	-	-	-	305 ≈ 328
		b	54	-	-	46	-	-	-	
305+369	40	a	77	-	-	-	22	-	-	305 > 369
		b	77	-	-	-	22	-	-	
305+354	50	a	70	-	-	-	-	30	-	305 > 354
305+372	50	a	88	-	-	-	-	-	12	305 > 372
		b	88	-	-	-	-	-	12	

Note: -, no reaction; a, pure culture; b, homogenate from a nodule

REFERENCES

Chanova DK (1982) Konkurentna sposobnost na stamove R. *leguminosarum*. Peti kongres po mikro-biologia, Vol. 3, p. 407-409, Ed. St. Nedyalkov. Sojus na nauchnite rabotnizi, Sofia.

Chanova DK (1982) Nitrogen-fixing and enzymatic activity of strains R. *phaseoli*. Reports, Bulgari-an Academy of Sciences (in press).

Clark FE (1965) Rhizobia. Methods of Soil Ana-lysis. CA Black, ed., part 2, p. 1487-1492, Madison, Wisconsin, American Society of Agronomy.

Fedorov MV (1941) Novaya modifikazia metoda sterilnoi i monobakterialnoi kulturi vishego rastenia, Mikrobiologia, Vol. X, No 1, 81-96.

Johnson HW, Means UM and Weber CR (1965) Com-petition for nodule sizes between strains of *Rhizobium japonicum* applied as inoculum and strains in soil, Agron. J. 57, 179-185.

Makarova NM (1974) Sravnenie dvuch metodov izuchenija konkurntnoi sposbnosti klubenkovih bakterii, Bull. VNIISHM, Leningrad, Vol. 3, No 16, 7-11.

Means UM, Johnson HW and Date RA (1964) Quick serological method of classifying strains of *Rhizobium japonicum* in nodules, J. of Bact. 87(3), 547-553.

Novikova AT and Vassilieva ND (1974) Konkurent-naja sposobnost klubenkovih bakterii soi i metodi ego izuchenia, Bull. VNIISHM, Leningrad, Vol.1, No 17, 5-11.

Novikova AT and Vassilieva ND (1978) Opredelenie konkurentnoi sposobnosti *Rhizobium*. Rol mikro-organismov v povishenii plodorodija pochv i uro-zaja kulturnih rastenii, Bull. VNIISHM, Leningrad, Vol. 47, 63-73.

Vyazov OE (1967) Laboratornaja imunologia, Medizi-na (Moskwa), 200-218.

PART VII

POLLUTION

THE IMPORTANCE OF FORESTS IN THE REGION OF THE PARDUBICE INDUSTRIAL AGGLOMERATION

Z. MRÁČEK (ČSVTS, Prague, Czechoslovakia)

1. INTRODUCTION

The importance of forest in the formation of a healthy environment in large towns has increased during the last 20 years. Sanitary and recreation functions have become primary in suburban and urban forests and decide the management form. Especially important is the mission of forest as green vegetation in regions of industrial agglomerations with disturbed environment. Forests favorably influence air temperature, moisture and ionisation, and the direction and power of wind streams. They reduce powder dispersion, lower noise, and intercept toxic gases. Also notable are the bactericide and bacteriostatic effects of phytocides.

The environment of the Pardubice industrial agglomeration is in the most exposed region of the Czech Socialist Republic. It may be expected that the construction of two large power plants in Chvaletice will further impair environment conditions. The limited area of forests near the town and the increase in town satellites will increase demands on the health and recreation functions of forests. Other green areas, such as meadows, tree lines and groups, riparian forests and remnants of lowland forests also have aesthetic value, and contribute to health and recreation.

2. DISCUSSION

2.1. Natural history of the Pardubice region

The landscape around Pardubice is typical of a region where settlement is associated with damage or even devastation. During 300 years, man has greatly changed both the nature and the original appearance of the landscape. The area of forests and water was reduced, as changing the local climatic conditions. The change of climate in turn unfavourably influenced agriculture, silviculture, and the environment of the whole region.

Although the area of forests and ponds has now been stabilized, a more dangerous environmental influence has appeared. This is the chemical industry, whose rapid development has caused further impairment of the environment of the satellite agglomeration. The maps (Fig. 1) indicate the main changes since 1688. During 250 years forests were reduced to about one third and ponds to one tenth of their former area. This state is permanent. Apart from that, the straightening of Labe River and the loss of large water surfaces have disturbed the water regime of the landscape, lowering the ground water table. In the Pardubice -Bohdaneč - Opatovice area alone, several thousand hectares of ponds disappeared lowering air moisture and increasing evaporation from the soil. Strips of lowland forests along the zigzag line of the Labe River were replaced by rows of birches and poplars, along the new stony river bed. Tree species of lowland associations of autochthonous type, such as oak, lem, linden, poplar, and hornbeam disappeared, and plantations of pine and spruce, with scattered acid oak forests, developed. Results of a study of the environment in 1975 show that Pardubice agglomeration is among the most unfavorable in our country. Air pollution due to aerosols of variable origin is on the average, very low in quality. On the degree scale we use, it is 4-5, which means at least 250-400 tons of particulates per square kilometer are deposited annually. Even worse, the sulfur dioxide content of the air is higher than 0.50 mg.m^{-3} and is classed as the worst (5th) degree in quality. Two thirds of the populations live in the area of Centrum, Dukla, Višňovka, Skřivánek, Na Spořilově, U kostelíčka and Na Drážce, where particulate fallout and SO_2 of the air content put the environment quality into the 5th degree. The 1975 evaluation shows that 69 percent of the inhabitants of Pardubice live in an environment strongly disturbed by negative effects (i.e., the

4th and 5th qualitative degree) on an area which
is 47 percent of the total town area, 2356 ha. A
more detailed qualitative evaluation of the agglomeration is given in Table 1.

The share of green vegetation per inhabitant of
the agglomeration is very low. The publicly
available green area,in the circle extending to
1 km from the agglomeration, ranges from 20 to 5
square meters per inhabitant in the densely popu-
lated parts of the town. The average is in the
3rd degree of quality, from 11 to 18 square meters
per inhabitant.

2.2. Forest management for recreation

Determination of the extent and treatment of recreation or park forests in the region of Pardubice
is based purely on the demand for short-term recreation by the present population, with consideration of its expected increase to 150 thousand.
This forest indicates all conditions of urban
and suburban forest. One condition is availability
of forest from town center within 20 to 30 min
by public transport. For the forest to be used
for recreation, however, it must have adequate
facilities for recreation and movement of visitors.
Management classifies the forest as recreation
forest or park forest.

2.2.1. Park forests. The area now being considered
park forest is a small forest unit of 19 ha
situated near Dukla, on the railway line from
Rosice to Chrudim and the road from Pardubice to
Přelouč. These forest stands are connected directly with the town of Dukla, and are subject to high
frequency of visits.

These forest stands are partly equipped with recreation facilities and in future should be
managed as park forest. For that reason they are
to be excluded from the State Forests, and will
become a part of the town vegetation.

2.2.2. Recreation forests. Areas considered recreation forest are mainly those immediately
connected with the town or with town satellites
under construction or being planned. Forests
which by their exceptional character serve for

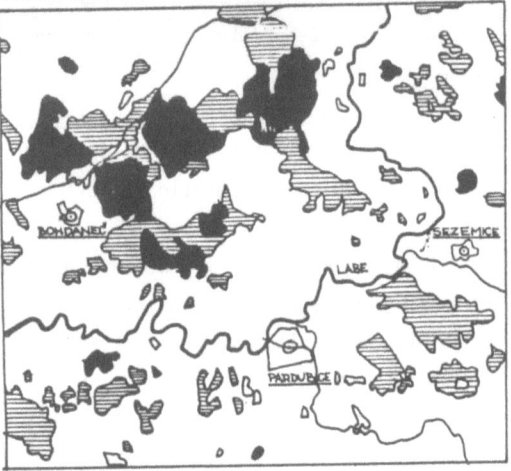

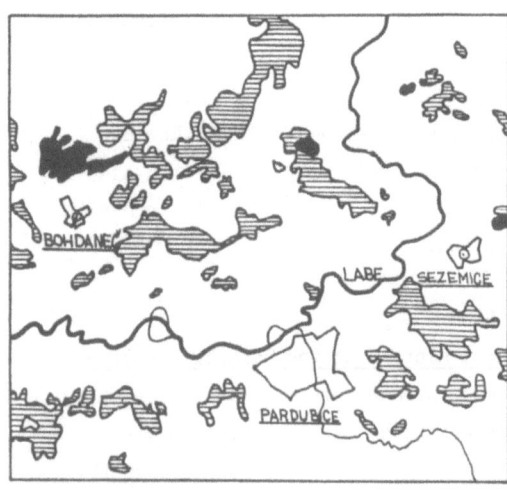

FIGURE 1. Forest and pond area in the Pardubice
region 1688 (a) 1850 (b) and 1950 (c)

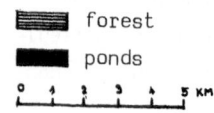

forest

ponds

TABLE 1. Quality evaluation of individual territorial wholes of Pardubice industrial agglomeration according to selected factors.

Territorial unit Name	inhabitants %	powder outfall $t.km^{-2}.year^{-1}$	SO_2 content in $mg.m^{-3}$ air	frequent noise level in streets in dB	industry intensity	public green in the circle 800 m m^{-2} per inhabitant
				resulting quality $\frac{\text{degree}}{\text{value}}$		
Centrum	19	5 / 0,25	5 / 0,25	5 / 0,25	4 / 0,32	3 / 0,30
Dukla Višňovka Skřivánek	29	5 / 0,25	5 / 0,25	5 / 0,25	5 / 0,40	5 / 0,50
Na Spořilově U Kostelíčka Na Drážce	20	5 / 0,25	5 / 0,25	4 / 0,20	3 / 0,24	2 / 0,20
Polabiny	18	3 / 0,15	5 / 0,25	4 / 0,20	2 / 0,16	3 / 0,30
Svítkov	5	4 / 0,20	5 / 0,25	5 / 0,25	4 / 0,32	1 / 0,10
Rosice	2	3 / 0,15	5 / 0,25	4 / 0,20	4 / 0,32	1 / 0,10
Rybitví Semtín	3	4 / 0,20	5 / 0,25	4 / 0,20	5 / 0,40	1 / 0,10
Uhrazenice	2	3 / 0,15	5 / 0,25	5 / 0,25	5 / 0,40	2 / 0,20
Staré Hradiště	1	2 / 0,10	5 / 0,25	3 / 0,15	5 / 0,40	5 / 0,50
Spojil	-	2 / 0,10	5 / 0,25	1 / 0,05	1 / 0,08	1 / 0,10
Černá za Bory	1	2 / 0,10	5 / 0,25	3 / 0,15	3 / 0,24	2 / 0,20

recreation are to be included (e.g., forest V olšině near St. Čivice).

To establish a basis of recreation forest, the
following forests or parts of them are to be
used (Fig. 2).

1. Studánka, a forest closely joined to the suburban area of the original Pardubice. It is 134
ha in area.

2. Bělobranská dubina, the western half bounded
by the road from Černá za Bory to Staročernsko
to Pardubice. The eastern part is to be left as
wood production forest. The areas for recreation
are to be 192 ha.

3. Zelenobranská dubina. In the management of
this forest for recreation, its importance for
health and landscape values is to be evaluated
in relation to neighboring factories. The area
totals 33 ha.

4. The whole forest near Polabiny, northeast
from the road from Trnová to Rosice and Labem
(the so-called green belt). The area is 112 ha.

5. The V olšině forest, the former game preserve
at Staré Čivice. The area is 58 ha.

The establishement of these recreation forests
would provide about 550 ha where management is
to be directed mainly to the fulfilment of
health and recreation functions.

2.3. Principles of management for intensive recreation

2.3.1. Park forests. A park forest is expressly
limited to a small part of the total forest area.
Its share of the total recreation forest is only
some tenths of a percent of the forest area. This
type is more park than forest.

Park forests are intensively used for recreation
and for that reason are usually located in close
vicinity to towns, spas, and recreation and
tourist centers. Frequency of visits is high
during the whole year.

The management plan requires special methods.
The cultivation of trees outside forest stands
is limited to the regeneration ensured by the
individual and group planting of stronger plants.

Attention is to be paid to coniferous and broad-
leaved species of native trees, suitable to the
site, including aesthetically effective exotic
tree species. Logging will be done only to remove
diseased, damaged and older trees. The same principle used in parks are valid for the protection
of the green of the park forest.

The pronounced characteristics of park forests are
grassland, forest meadows, and seed trees and their
share of the area is usually higher than that
covered by individual trees and groups.

All costs connected with the management of park
forests are provided by the whole society.

2.3.2. Recreation forests. This category includes
those parts of the forest where the main stress is
on recreation facilities. Here up to 10 percent of
the total area of the forest is used directly for
recreation. The important feature is removal or
thinning of the selected forest stands for establishment of meadows, play and rest grounds, and
other recreation facilities. It is therefore
necessary to exclude a certain part of the land
from production of wood and to limit choice of
management actions.

The forests of this group are intensively visited
mainly in one season, mostly in spring. The various
recreation facilities are connected by forest roads
with areas for parking. Tourists are offered forest
paths, but they may walk also in the forest itself.
In park forests movement and stay is limited to
roads, paths, and certain plots. The area of these
forests amounts as a rule to hundreds of hectares.

The target of forest management in this group is
to establish an aesthetically effective forest
environment, mainly around roads and recreation
facilities. The larger area of these forests
enables the use of small area cutting systems for
the opening of views into the forest and surrounding landscape. Felling quantity is determined from
the same principles that apply to normal forests
for wood yield. The regeneration period is longer
by 20 to 40 years. In this way it is possible to
create a more complicated and more aesthetically
effective forest structure. Forest regeneration

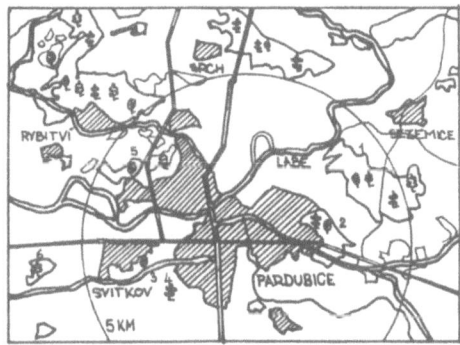

FIGURE 2. The areas of recreational forest (1-6) in the Pardubice region.

methods and processes of the usual type will be
used, based on site conditions and management
targets. A special feature of recreation forests
is to be based on protection of plantings and of
regeneration against damage by visitors. Roughly
about 10 to 15 percent of the area in forests
of this group will be fenced and cannot be used
directly by visitors. To increase aesthetic
effect , exotic tree species may be introduced,
but mainly on localities near frequently visited
roads and recreation facilities. In the introduction of these exotic trees, increase in wood
production is also to be taken into account;
since, in this forest type, wood production
is also a constant task of forestry.
Costs are increased because logging operations are
relatively less concentrated and there are smaller outputs by the machines used, more careful
skidding and haulage of wood, removal of felling
residues, and reduced or no use of chemicals
for forest protection. Expenses are also higher
due to complete protection of plantations and
intensive protection measures aimed at providing
complete recreation opportunities for inhabitants.
Costs are to be covered mainly from funds provided
by the whole society.

3. CONCLUSION

In the urban and suburban forests of the industrial agglomeration of Pardubice, health and recreation functions are more and more important.
For that reason these forests are to be divided
from management based on industrial wood production and managed as recreation and park forests.

MODIFICATIONS OF POLLUTANT EFFECTS BY ECOLOGICAL FACTORS [+]

Wolfgang PUNZ (Vienna, Austria)

1. INTRODUCTION

The number of papers dealing with the effects of pollutants on plants during the last years can be quoted as nearly unnumerable, among them being a lot of publications concerning heavy metal influences, especially lead. Nevertheless, among these there is only a small percentage revealing the problems of pollutant interactions - among one another and with ecological factors as well. For the first problem - efficiency of pollutant combinations on plants - might be cited Reinert et al. (1975), Burian et al. (1981). Interactions of pollutants with ecological conditions (the second problem) are usually reviewed and discussed implicitly in respective papers but rarely separately focused. Most of these papers (e.g. Heck et al.,1965, Davis, Wood, 1973, Guderian, 1977, Heck, Dunning, 1978, Noland, Kozlowski, 1979, Bell, 1980, Norby, Kozlowsky, 1981a,b) deal with sulfur dioxide effects, being of more economic interest as well as possible salt tolerance affected by environmental changes (e.g. Hoffman et al., 1978). But though an obvious damage of plant growth by lead can rarely be observed because of the buffering capacity of soil for heavy metal ions (Ernst, 1976, Sharp, Denny, 1976, Koeppe, 1977, Lane, Martin, 1977, Engenhart, 1980a,b), a conside-

rable effect of lead contamination on higher plants (Davis, Barnes, 1973, Bazzaz et al., 1974, Carlson et al., 1975, Hassett et al., 1976, Judel, Stelte, 1977, Walker et al., 1977, Flückiger et al., 1978, Raghi-Atri, 1978, Altgayer, 1979, Maier, 1980a,b, Maier et al., 1981) must be quoted evident.

Our own investigations focused on the modifying effects of environmental conditions on pollutant stresses to plants, summarized from several experiments and exemplified for lead as a prominent representative of heavy metals. Special attention has been given to aspects of pollutant monitoring as well as damage assessment as far as concerned by our results.

2. MATERIALS AND METHODS

Experiments with Hypogymnia physodes (L.) Nyl.: Methods are described in Punz (1979a,b,c).

Experiments with Zea mays L.: Corn was cultivated in single pots in soil (ED 63), light 12:12 hrs, 78 μmole.m^{-2}.s^{-1}; temperature as described in Tab. 1 below. Lead application simultaneous with temperature change after three weeks, offered as lead nitrate ($Pb(NO_3)_2$) via soil (7500 and 15000 ppm with respect to soil weight). Total length of corn leaves was determined at the end of week III, IV, V for each plant; results from week IV and V were expressed in per cent of week III. Statistical treatment: $\bar{x} \pm$ s.t (t-tested standard error of mean values).

[+])Publication Nr. 44 of the Vienna MAB project group 'urban ecology'

292

Experiments with Phaseolus vulgaris var.
nanus L. 'Saxa': Beans were cultivated
in single pots in soil (ED 63), light
12:12 hrs, 78 μmole.m^{-2}.s^{-1}; temperature 24 ^{O}C, after lead application 16
or 32 ^{O}C. Lead application via soil
(75oo ppm with respect to soil weight)
as lead nitrate (Pb(NO$_3$)$_2$). Photosynthesis measured by an URAS I (Fa. H&B)
described by Punz (1979a). Transpiration and water saturation deficit
measurement described by Schinninger-
Rothschedl (1976). Starch was determined using the methods described by
Albert, Falter (1979). Chlorophyll was
determined according to Schopfer (197o).
Optimum temperature was formally calculated from temperature coefficient
of photosynthesis (Q_5) described by
Punz (1980). Control series were grown
with aqua destillata and sodium nitrate.

3. RESULTS

3.1. Hypogymnia experiments

In experiments with Hypogymnia formerly
published (Punz, 1979a,b,c, 1980) it could
be shown that pollutant influence produced
a kind of shift in photosynthesis optimum temperature. From our experiments
we assumed that it depends probably on
temperature at the time of pollution which
kind of shift - to lower or higher temperatures - would occur. An example is
given in Fig. 1. Photosynthesis optimum
temperature of lead treated lichens is
lowered. This could be of considerable importance e.g. if lichens are used for biological monitoring of air pollution in an
uncritical way.   As it can be seen
from the figure, too, isolated measurement of photosynthesis at two different
temperatures would provide opposite results.

TABLE 1. Temperature conditions of
corn during the experiments (control
and lead treated plants as well)

week	1.	2.	3.	4.	5.
	24	24	24	16	16
	24	24	24	16	32
	24	24	24	32	32
	24	24	24	32	16

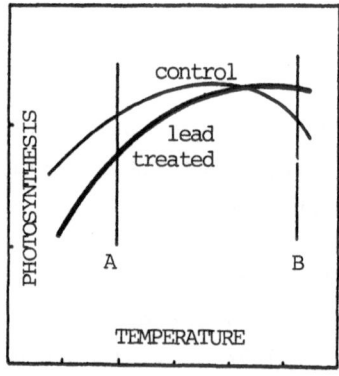

FIGURE 1. Photosynthesis as a function of
temperature with lead treated and untreated (control) plants. Isolated measurement only at temperature 'A' or 'B'
gives opposite results (from Punz, 1979c)

3.2. Zea experiments

3.2.1. Preliminary remarks

The concept of these experiments was simply to answer the question whether temperature would influence lead effects on plant production. Focusing on this question we made at first the following assumptions: 1) a relative high lead concentration was used in our experiments to approach an answer in principle as soon as possible 2) the complex system lead-soil-plant (e.g. John, 1972, Mikkelson, 1974, Tyler et al., 1974, Hildebrand, Blum, 1975, Zimdahl, Skogerboe, 1977, Ernst, 1978, Doelman, Haanstra, 1979a,b,c, Maier et al., 1981) was methodically considered as a 'black box'; we registered only the input (lead) and the output (plant growth modification) 3) lab experiments were preferred for this step 4) growth period investigated was minimized to avoid complex phenomena as described by Altgayer (1979) and explained according to the results of Malone et al. (1974), Jarvis et al. (1977), Lane, Martin (1977), Engenhart (1980a,b); from these results, a periodic inhibition of lead uptake resp. immobilisation of lead ions in batches can be derived, and periodic changes of metabolic activations and inhibitions could result.

3.2.2. Results (summarized in Tab. 2)

Corn treated at 16 $^{\circ}$C showed unsignificant increase in length growth (this and all following results always compared with control); after a week at 16 $^{\circ}$C and a subsequent week at 32 $^{\circ}$C a significant reduction of length growth could be observed.

Maize plants treated at 32 $^{\circ}$C showed significant increased length growth becoming unsignificant after changing the conditions to 16 $^{\circ}$C in the subsequent week. Using the double lead concentrations growth was always smaller than with control plants but because of a greater deviation no significant results could be obtained so far.

To explain the results we used a modified conception of Maier (1979, 1980a,b). It was derived from Maier's experiences with lead effects on enzymes in plants but we think it is a good and clear concept to explain some formerly misunderstood results in pollutant investigation. The principles of this model can be demonstrated in Fig. 2: Maier's conception is to see the lead influence in a dynamic way. The abscissa represents increasing (experimental) lead concentrations, the ordinate the activity of the investigated enzyme. + represents stimulating, - inhibiting effects. In a more general view we think we can consider plant metabolism as a 'black box' with a complex of interacting metabolic processes. But in any case response of plants seems to follow the same pattern as it is described for this model experiment: Primary pollutant effects usually produce negative responses (often the results of experiments on a cellular or subcellular level); secondary pollutant effects seem to possess at first stimulating, with increasing concentrations inhibiting quality; and the result from our black box is therefore the difference between both, usually at first +, then -. This might explain why 'zero response' to considerable pollutant doses can often be observed. Besides the results of Maier and our own results, the results of Gschliffner (1976) follow this scheme accurately, so we believe that it might be used easily for a first classification of pollutant effects.

TABLE 2. Length growth of lead treated corn (Zea mays L.) in relation to control (summarized)

Temperature of 4th week	Deviation from control	p	Temperature of 5th week	Deviation from control	p
16 °C	±(a)	-	16 °C	+(b)	-
			32 °C	-(c)	o.o5
32 °C	+(d)	o.o5	32 °C	+(e)	o.1
			16 °C	+(f)	-

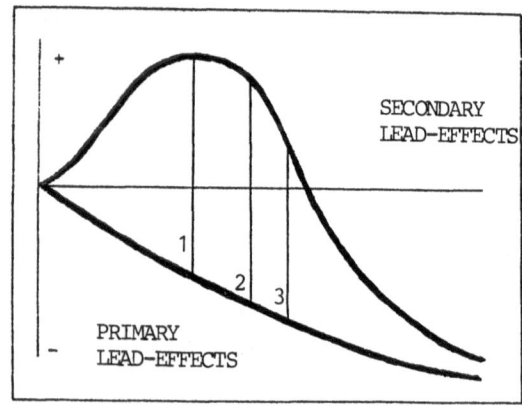

FIGURE 2. Scheme proposed by Maier (1979, 198oa,b) to explain the cooperation of primary and secondary lead effects for enzymes; altered. 1...positive effects predominant, 2...ambivalence, 3...negative effects predominant

According to this scheme we assume that lead might have at first (i.e. at lower concentrations) a stimulating effect, this being the reason for the results marked with a)b)d)e) in Tab. 2. The shift to inhibiting effects can be reached either by increasing concentrations (not mentioned in the Tab. 2) or by changing ecological conditions to which the plant has been adapted: a significant negative response in case of c)(former state: unsignificant positive resp. equal to control) or a shift from significant (d)e)) to unsignificant increase in length growth (f).

In the first case, + and - effects of lead are at first in an equilibrium as they are at mark 2 of Fig. 2, this equilibrium being disturbed by changing temperature regime; in the second case, plant state has just reached the stimulating phase of the scheme (mark 1), and a change of temperature produces an approach to mark 2. Anyway, a change of the very temperature plants are adapted at, seems to produce the worst results with lead treated plants. This might be the most interesting result - apart from all functional interpretation - because pollutant experiments rarely take into account that pollutant effects might also vary according to temperature before, during and after intoxication. A functional interpretation is to be given by further work but might be suspected in the field of enzymatic action, uptake control, or other effects as well.

3.3. Phaseolus experiments

Our investigations with beans focused at first on one question: a possible alteration of photosynthesis after lead treatment as influenced by temperature and light. The results can shortly be summarized as follows: As it can be seen from Fig. 3 and Tab. 3 temperature after lead application causes a shift in photosynthesis optimum temperature

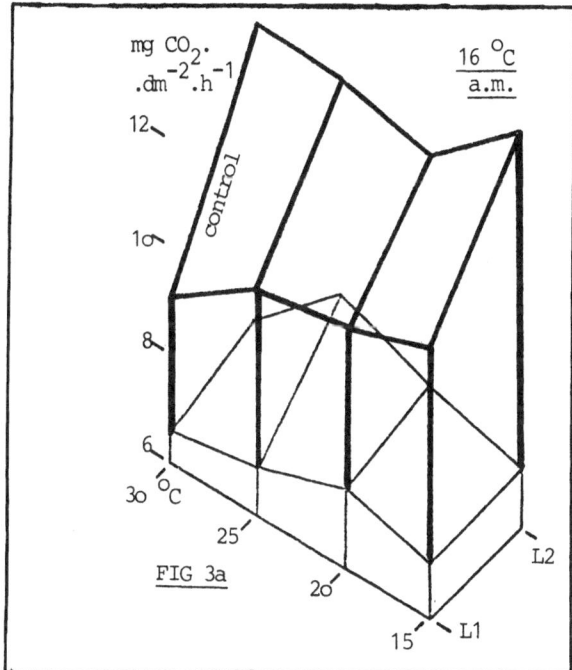

FIG 3a

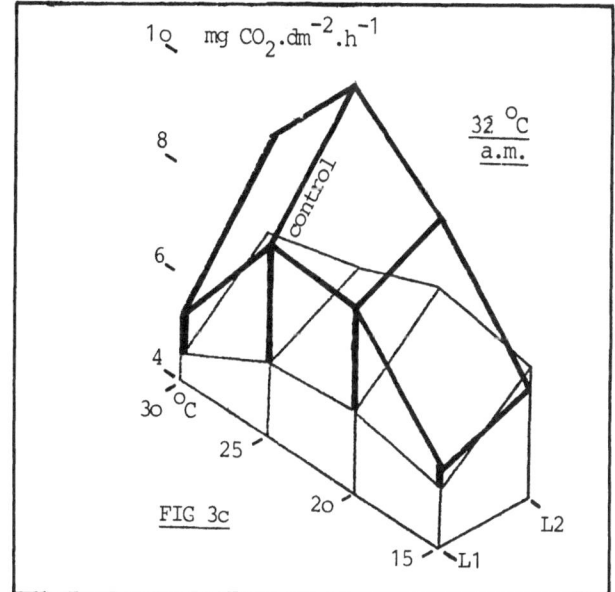

FIG 3c

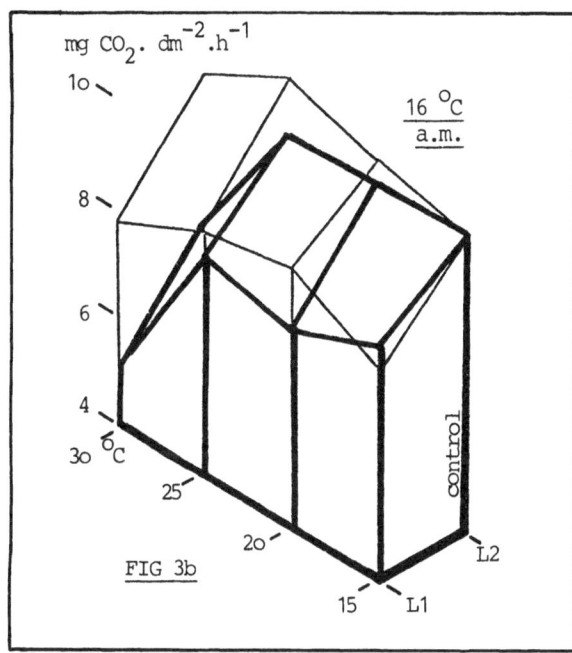

FIG 3b

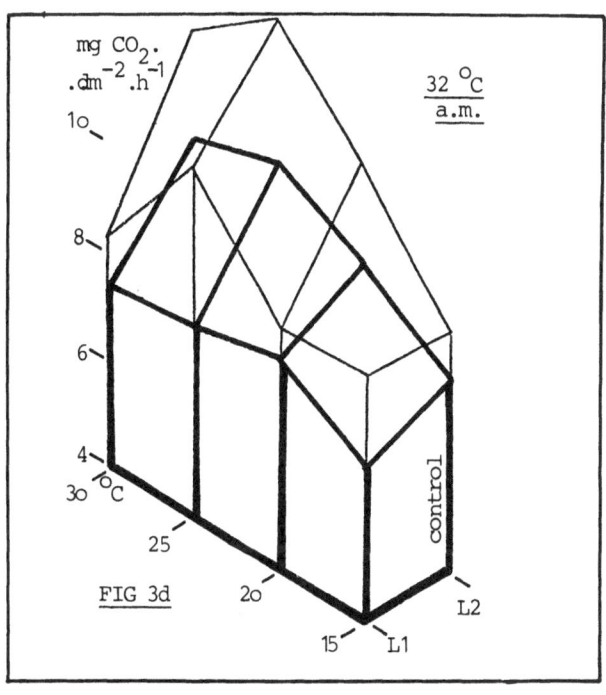

FIG 3d

FIGURE 3. Net photosynthesis of lead treated and untreated beans (Phaseolus vulgaris var. nanus L.) as a function of temperature and light intensity; photosynthe-
sis in mg $CO_2.dm^{-2}.h^{-1}$, temperature in ^{o}C,
Light L1...98 μmole.m^{-2}.s^{-1}, L2...195 μmole
.m^{-2}.s^{-1}; lead treatment: 3a and 3b at 16°C
3c and 3d at 32 ^{o}C; 3a and 3c measured at
1o^{30}a.m.; 3b and 3d measured at 15^{30}(p.m.)

TABLE 3. Shift of bean (Phaseolus vulgaris
var. nana L.) photosynthesis optimum temperature compared with control (summarized
from Fig. 3)

light intensity	lead applied at a temperature of 16 ^{o}C	32 ^{o}C
low	raised	lowered
high	raised	lowered

296

the latter being lowered after lead application at high temperatures and vice
versa. This is again of considerable relevance for bioindication research because of the reasons described above
(3.1.): A simple measurement of damaged
plants at only one temperature might be
insufficient and lead to wrong interpretations. A functional explanation of the
phenomenon should be approached by further experiments but is suspected probably in the field of enzymatic phenomena.

Measurements during the whole light period showed a specific effect that can
be seen in Fig. 4: During the first hours
of the light period photosynthesis of
lead treated plants is lower than control
plant photosynthesis, but in the second
half of the light period the inverse effect could be observed. This might be of
a similar meaningfulness for bioindication showing the importance of light period for pollution effects but might be
a singular effect with beans only because
it is known that beans may have daily
changes in photosynthetic rate and chloroplast ultrastructure (Wieckowsky, 1967,
1968) at a special ontogenetic stage.
Nevertheless we made the attempt to clear
up these phenomena; it is known from
Engenhart (1980a,b) and Schinninger (1979,
1981a,b) that lead reduces transpiration
of plants and levels diurnal period of
plant root water flow capacity. We observed lower transpiration rates with lead
treated plants but without any diurnal
change; the same results could be gath-
ered for water saturation deficit measurements. Chlorophyll analysis showed no
significant results (a slight decrease
in a/b ratio at the beginning of the
light period as well as an increase after

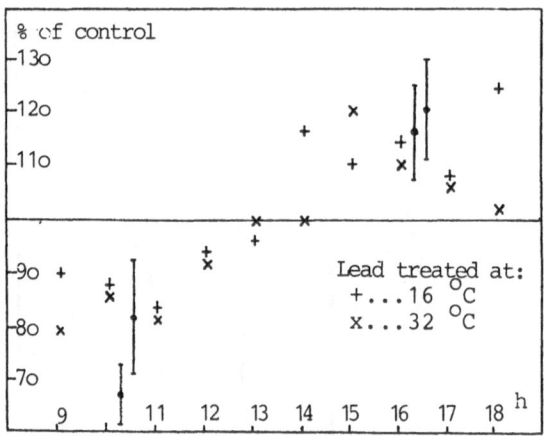

FIGURE 4. Net photosynthesis in per cent
of control during the day, measured at
24 ^{o}C, 195 μmole.m^{-2}.s^{-1}; mean values $\pm$
deviation from Fig. 3 (all values from a)
b)c)d) respective)

some hours could be observed with lead treated plants in relation to control plants.

A marked increase in starch content of control leaves up to 300 per cent above lead
treated plants could be observed (Fig. 5);
this might indicate negative response of
photosynthesis to high starch levels, but
this hypothesis is not accepted everywhere
(review by Neales, Incoll, 1968). Beside
this possibility for an explanation (cf.also
Austin, McLean, 1972) we cannot give a sufficient interpretation for this phenomenon.

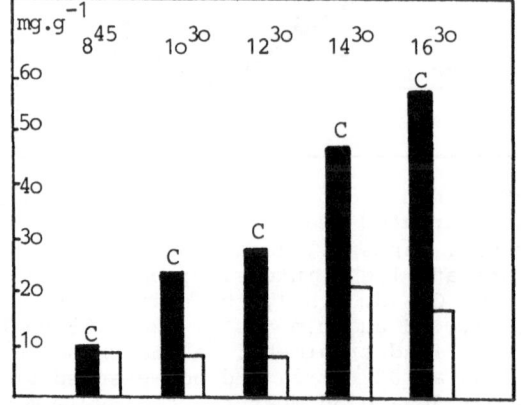

FIGURE 5. Starch content in leaves of lead
treated and untreated (C...control) plants
during the day (see Fig. 4)

4. DISCUSSION

The discussion of the whole subject should focus on two aspects of the results: On the one hand, there is the functional and physiological interpretation of the observed phenomena - according to the data obtained. On the other hand it must be stressed, however, that even without a complete functional explanation of the data they may serve to improve our knowledge about possible pollutant reactions and to increase caution in setting experiments as well. A complete functional interpretation of the temperature dependence of lead effects seems to be impossible using only the results obtained. But we consider it to be proved that temperature before, during, and after application of pollutants may cause severe changes in photosynthesis and production of plants. It is not possible therefore to describe a single pollution response of a plant to a special pollutant. The whole pattern of pollutant effects must be described with respect to ecological conditions, at least with respect to temperature. As it could be proved with lichens as well as with maize, complete inverse effects can occur if various temperature regimes are chosen for measurements. This is also according to the results of Maier (1977, 1979) and other works. The results of the maize experiments described above (i.e. at first stimulation, then inhibition of growth by increasing lead concentrations) may be parallelized with the data given by Stoklasa (1913), Stutzer (1916), Bonnet (1922), Keaton (1937), Zegers et al. (1976), John (1977), Lane et al. (1978). But in difference with the results of Norby, Kozlowsky (1981a,b) - working on a similar topic but with sulfur dioxide -

we found no special increase of lead caused damage at low temperatures but observed that in general a drastic change in temperature conditions produced the worst results (in the case of maize production). Anyway, the important question of plant productivity affected by air pollution (cf. Laurence, Weinstein, 1981) seems to need far more and even more sophisticated investigations than have been done so far. The results obtained with beans during the light period do not seem to be a general phenomenon and might be a special problem of this species; nevertheless it should be expected that possibly other plants might produce similar problems and therefore it should be taken care before scientific experiments are generalized for practical use.

ACKNOWLEDGEMENT

I am indebted to the Austrian Academy of Sciences (Österreichische Akademie der Wissenschaften) and to the Municipality of Vienna (Magistrat der Stadt Wien) for their financial supply. I should like to thank Prof. Burian for his aid and advice.

REFERENCES

Albert R and Falter J (1979) Stoffwechselphysiologische Untersuchungen an Blättern streusalzgeschädigter Linden in Wien II. Stickstoff- und Kohlehydratstoffwechsel, Phyton 19(3-4), 141-162.

Altgayer M (1979) Morphologische und physiologische Veränderungen von Pflanzen unter Bleieinfluß. Diss.Univ.Wien.

Austin RB and McLean MSM (1972) Some effects of temperature on the rates of photosynthesis and respiration of Phaseolus vulgaris L., Photosynthetica 6(1), 41-5o.

Bazzaz FA, Rolfe GL and Windle P (1974) Differing sensitivity of corn and soybean photosynthesis and transpiration to lead contamination, J.Environ.Quality 3(2), 156-157.

Bell JNB (198o) Response of plants to sulphur dioxide, Nature(London) 284, 399-4oo.

Bonnet E (1922) Actions des sels solubles de plomb sur les plantes, C.R.Acad. Sci.174,488-491.

Burian K, Punz W and Schinninger R (1981) Efficiency of pollutant combinations on plants. Methods of indication of air pollutants by plants, Symposium Osnabrück 1981, in press.

Carlson RW, Bazzaz FA and Rolfe GL (1975) The effects of heavy metals on plants II. Net photosynthesis and transpiration of whole corn and sunflower plants treated with Pb, Cd, Ni and Tl, Environ.Res. 1o, 113-12o.

Davis JB and Barnes RL (1973) Effects of soil applied fluoride and lead on growth of loblolly pine and red maple, Environ.Pollut.5(1), 34-45.

Davis DD and Wood FA (1973) The influence of environmental factors on the sensitivity of Virginia pine to ozone, Phytopathology 63, 371-376.

Doelman P and Haanstra L (1979a) Effects of lead on the soil bacterial microflora, Soil Biol. 11, 487-491.

Doelman P and Haanstra L (1979b) Effects of lead on the decomposition of organic matter, Soil Biol. 11, 481-485.

Doelman P and Haanstra L (1979c) Effects of lead on soil respiration and dehydrogenase activity, Soil Biol. 11, 475-479.

Engenhart M (198oa) Einfluß von Blei auf die Wasserpermeabilität. International workshop on problems of bioindication Kongreß- und Tagungsberichte der Martin-Luther-Universität Halle-Wittenberg; Wissenschaftliche Beiträge 25(P9),Bd.2, 77-82.

Engenhart M (198ob) Hemmung des Wasserdurchsatzes in Wurzelsystemen von Phaseolus vulgaris L. unter Bleieinfluß. Diss. Univ.Wien.

Ernst W (1976) Physiological and biochemical aspects of metal tolerance. In Mansfield TA, ed. Effects of Air pollutants on plants, 115-133, Soc.Exp.Biol., Seminar Ser.1, Cambridge.

Ernst W (1978) Chemical soil factors determining plant growth, Verh.Konink. Nederl.Akad.Wetensch., 155-187.

Flückiger W, Flückiger-Keller H and Oertli JJ (1978) Der Einfluß verkehrsbedingter Luftverunreinigungen auf die Peroxidaseaktivität, das ATP-Bildungsvermögen isolierter Chloroplasten und das Längenwachstum von Mais, Z.Pflanzenkrankh. 85(1), 41-47.

Gschliffner Ch (1976) Die Wirkung einzelner und kombinierter Umweltgifte auf die Weitungsfähigkeit und Beweglichkeit von Spaltöffnungen des Amaryllis-Typs. Diss.Univ.Wien.

Guderian R (1977) Air pollution. Springer Verlag, New York.

Hasset JJ, Miller JE and Koeppe DE (1976) Interaction of lead and cadmium on maize root growth and uptake of lead and cadmium by roots, Environ.Pollut. 11, 297-3o2.

Heck WW and Dunning JA (1978) Response of oats to sulfur dioxide: interactions of growth temperature with exposure temperature or humidity. J.Air Pollut.Control Assoc. 28, 241-246.

Heck WW, Dunning JA and Hindawi IJ (1965) Interactions of environmental factors on the sensitivity of plants to air pollution, J.Air Pollut.Control.Assoc. 15, 511-513.

Hildebrand EE and Blum WE (1975) Fixation of emitted lead by soils, Z.Pflanzenernährg. Bodenkde. 138, 279-294.

Jarvis SC, Jones LHP and Clement CR (1977) Uptake and transport of lead by perennial ryegrass from flowing solution culture with a controlled concentration of lead, Plant and Soil 46, 371-379.

John MK (1972) Lead availability related to soil properties and extractable lead, J. Environ.Qual. 1(3), 295-298.

John MK (1977) Varietal response to lead by lettuce, Water, Air and Soil Pollution 8(2), 133-144.

Judel GK and Stelte W (1977) Gefäßversuche mit Gemüsepflanzen zur Frage der Bleiaufnahme aus dem Boden, Z.Pflanzenernährg. Bodenkde. 14o, 421-429.

Keaton CM (1937) The influence of lead compounds on the growth of barley, Soil Sci. 43, 4o1-411.

Koeppe DE (1977) The uptake, distribution and effect of cadmium and lead in plants, Sci.Total Environment 7, 197-2o6.

Lane SD and Martin ES (1977) A histochemical investigation of lead uptake in Raphanus sativus, New Phytol. 79, 281-286.

Lane SD, Martin ES and Garrod JF (1978) Lead toxicity effects on Indole-3-Ylacetic Acid-induced cell elongation, Planta 144, 79-84.

Laurence JA and Weinstein LH (1981) Effects of air pollutants on plant productivity, Ann.Rev.Phytopathol. 19, 257-271.

Maier R (1977) Der Einfluß von Blei auf die Aktivität der Esterase und ihrer multiplen Formen, Biochem.Physiol.Pflanzen 171, 455-468.

Maier R (1979) Zur Bioindikation von Bleiwirkungen in Pflanzen über Enzyme, Verh.Ges. Ökol. VII, 315-322.

Maier R (198oa) Nachweis von Bleieffekten in Pflanzen mit Hilfe der Gelelektrophorese von Enzymen. International workshop on problems of bioindication/ Kongreß- und Tagungsberichte der Martin-Luther-Universität Halle-Wittenberg; Wissenschaftliche Beiträge 25(P9) Bd.2, 89-95.

Maier R (198ob) The influence of lead on the activity of enzymes and their multiple forms in plants. Symposium Santiago/Spanien.

Maier R, Maier F and Maier G (1981) Aktivität TTC-reduzierender Enzyme in Boden und Pflanze unter dem Einfluß von Blei im Boden, Ber.Deutsch.Bot.Ges. 94, 7o9-718.

Malone C, Koeppe DE and Miller RJ (1974) Localization of lead accumulated by corn plants, Plant Physiol. 53, 388-394.

Mikkelson JP (1974) Effects of lead on the microbiological activity in soil, T.Plantes 78, 5o9-516.

Neales TF and Incoll LD (1968) The control of leaf photosynthesis rate by the level of assimilate concentration in the leaf: A review of the hypothesis, Bot.Review 34(2), 1o7-125.

Noland TL and Kozlowski TT (1979) Effects of SO_2 on stomatal aperture and sulfur uptake of woody angiosperm seedlings, Can.J.Forest Res. 9, 57-62.

Norby RJ and Kozlowski TT (1981a) Response of SO_2-fumigated Pinus resinosa seedlings to postfumigation temperature, Can.J.Bot. 59, 47o-475.

Norby RJ and Kozlowski TT (1981b) Interactions of SO_2-concentrations and postfumigation temperature on growth of five species of woody plants, Environ. Pollut. 25, 27-39.

Punz W (1979a) Der Einfluß isolierter und kombinierter Schadstoffe auf die Flechtenphotosynthese, Photosynthetica 13(4), 428-433.

Punz W (1979b) Beiträge zur Verwendung von Flechten als Bioindikatoren: Der Einfluß von Schadstoffkombinationen, Sitzg. ber.Österr.Akad.Wiss., mn Kl. 188(1), 1-25.

Punz W (1979c) Beiträge zur Verwendung von Flechten als Bioindikatoren II. Mögliche Einflüsse von Temperatur und Jahreszeit, Sitzg.ber.Österr.Akad.Wiss., mn Kl. 188(4), 63-85.

Punz W (198o) Pollutant combinations and their effects on some aspects of lichen physiology. International workshop on problems of bioindication/ Kongreß- und Tagungsberichte der Martin-Luther-Universität Halle-Wittenberg; Wissenschaftliche Beiträge 25(P9), Bd.2, 26-32.

Raghi-Atri F (1978) Einfluß von Schwermetallen (Pb, Hg) im Substrat auf Glyceria maxima (Hartm.)Holmbg. I.Blei, Angew. Botanik 52, 185-192.

Reinert RA, Heagle AS and Heck WW (1975) Plant responses to pollutant combinations. In Mudd JB and Kozlowski TT eds. Responses of plants to air pollution, 159-177, Academic Press, New York-San Franzisco-London.

Schinninger-Rothschedl R (1976) Wasserhaushalt, Transpiration und Austrocknungsresistenz verschiedener Ruderalpflanzen nach Einzel- und kombinierter Schädigung durch Umweltgifte. Diss.Univ.Wien.

Schinninger R (1979) Der Einfluß isolierter und kombinierter Schadstoffe auf Austrocknungsresistenz und Transpiration bei Festuca rubra L., Z.Pflanzenphysiologie 94, 351-362.

Schinninger R (1981a) Der Einfluß isolierter und kombinierter Schadstoffe auf Austrocknungsresistenz und Transpiration bei Amaranthus chlorostachys Willd., Flora 171, 187-198.

Schinninger R (1981b) Der Einfluß isolierter und kombinierter Schadstoffe auf Austrocknungsresistenz und Transpiration bei Trifolium repens L., Phyton 21(2), 245-259.

Schopfer P (197o) Experimente zur Pflanzenphysiologie. Freiburg, Rombach-Verlag.

Sharp V and Denny P (1976) Electron Microscope studies on the absorption and localization of lead in the leaf tissue of Potamogeton pectinatus L., J.Exp.Bot. 27(1o1), 1155-1162.

Stoklasa J (1913) De l'influence de l'uranium et du plomb sur la vegetation, C.R. Akad.Sci.156, 153-155.

Stutzer A (1916) Die Wirkung von Blei als Reizstoff für Pflanzen, J.Landwirtschaft 64, 1-8.

Tyler G, Mornsjo B and Nilsson B (1974) Effects of cadmium, lead and sodium salts on nitrification in a mull soil, Plant and Soil 4o, 237-242.

Walker WM, Miller JE and Hassett JE (1977) Effects of lead and cadmium upon the calcium, magnesium, potassium and phosphorus concentration in young corn plants, Soil Sci. 124, 145-151.

Wieckowski S (1967) Chloroplasts in growing bean leaf, Acta Soc.Bot.Polon. 36, 161-169.

Wieckowski S (1968) Daily changes in the photosynthetic rate and chloroplast ultrastructure in growing bean leaf, Photosynthetica 2(3), 172-177.

Zegers, PV, Harmet KH and Hanzely L (1976) Inhibition of IAA-induced elongation in Avena coleoptile segments by lead: a physiological and an electron microscopic study, Cytobios 15, 23-25.

Zimdahl RL and Skogerboe RK (1977) Behaviour of lead in soil, Env.Sci.Techn. 11, 12o2-12o7.

ADDENDUM

Hoffman GJ, Jobes JA, Hanscom Z and Maas EV (1978) Timing of environmental stress affects growth, water relations and salt tolerance of pinto bean, Transactions ASAE 21(4), 713-722.

TREE BARKS AND POLLUTANT ZONING

Wolfgang PUNZ & Rosemarie SCHINNINGER (Vienna, Austria)

1. INTRODUCTION

For urban planning as well as for health affairs it is absolutely necessary to have an adequate knowledge about the regions of different air pollution. Chemical and physiological analysis provides precise but quite expensive systems for this purpose; bioindication (i.e. biological monitoring of pollutants) gives integrative values but is rarely specific for single pollutants, by no means in a quantitative manner. Tree bark as a monitor of air pollution seems to be a simple, cheap and accurate method. Tree bark can be considered as a natural sink for all gaseous, aerosolic or rain-solved pollutants.

Investigations described here are part of a project to test tree bark as indicator of sulfur dioxide and dust in an urban area of Austria. The project is to answer the question whether bark analysis can provide a supplement to chemical and physiological methods of air pollution monitoring, and which tree species are appropriate for this purpose.

2. MATERIALS AND METHODS

(Methods were chosen using especially the results of Härtel, Grill, 1972, Kienzl, 1978, Kienzl, Härtel, 1979) Tree bark is able to absorb air pollutants (the process is not completely clear so far); this absorption increases the electrolytic conductivity of bark extract(=ELB)(see Härtel, Grill, 1972, Köhm, 1976). The extent of the ELB-increase provides a measure for the in-

tensity of preceding immissions. ELB integrates sulfate conductivity and the conductivity of non sulfatic components (esp. dust). To discern these two components it is necessary to determine sulfate concentration in bark extract. We used a turbidimetric method: soluted sulfate is precipitated by acid bariumchloride and its extinction is measured at 575 nm using a spectrophotometer. For calibration, solutions of known sulfate concentration ($CaSO_4$) are used. From the measured conductivity of calibration solutions a relation between the conductivity of 'pure' sulfate solutions and sulfate concentrations can be derived. If the fictitious sulfate conductivity is compared with the actually measured values of bark extract, the 'rest conductivity' i.e. the conductivity of the non sulfate component of the extract can be determined. From this, the amount of the non sulfate components of air pollution can be estimated. The pH of the extract can also be measured but its role as a monitor of air pollution is inconsistent (Härtel, Grill, 1972, Kienzl, Härtel, 1979). Tree bark values reflect the immission situation four weeks before sampling; they are influenced by rainfall and high air humidity periods (Kienzl, Härtel, 1979). It must be stressed, however, that data from bark analysis can always be only relative values: The immission load of single sites can be compared with others giving a relative pattern of more or less polluted regions; no absolute values of air pollution can be derived from bark analysis.

Bark samples were taken from Picea abies (L.)Karst., Acer platanoides L., Tilia cordata Mill. bzw. Tilia platyphyllos Scop. and Aesculus hippocastanum L. at a maximum thickness of ca. 3 mm. Samples were taken up to six times (August 1980/ December 1980/ March 1981/ July 1981/ September 1981/ December 1981). Air dried samples were crushed into pieces not more than 30 mm^2 and extracted in aqua destillata for 24 hours. The extract was filtered, electolytic conductivity (=ELB) pH and sulfate concentration were determined using a conductivitymeter L 17 (Fa. Hoelzle&Chelius), a pH-meter (Fa. Beckmann) and a spectrophotometer MONO-SPAC 103 (Fa. Jobin-Yvon). Calculations were made with an HP 41CV using the HP STAT 1 pack.

The four tree species were chosen with respect to their abundance in the investigated area and for comparisons of coniferes and deciduous trees. As far as possible, tree samples were taken from sites that were not exposed to direct pollutant influence (esp. traffic).

3. RESULTS
3.1. Preliminary investigations

It is known from literature that bark of Picea is appropriate for monitoring air pollution (Grill, Hofer, 1979, Grodzinska, 1971, Härtel, 1977, Härtel, Grill, 1972, Härtel, Grill, Krzyscin, 1980, Hofer, 1979, Kienzl, Härtel, 1979, Köhm, 1976, Maier et al., 1980, Świeboda, Kalemba, 1979). The first step of our investigations was, therefore, to compare the results of analysed bark of Picea and three species of deciduous trees (Acer, Aesculuc, Tilia). It was expected that deciduous trees might replace Picea for monitoring air pollutant

zoning.

The sulfate concentration of bark extract calculated from about 25 values respectively is summarized in Tab. 1. From this table it can be seen that Aesculus and Tilia show no significant correlation with Picea. The correlation for Acer is better (0.82).

TABLE 1. Sulfate content of bark extract of Picea correlated with the values of Acer, Aesculus and Tilia

	r
Picea - Acer	0.82
Picea - Aesculus	0.39
Picea - Tilia	0.52

The relation between particular sites can be drawn from Fig. 1. The investigated area was divided into 25 squares. The values of bark sulfate concentrations were classified into three categories (0-300/ 301-600/>600 mg $SO_4.1^{-1}$). All given data are from one simultaneous bark sampling in March 1981 representing winter values (sulfur dioxide concentration in urban areas being usually higher in winter than in summer).

Classified results of deciduous trees were again compared with those of Picea. From a total of 13 squares we can find that only four squares show the same class for Picea and deciduous trees as well. Eight squares show a difference of one degree with respect to at least one deciduous tree. In one case, a difference of two degrees can be observed.

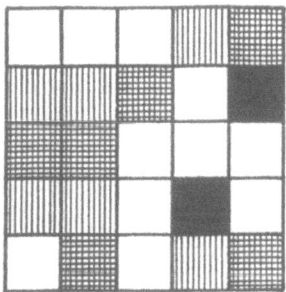

FIGURE 1a. Picea.

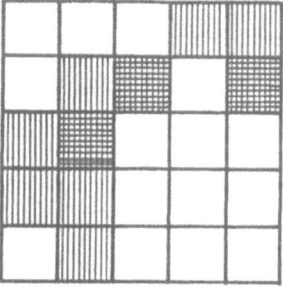

FIGURE 1b. Acer.

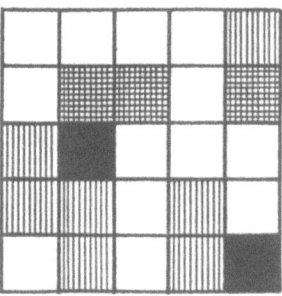

FIGURE 1c. Aesculus.

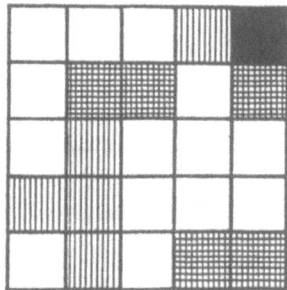

FIGURE 1d. Tilia.

FIGURE 1a-d. Classified sulfate concentration of bark extract (investigated area divided into squares; white squares no data)

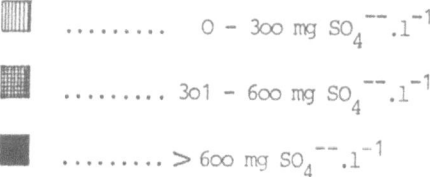

It might be argued that classification for Picea is not convenient for deciduous trees. This can be denied for Aesculus and Tilia as all three degrees can be found but might be true for Acer. The results given so far made us exclude Aesculus and Tilia from further investigations. In spite of its good correlation values (Tab. 1) Acer was also excluded especially for reason of its bark consistence (it is quite difficult to collect Acer bark particles of 3 mm thickness as required in the prescription).

3.2. Picea results
3.2.1. Seasonal variation of bark sulfur content

The results given below reflect continuous sampling during a period of one and a half years (sampling quarterly) at seventy sites in an urban area of Austria.

For each site the course of bark extract sulfate concentration during the whole investigated period is shown. The diagrams reflect the seasonal different immission load as well as local phenomena of each site.

304

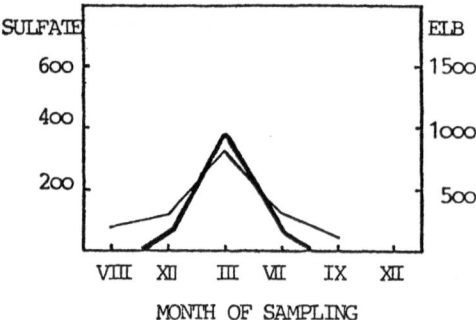

FIGURE 2. Picea. Sulfate content (mg SO$_4$. .1^{-1}) and electrolytic conductivity (ELB; μS.cm^{-1}) of bark extract during the whole investigated period. Winter peak. Full line: Sulfate

Fig. 2 represents a typical pollutant situation in an urban area. Sulfate concentrations are distinctly increased in winter compared with those in summer. This type could be observed frequently.

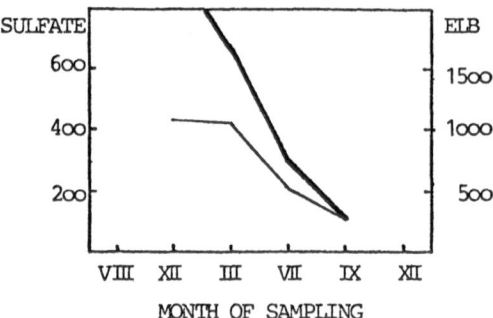

FIGURE 4. Picea. Sulfate content (mg SO$_4$. .1^{-1}) and electrolytic conductivity (ELB; μS.cm^{-1}) of bark extract during the whole investigated period. Decrease. Full line: Sulfate

The site described by Fig. 4 shows a continuous decrease of sulfate concentration. The reason for this phenomenon is shrubbery surrounding the investigated tree. The increasing foliage of these bushes protects the bark from air pollution.
An opposite result can be seen in Fig. 5: Responsible for the increasing sulfate concentration during the investigated period was a building site.

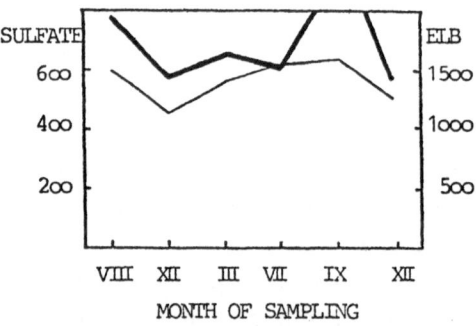

FIGURE 3. Picea. Sulfate content (mg SO$_4$. .1^{-1}) and electrolytic conductivity (ELB; μS.cm^{-1}) of bark extract during the whole investigated period. Continuously high. Full line: Sulfate

In some cases, however, high sulfate concentrations were observed during the whole period (Fig. 3).

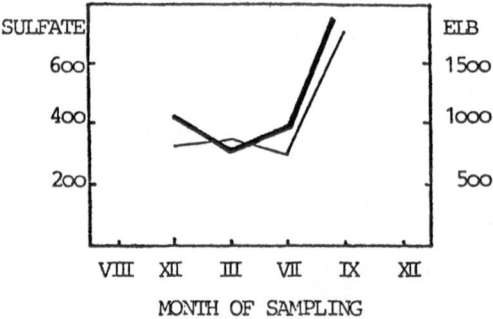

FIGURE 5. Picea. Sulfate content (mg SO$_4$. .1^{-1}) and electrolytic conductivity (ELB; μS.cm^{-1}) of bark extract during the whole investigated period. Increase. Full line: Sulfate.

3.2.2. Significance of pH-value

pH-values of bark extracts were measured
and correlated with sulfate concentration. No significant correlation could
be found (r=o.19; see Fig. 6). For this
reason, pH-values were considered not to
be appropriate for the referred investigations.

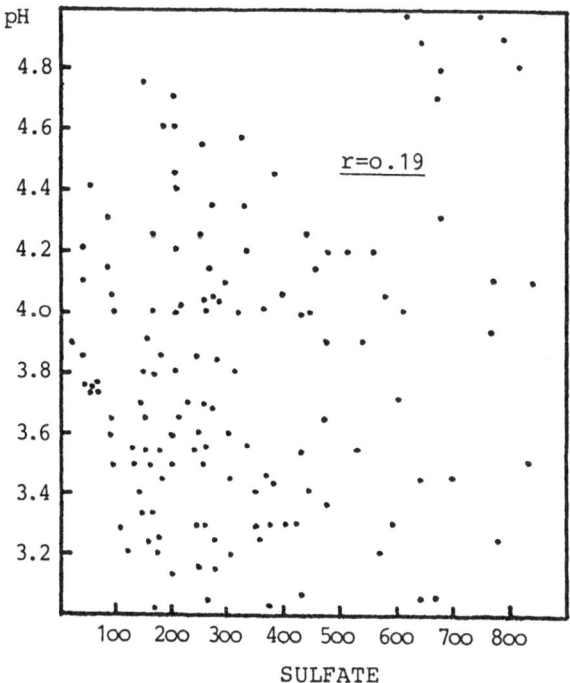

FIGURE 6. Picea. Correlation "sulfate
content - pH" of bark extract.

3.2.3. Bark transplantations

It is possible that neither Picea nor
other trees appropriate for pollutant
monitoring by means of bark analysis
are available. In this case transplants
of Picea bark can be used instead of
(Hofer, 1979, Grill, Hofer, 1979).
This method allows zoning of air pollution in the same way as the use of Picea
trees growing in the investigated region
(unpublished results; Grill, Hofer, 1979).

4. CONCLUSIONS

+ In our investigations it could be shown
that sulfate concentration of Picea bark
extracts provides a good mean for mapping
regions of different air pollution.(These
results are in accordance with Härtel,
Grill, 1972, Kienzl, Härtel, 1979, and
others). pH-values of Picea bark extracts
proved to be unsuitable for this purpose.

+ From the data obtained resulted the exclusion of the three deciduous tree species
(Acer, Aesculus, Tilia) from our further
investigations.

+ The course of bark extract sulfate concentrations during the whole investigated
period reflects the changes of the ambient
SO_2-pollution of each site in a suitable
way.

+ The lack of Picea trees in the investigated area can be compensated by Picea
bark transplants.

+ It must be stressed that data from bark
analysis can always be determined as relative values only.

+ The prescription for this method must be
strictly observed (bark thickness 3mm/
sites without dust and trees not sheltered)

+ A special advantage of the method referred
to is its simplicity, its cheapness and its
easy handling.

ACKNOWLEDGEMENTS
We should like to thank Prof.Dr.Karl BURIAN
for his aid and advices. We also thank Prof.
Dr.Otto HÄRTEL and Univ.Doz.Dr.Dieter GRILL
for their encouragement to perform these

306

investigations and their useful counsels. We are indebted to the Austrian Federal Ministry for Health and Environment (Österreichisches Bundesministerium für Gesundheit und Umweltschutz) and to the Austrian Academy of Sciences (Österreichische Akademie der Wissenschaften) for their financial support. The investigations are part of a project of the Ministry as well as of the Vienna MAB project group 'urban ecology'.

REFERENCES

Grill D and Hofer H (1979) Der Borken-test als Hilfsmittel zur Rauchschadens-kartierung. Zbornik Vol.1-39o, 283-288, Ljubljana.

Grodzinska K (1971) Acidification of tree bark as a measure of air pollution in Southern Poland. Bull.Acad.Polon.Sci. Sér.biol.Cl II 19(3), 185-195.

Härtel O (1977) Fichtenborke gibt Aus-kunft über die Luftgüte. Umschau 77(1o), 3o8-3o9.

Härtel O and Grill D (1972) Die Leitfä-higkeit von Fichtenborken-Extrakten als empfindlicher Indikator für Luftverun-reinigungen. Eur.J.Forest.Pathol. 2, 2o5-215.

Härtel O, Grill D and Krzyscin F (198o) Die Abgasbelastung im Raum Voitsberg-Köflach - Die Verwendung von Fichten-borke zur Indikation. Mitt. FBVA Wien 97, 367-386.

Hofer H (1979) Prüfung der Brauchbarkeit von Borkenexplantaten für den Borkentest. Hausarbeit Universität Graz.

Kienzl I (1978) Baumborke als Indikator für SO_2-Immissionen. Diss.Univ.Graz.

Kienzl I and Härtel O (1979) Die Luft-verunreinigungen im Stadtgebiet von Graz dargestellt anhand von Borkenuntersuchun-gen. Mitt.naturwiss.Ver.Steiermark 1o9, 113-135.

Köhm H-J (1976) Indikatoreigenschaften der Baumborke. Dissertationes botanicae 32, 91 S, J.Cramer(Vaduz).

Maier R, Sieghardt H, Punz W, Slad H, Engen-hart M, Domschitz E and Nagl A (198o) Öko-physiologische Untersuchungen in industriell belasteten Pflanzenbeständen im Raume Gai-litz, Kärnten. Carinthia II, 17o./9o.Jg., 279-299.

Świeboda M and Kalemba A (1979) The bark of Scots pine (Pinus sylvestris L.) as a bio-logical indicator of atmospheric air pol-lution. Acta Soc.Bot.Pol. 48(4), 539-549.

HEAVY METAL UPTAKE BY RADISH IN RELATION TO SOIL FERTILITY AND CHEMICAL EXTRACTABILITY OF METALS

DAVIES, BRIAN E. AND LEAR, JEAN M. (Department of Geography, University College of Wales, Penglais, Aberystwyth, Dyfed, SY23 3DB, Wales)

ABSTRACT

Heavy metal pollution of agricultural and horticultural land is a significant modern problem. There is an urgent need to develop reliable techniques to predict the availability of soil metals to plants and to assess how agronomic practices may limit uptake. This paper reports the results of growing radish (Raphanus sativus, L.) in metal mining areas. The Pb, Zn, Cu and Cd contents of the plants were compared with results from extracting the soils with several extractants for heavy metals and major nutrients. The data were examined statistically and the usefulness of the extractants assessed together with a consideration of the contribution of pH, humus and P contents in modifying metal uptake.

INTRODUCTION

Metals whose densities exceed 6 g/cc are collectively described by the convenient name 'heavy metals'. Some are essential for life processes (Co, Cr, Cu, Fe, Mn, Mo, Sn, V, W, Zn) whereas others (e.g., Cd, Pb, Hg) have no known beneficial biological role; all of them are injurious to plants or animals at higher concentrations. Plants absorb metals from the soils in which they root and there is a broad, direct relationship between the concentration of a given metal in a plant and in the soil. But availability to the root is influenced by soil factors such as soil reaction, redox potential or solubility and also by antagonisms or synergisms at the root-soil interface. Uptake of metals is influenced by the plant's metabolism and there are not only differences in metal uptake between species but also between varieties of the same species.

Very high concentrations of heavy metals in the soil can cause it to be sterile so far as higher plants are concerned. More commonly plants will adapt to the hostile environment, an effect which is readily observed in and around metal mine tailings areas. Some plants are more tolerant than others of high metal concentrations and therefore appear to thrive in inimical environments. For example Minuartia verna (vernal sandwort) is a Pb-tolerant calcicole which grows profusely on calcareous mine tailings in England and Wales. Other plants respond by evolving tolerant strains: Agrostis tenuis (Common Bent grass) is a good example since about 4 in 1000 seedlings will continue growing on mine spoil (Bradshaw, McNeilly, 1981) and can be shewn to be tolerant to specific metals. At lower concentrations many plants will grow normally yet their tissues will contain relatively high concentrations of heavy metals and thus pose a health threat to men or animals when eaten.

The evaluation of the plant-available heavy metal status of soils is an essential research need in several disciplines and chemical extraction rather than bioassay procedures is the most common choice of technique. Because the processes whereby metals are absorbed by plant roots are only poorly understood the approach to chemical extraction is largely empirical. Most extractants have been accepted for general use because they satisfactorily predict when crop responses will be observed following the application of nutrient elements in fertilizers. It is often, but erroneously, assumed that these same extractants will predict uptake of metals both at low or high soil concentrations.

There is an urgent need to develop reliable extraction techniques. This paper recounts the results of calibration investigations at Aberystwyth for Cd, Cu, Pb and Zn in soil using radish as a test plant.

FORMS OF HEAVY METALS IN SOILS AND THE CHOICE OF EXTRACT

The forms of heavy metals in soil can be considered as comprising several fractions or compartments which act as reservoirs of available metals. No one chemical reagent selectively and uniquely dissolves metals from a compartment and the compartments are to some extent differentiated by solubility. An extractant designed to extract metals at a given level of solubility necessarily also extracts those which are more soluble. In Table 1 some common soil extractants are classified in terms of the soils fractions they are thought to attack.

TABLE 1. Some common soil extractants and the fractions they are thought to extract. (EDTA = ethylene diamine tetraacetic acid; EDDAH = ethylene diamine di-o-hydrophenyl acetic acid; DTPA = diethylene triaminepenta acetic acid; TEA = triethanolamine)

Soil fraction	Common extractant
Soil solution	H_2O; 0.01M $CaCl_2$
Exchangeable	0.5M CH_3COONH_4; 0.2M $MgSO_4$; 0.1M NH_4Cl 1M NH_4NO_3
Adsorbed	0.5M CH_3COOH; 0.1M HCl; 0.1M HNO_3
Complexed	0.05M EDTA (diammonium or disodium salt) 0.05M EDDAH; 0.005M DTPA/0.1M TEA/0.05M $CaCl_2$ @ pH 7.3
Oxide bound	1M CH_3COONH_4/0.002M $C_6H_5(OH)_2$; 0.2M $(COO)_2(NH_4)_2$; 0.15M $(COOH)_2$ @ pH 3.3
Total	HF; $HNO_3/HClO_4$; HNO_3; fusion mixtures.

The soil solution is believed to be the immediate source of metals for plants and the soluble metals are present as organic and inorganic complexes: Sposito (1981) has described a computer program (GEOCHEM) which may be used to predict the ion species and complexes. In this investigation soils were extracted with 0.01M $CaCl_2$ but many of the levels were below the detection limit of flame atomic absorption spectrometry and the results are not therefore reported here.

Ions may be held on the charged surfaces of clay minerals and humus by adsorption forces and may be displaced from these sorption sites by simple salt solutions. Molar ammonium nitrate was chosen in this investigation since it is the extractant preferred by the official extension service (ADAS) for estimating available potassium and magnesium which were also determined.

Acetic acid is commonly used to extract both specifically adsorbed cations and precipitated or other slightly soluble forms. It has also been widely used to estimate plant-available phosphorus. Soil humus plays an important part in retaining metals for release to the soil solution in addition to its contribution to the exchange complex. It holds metals by the formation of complexes, especially chelates, and the metals can be decomplexed by chelating agents. EDTA was first proposed by Viro (1955) to predict Cu and Zn deficient soils and has been widely used for Cu since then.

In this work oxide bound metals were not estimated. Total metal content is useful for identifying polluted soils although its ecological or agronomic value is limited. A complete dissolution of the soil may be achieved using HF but the procedure is too slow for processing large numbers of soils. Accordingly, hot concentrated HNO_3 was used to extract heavy metals.

METHODS

Radish (Raphanus sativus l. var. Webb's French Breakfast) was sown direct in drills 1-2 m long in private vegetable gardens in the old lead mining areas of Derbyshire, Clwyd and Dyfed in England and Wales. The plots were dug and weeded prior to sowing in 1978, no fertilizer was used but the

plants were watered using a local supply. Soil samples were taken (0-12 cm) with a stainless steel trowel. At harvest, some 6-8 weeks after sowing, mature radish plants were removed for analysis from 40 plots. In the laboratory leaves and edible roots were separated, washed in distilled water, dried at $100^{o}C$ and milled. Soils were air dried and disaggregated and passed through a sieve of 2 mm aperture.
The dry plant material (2-5 g) was ashed for 24 hours at $430^{o}C$ and extracted into 0.1M HNO_3. Soils were analysed for pH (equilibration in 0.05M $CaCl_2$) and organic content was estimated from the loss-on-ignition at $430^{o}C$. Soils were extracted thus: total metals – 10 g soil/20 cc conc. HNO_3/simmer 60 minutes/take to near dry and extract into 0.1M HNO_3; EDTA – 15 g soil/ 75 cc 0.05M EDTA-$(NH_4)_2$ at pH 7/shake 60 minutes/ filter; acetic acid – 5 g soil/100 cc 5% (v:v) CH_3COOH/shake 30 minutes/filter; ammonium nitrate – 10 g soil/100 cc 1M NH_4NO_3/shake 60 minutes/filter. Cation exchange capacity was determined using Na as the displacing ion at pH 7. All metals were determined by atomic absorption flame spectrometry using a double beam instrument with background correction. Phosphorus was determined colorimetrically through formation of the blue phospho-molybdate complex.

RESULTS AND DISCUSSION
Soil heavy metal values are summarised in Table 2.
Although the experimental plots were situated in productive, private vegetable gardens they were all located in old mining areas where there is widespread metal pollution (Davies, Roberts, 1975; Davies, White, 1981) and the metal values reveal a range from normal to heavily contaminated. In some of the gardens with very high zinc contents vegetable leaves were observed to be chlorotic and some sensitive plants, e.g., spinach do not grow well.

TABLE 2. Summary data for the heavy metal content of the 40 experimental soils

| | mg/kg dry soil | | | |
	Pb	Zn	Cu	Cd
TOTAL				
Mean	1394	1019	61	7.2
Minimum	25	57	10.5	0.05
Maximum	8355	5929	346	62
EDTA				
Mean	337	210	22	4.2
Minimum	5.5	3.5	2.3	0.05
Maximum	900	950	75	24
ACETIC ACID				
Mean	103	372	3.2	2.6
Minimum	2.0	2.0	2.0	0.05
Maximum	720	2000	16.0	23
AMM. NITRATE				
Mean	2.6	13.5	1.0	0.18
Minimum	1.0	1.0	1.0	0.05
Maximum	39	95	1.0	1.30

The mean metal values for the different extractants were compared with their respective total values. Except for Cd less than half the total metal content could be extracted by the various extractants. Thus, the maximum extracted (as % total metal) was by EDTA for Pb (24.2%), Cu (36.1%) and Cd (58.3%) and by acetic acid (36.5%) for Zn (EDTA-Zn = 20.6%). This suggests that humus represents a significant storage compartment for Cd, Cu and Pb but less so for Zn and the contrast between Zn and Cd in this respect is interesting in view of their general chemical similarity. Acetic acid extractably Cd was comparable with Zn (36.1%) but it was a poor extractant for Pb (7.4%) and Cu (5.2%). Readily exchangeable Pb, as measured with ammonium nitrate was a negligible proportion of total (0.2%) and proportions for Zn (1.3%) and Cd (2.5%) although higher were also small.
Summary values for major nutrients and fertility parameters are presented in Table 3.
The data in Table 3 reveal that the gardens varied considerably in overall fertility, from very acid to mildly alkaline, overall moderately high organic contents and a wide range of major nutrient status.

TABLE 3. Soil fertility parameters

	pH	% Humus	Me/100 g Cation Exchange Capacity	mg/kg dry soil		
				Mg*	K*	P***
Mean	5.9	10.2	27.8	129	293	266
Minimum	3.9	4.9	11.8	49	55	1.8
Maximum	7.1	20.0	48.9	310	1000	3625

*Ammonium nitrate extract
**Acetic acid extract

Table 4 summarises the radish metal contents.

TABLE 4. Heavy metal contents of radish leaves and edible root

		mg/kg dry matter		
		Mean	Minimum	Maximum
Pb	leaves	51	2.3	233
	root	24	2.2	162
Zn	leaves	416	52	2717
	root	313	41	1871
Cu	leaves	6.1	2.9	15.1
	root	3.8	1.8	12.5
Cd	leaves	2.6	0.09	20
	root	0.90	0.10	3.9

The values in Table 4 indicate a marked tendency for heavy metals to accumulate in radish leaves, even for Pb which, more generally, tends to concentrate in roots of vegetables (Davies, White, 1981). In Britain it is an offence to sell vegetables containing more than 1 mg Pb/kg fresh weight and this can be equated with 10 mg Pb/kg dry weight. Both the mean and the maximum radish root (edible) lead values exceed the legal limit and overall it was exceeded by 22/40 samples. This problem of high metal contents in vegetables grown in these mine contaminated areas has been reported before (Davies, White, 1981) and in that paper it was suggested that 0.2 mg Cd/kg fresh weight or 2 mg Cd/kg dry weight could be taken as a useful threshold by which to judge the health implications of Cd in food. The mean and maximum root Cd levels exceed this limit but only 6/40 samples were in excess. The picture that emerges from Tables 2-4 is of a group of fairly typical private vegetable gardens in Britain but having the special problems of heavy metal contamination so that plants grown in them range from acceptable, unpolluted vegetables to a proportion containing excess Cd and Pb. In none of the gardens was the Zn content high enough to preclude growth but some of the radish leaves tended to be chlorotic. The data were next examined by correlation analysis to investigate the relationships between plant metal contents and soil values according to the several extractants used. Although there was a tendency to positive skewness in several of the data sets the populations were not normalised on this occasion by, e.g., converting to $\log_{10}$ values. The results of the correlation analysis are presented in Table 5. Any correlation coefficient greater than 0.5 is significant at $p = 0.001$ for the forty radish-soil pairs represented in Table 5. It is interesting that all four extractants yielded satisfactory correlations for plant Pb except for EDTA/root Pb and this is in contrast with other work in this laboratory where EDTA has been found to be a useful predictor of Pb adsorption by radish. E.g., in an investigation of London soils (Davies, et al., 1979) a correlation coefficient of 0.87 ($p = 0.01$) for radish root Pb and EDTA extractable Pb was obtained. The correlations with total Pb are particularly noteworthy since they suggest this is as good a predictor as any whereas for most trace elements the total concentration has only limited ecological or agronomic significance. The behaviour of zinc is very different and ammonium nitrate is the only extractant to yield satisfactory correlations for both leaf and root zinc. This implies that the readily exchangeable Zn fraction, i.e., the Zn adsorbed on to clay and humus exchange sites, represents the main pool of plant-available Zn

TABLE 5. Correlation coefficients for heavy metals in soils and leaves and edible roots of radish. Correlations underlined are significant at p = 0.001

	TOTAL	EDTA	HAC	NH$_4$NO$_3$	TOTAL	EDTA	HAC	NH$_4$NO$_3$
	Pb				Zn			
Leaves	0.828	0.653	0.804	0.587	0.404	0.198	0.582	0.891
Roots	0.743	0.408	0.662	0.737	0.283	0.146	0.458	0.864
	Cu				Cd			
Leaves	0.413	0.606	0.516	-	0.610	0.772	0.665	0.274
Roots	0.373	0.317	0.134	-	0.700	0.854	0.773	0.316

whereas Zn complexed in humus (EDTA extractable) has little importance. Acetic acid performed satisfactorily for leaf Zn which is consistent with this reagent's properties of extracting both specifically and non-specifically sorbed metals but not, to any great extent, organically complexed metals. The cadmium data are interesting. Symeonides, McRae (1977) concluded that 1M NH$_4$NO$_3$ was the most useful extractant from a range of reagents they investigated and they obtained a correlation coefficient for soil Cd and radish leaf cadmium of 0.97 (p = 0.001). In this investigation the same extractant performed badly and, overall, EDTA proved the most useful. Here also there is an intriguing dissimilarity with Zn considering that the chemistry of the two metals is so often similar. In the case of Cu none of the extractants investigated was apparently very useful, including EDTA, which is surprising considering the widespread use of EDTA for identifying Cu deficient soils.

According to reports in the literature, soil pH, humus content, cation exchange capacity or phosphorus content have each been identified as modifying the uptake of heavy metals. The data were therefore examined by partial correlation analysis to establish whether the correlations could be improved by controlling for pH, organic content, cation exchange capacity and extractable phosphorus. This approach was disappointing.

For the highest correlations in Table 5 none was improved by partial correlation. The only significant improvement was found for EDTA-extractable Zn where controlling for pH increased the size of the correlation coefficient:

Radish leaf Zn/EDTA extractable Zn: r = 0.198
Radish leaf Zn/EDTA Zn + pH: r = 0.381
Radish root Zn/EDTA extractable Zn: r = 0.146
Radish root Zn/EDTA Zn + pH: r = 0.452

The corresponding multiple regression equations were:

Lead-Zn = 0.37 EDTA-Zn - 0.37 pH + 1827
Root-Zn = 0.37 EDTA-Zn - 0.48 pH + 1569

Although many of the metal-plant correlations are highly significant the data tend to be scattered which reflects the fact that the metal content of a plant depends not only on the supply but on other controls of uptake. Furthermore, the data do not indicate that a single extractant will satisfactorily predict uptake of all metals. EDTA emerges as suitable for Cd and Pb and perhaps for Zn if account is taken of soil pH.

In the introductory section it was pointed out that for the most part soil extractants are used in agronomy to identify responsive soils, i.e., nutrient deficient situations in which the application of fertilizers will increase crop production. Can the same extractants identify soils

312

where excess of an element will either diminish crop yield or cause an outright failure of growth? A subsequent radish growth experiment (1980) in the same areas was helpful in answering this question. Radish was grown in soils with a wider range of metal contents and in several cases there was either a tiny yield of immature plants (near failure) or growth was very unsatisfactory. A note was made of the near failure sites and also of those plots where leaves were chlorotic and formation of the radish root was unsatisfactory, leading to small or misshapen bulbs. It was assumed that the phytoxicity was due to Zn and the three groups of radish (near failure, small bulbs, satisfactory growth) were separated and the corresponding EDTA-extractable soil Zn values noted. These are summarised in Table 6.

TABLE 6. Summary data for soil Zn for three categories of radish growth

	mg Zn/kg dry soil (EDTA extract)			
Growth class	Mean	Median	Minimum	Maximum
Near failure (7)	642	430	272	1331
Small (8)	267	257	85	560
Satisfactory (14)	147	117	39	389

Although the groups are separated by both the mean and median values there is a considerable overlap of the ranges. This is not surprising since the plant's sensitivity to zinc must depend on many interacting factors. Therefore a simple unequivocal toxic limit cannot be established. For radish, from these data, it lies somewhere in the range 642 - 1331 mg Zn/kg soil.

CONCLUSIONS

It is concluded from this work that, within the total metal content of the soil, Cd, Cu and Pb are held primarily in the humus fraction whereas for Zn the exchangeable and specifically sorbed forms are more important. The success of the extractants in predicting metal uptake is consistent with this except for Cu where EDTA proved disappointing. Exchangeable Zn correlated well with leaf and root Zn and EDTA was a successful extractant for Cd and Pb. The performance of EDTA for Zn was improved when allowance was made for soil pH. In general EDTA approximates best to a wide spectrum extractant for predicting metal uptake. It may also be used for diagnosing soils where available Zn may severely restrict growth: for radish this toxic threshold lies in the range (approximately) 650-1300 mg Zn/kg soil.

ACKNOWLEDGEMENTS

We are grateful for financial support from the Welsh Office and the cooperation of many gardeners.

REFEERENCES

Bradshaw AD and McNeilly (1981) Evolution and pollution. Studies in Biology No. 130. London. Edward Arnold.
Davies BE and Roberts LJ (1975) Heavy metals in soils and radish in a mineralised limestone area of Wales, Great Britain, Science Total Environment 4, 249-261.
Davies BE and White HM (1981) Trace elements in vegetables grown on soils contaminated by base metal mining, J. Plant Nutrition 3, 387-396.
Sposito G (1981) Trace metals in contaminated waters, Environ. Sci. Technol. 15, 396-403.
Symeonides C and McRaw SG (1977) The assessment of plant-available cadmium in soils, J. Environ. Qual. 6, 120-123.

ADAPTATION OF SOIL MICROORGANISMS TO DECOMPOSITION OF SOME HERBICIDES

D. I. Bakalivanov (N. Poushkarov Institute of Soil Science and Yield Programming, Sofia, Bulgaria)

1. INTRODUCTION

The adaptation of soil microorganisms to the triazine herbicides atrazine and simazine as a factor influencing their degradation is very important agronomically. In crop rotation, some crops are often damaged by the same herbicides, due to their slower detoxification in the soil (Pilnik, Tarassov, 1965). On the other hand, atrazine and simazine as herbicides have a good effect on many weedy grasses and show low toxicity to mammals. That is why they are widely used in agriculture and every factor affecting their detoxification should be investigated, so as to improve their degradation. Soil microorganisms are the major avenue for decomposition (Alexander, 1967), and the adaptation of many soil microbes to these herbicides is a significant factor in detoxification (Bakalivanov, Popova, 1977).

The response of soil microorganisms to the addition of pesticides is dependent oh their enzymatic ability to react usefully with the pesticide molecules and to proliferate in the soil. Usually after a period of exposure to the pesticide, certain microorganisms acquire the complement of enzymes need to decompose the pesticides in such manner that they derive a growth advantage from this substrate. This lag period (lag phase), during which little or no appreciable loss occurs, is followed by a rapid decomposition phase (Holly, Roberts, 1963; Burnside, 1965). The duration of this lag phase depends on the chemical composition of the pesticides and on soil type, with respect to the soil microorganisms, and can continue several days to several months. This is a period of adaptation for the soil microorganisms, during which some of them synthesize specific enzymes, degrating the molecules of the pesticides (Sheets, 1970; Kruglov, 1976). But Sikka and Davis (1966) present data about decomposition of atrazine with-

out lag phase. In that case presumably the soil microorganisms have been already adapted to the herbicide by previous treatment, or possess a greater natural ability to decompose it.

2. MATERIAL AND METHODS

The herbicides used-atrazine (-2chlor-4-ethylamino-6-isopropyl -amino-s-triazine) and simazine (2-chlor-4,6-bis/ethylamino/-s-triazine) are products of Agria (Bulgaria) as preparation called respectively zeazine and herbazine, with 50% active ingredient. The same herbicides were also used in the experiment as substances with 100% active ingredient with pure cultures of microorganisms. The two soils used were alluvial, from the vicinity of Kostinbrod, and smolnitza, from the vicinity of Bojouriste. Pure cultures of soil fungi were isolated from these soils, from field experiments with the same herbicides. After that the microbes were additionally adapted to atrazine and simazine by several passages on Kaufman's medium (Kaufman et al., 1965) containing 5 mg per liter herbicide. The mycelia and cultural liquids of adapted fungi were homogenized (in a cooling bath) and centrifuged at 6000 rpm for 30 minutes. The supernatants were dialysed and exposed to the herbicides, in concentrations of 25 $mg.l^{-1}$ for 5 hours. The residual quantities of herbicides were analyzed by the method of Mattson et al., 1970. The natural acclimation of the soil microbes to the decomposition of the herbicides was demonstrated by several treatments of the soils with atrazine and simazine under the conditions of a model and a field trial.

3. RESULTS AND DISCUSSION

Our data for adaptation of some soil fungi, cultivated with 5 $mg.l^{-1}$ and 10 $mg.l^{-1}$ atrazine added to the medium, are presented in Table 1. They show that *Penicillium cyclopium* has good ability to

TABLE 1. Detoxification of atrazine and simazine herbicides by non-acclimated and acclimated soil fungi, at the level of adaptation treatment.

Soil fungi	Origin of soil	by non-acclimated fungi atrazine	simazine	by acclimated fungi atrazine 5 mg	10 mg	simazine 5 mg	10 mg
Penicillium lilacinum	Sopol	12	10	27	0	27	0
P. lilacinum	Slivnitza	12	16	27	0	24	0
P. cyclopium	Kostinbrod	15	15	30	0	37	0
P. cyclopium	Prague	0	24	24	12	18	0
P. terrestre	Prague	0	15	27	0	18	0
P. citrinum	Samokov	12	0	34	0	0	0
P. citrinum	Prague	0	0	24	0	0	0
P. funiculosum	Lavino	0	0	0	0	0	0
P. rubrum	Bojouriste	0	0	0	0	0	0
P. expansum	Prague	0	0	0	0	20	0
P. citreo-roseus	Borovetz	12	0	15	0	6	0
P. ochraceus	Bojouriste	0	11	0	0	16	0
Aspergillus niger	Belogradchik	0	0	12	0	0	0
A. niger	Borovetz	12	10	18	10	15	0
A. terreus	Bojouriste	0	0	18	0	0	0
A. sulfureus	Gen. Toshevo	0	0	0	0	30	22
A. glaucus	Bojouriste	12	11	20	0	20	0
Paecilomyces varioti	Grigorevo	12	10	18	12	30	12
P. varioti	Prague	0	0	18	12	0	0
Trichoderma viride	Bojouriste	7	10	10	0	23	0
T. viride	Prague	0	0	0	0	12	0
Fusarium roseum	Belogradchik	12	0	20	0	0	0
Cephalosporium curtipes	Prague	12	6	16	0	15	0
Culvularia lunata	Prague	9	6	12	0	27	0
Stachibotris atra	Riverside	0	0	0	0	0	0
Alternaria tenuis	Trastenik	5	12	10	0	16	0
Cladosporium herbarum	Trastenik	0	0	10	0	6	18

TABLE 2. Influence of the adaptation of soil microbes in alluvial soil on decomposition rate of atrazine (Model experiment)

Variant	Rate $(mg.kg^{-1})$	Residue of herbicide after 90 days (%)
Without previous treatment with atrazine	10	36
	5	30
	2,5	17
With previous treatment with atrazine	10	34
	5	29
	2,5	9
With two previous treatments with atrazine	10	34
	5	28
	2,5	8

Note: Mean errors are ±0,3-0,5%

TABLE 3. Influence of the adaptation of soil microbes in smolnitza on the decomposition rate of simazine (Field experiment)

Variant	Residue of herbicide (%) after 20	90	150 days
Without previous treatment (control)	93	35	22
With one previous treatment (during the previous year)	84	31	19
With 10 previous treatments (during 10 years)	82	31	14

Note: Herbicide used at a rate of 400 $g.dk^{-1}$. Mean errors are ±0,4-1%.

adjust. This fungus increases the amount of decomposed herbicide from 15 to 30% by cultivation with 5 mg herbicide. However, the rate of 10 mg atrazine is toxic for it and stops the decomposition. *Penicillium lilacinum* also has ability to acclimate to atrazine, that is, without acclimation it decomposes only 12%, but after acclimation with 5 mg it increases that percentage to 27. *Paecilomyces varioti* also increases decomposition from 12 to 18% by the same acclimation, but it decreases the amount of decomposed herbicide to 12% when cultivated with 10 mg. This evidence shows that a high rate of the herbicide has a toxic effect. Some other fungi, namely *Aspergillus terreus, A. niger* and *Penicillium terrestre* do not decompose atrazine without acclimation, but they decompose it after cultivation with 5 mg of the same, *A. terreus* up to 18%, *A. niger* up to 12% and *Penicillium terrestre* up to 27%. Degradation stops, if there are 10 mg herbicide in the medium.

A similar ability to adapt to simazine is demonstrated by the investigated soil fungi after cultivation with 5 and 10 mg of the herbicide. Both herbicides have very similar chemical composition and that is why the acclimation of the fungi is like that to atrazine. Therefore it is very possible that the enzymes decomposing these herbicides are the same, or similar. The data in Table 1 show that *Penicillium cyclopium* has a high level of ability to adapt to simazine (as well as to atrazine). It decomposes 15% of simazine without acclimation and 37% after cultivation with 5 mg herbicide. *Paecilomyces varioti* and *Penicillium lilacinum* also have high ability in this respect. They increase decomposition respectively from 10 to 30% and from 10 to 27%. Increase of the simazine rate to 10 mg.l^{-1} is also toxic for all the investigated fungi; they decrease or completely stop their decomposing activity.

The natural acclimation of some species of the soil microbial cenosis to the pesticides is speeded by previous treatment of the soil with herbicides during one or several years.

The data from our model experiment with alluvial soil repeatedly treated with atrazine show a well-expressed adaptation of soil microorganisms to the degradation of the herbicide, with an increase in their decomposing activity (Table 2). As a result, the residues of atrazine are smaller. The smallest residue of the herbicide is attained by triple treatment with atrazine. The data in the same table show that the decomposition of the herbicide is faster in treatment variants with lower rates of herbicide application, demonstrating better acclimation of the microorganisms to these lower rates. The unsatisfactory action of the high rate could be related to its toxic effect on the microorganisms. The data from our field experiment (Table 3) confirm the results of the model trial. They show that the previous treatment of the soil with simazine increases its decomposition.

In this respect the investigations of Fryer and Kirkland (1966) and Kirkland (1967) are very interesting. These authors established a faster degradation of MCPA in soil that was treated over a 4-year period with the same herbicide. The studies of Suess (1967) and Fryer et al. (1980) with simazine and Ahrens (1967) with 2,4-D show a similar tendency. Kaufman (1964) found a reduction of the lag period in the decomposition of dalapon when soil was re-treated with the same herbicide. According to Hiltbolt (1974) decomposition of pyrazon was preceded by a lag phase of about 6 weeks after the initial treatment, but decomposition began immediately upon renewed addition of pyrazon. Degradation of pyrazon was developed rapidly in soil without such treatment when it was inoculated with a small amount of soil previously treated with the same preparation, as a source of acclimated microorganisms.

The presented data (Figure 1) of the quantity of some microbe populations are a direct proof of acclimation of soil microorganisms to atrazine. They show the number of soil bacteria, actinomycetes and fungi in smolnitza soils, where the field experiment was carried out.

The curves of development of these microorganisms show significant differences in their quantity that is, in the depressive effect of the herbicide at

316

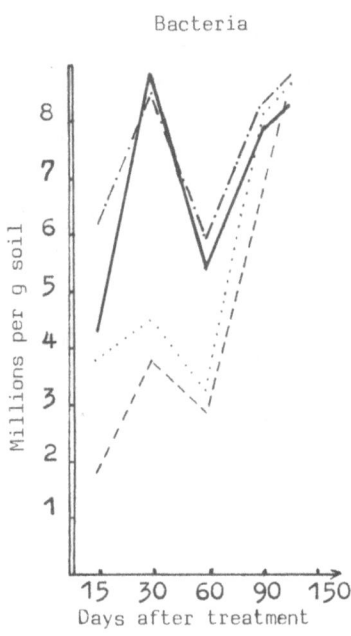

Bacteria

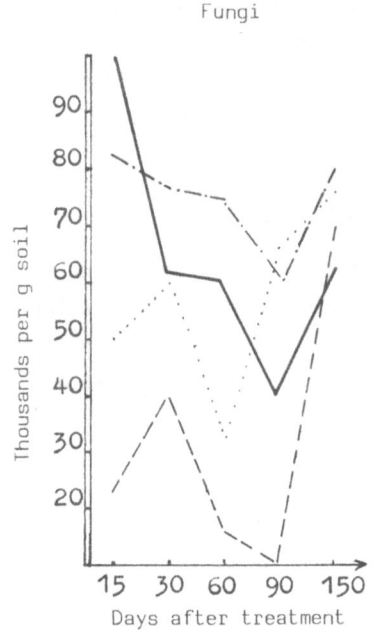

Fungi

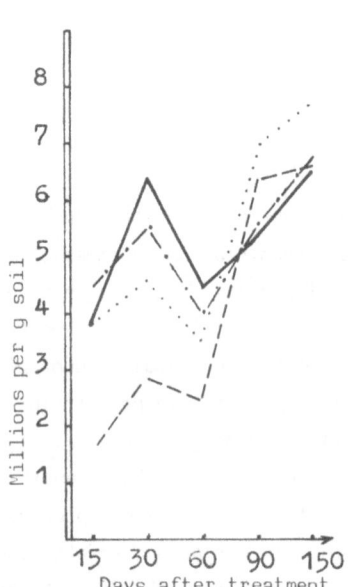

Actinomycetes

Key

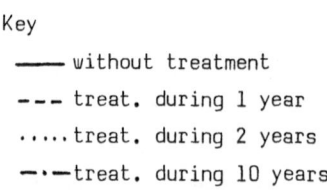

——— without treatment

– – – treat. during 1 year

·····treat. during 2 years

—·—treat. during 10 years

FIGURE 1. Quantity of soil microorganisms as affected by frequency of treatment with simazine.

the beginning of the experiment. It goes down after 1-3 months and practically stops towards 150 days after the treatment. This negative effect is the strongest during the first year and the weakest in the variant of 10 years treatment of soil with simazine. The diminution of the effect of the herbicide on the investigated microorganisms can be attributed to excretion of adaptive enzymes from soil microorganisms decomposing the herbicide, and to the adaptation of the soil microflora to proliferate in the presence of the same herbicide.

4. CONCLUSIONS

1. The acclimation of soil microbes to atrazine and simazine is an important factor for the detoxification of these herbicides.

2. Repeated treatments with these herbicides increase decomposition of the herbicides as a result of that acclimation.

3. Soil microorganisms are less sensitive or are insensitive to the toxic effect of atrazine and simazine when the soil is retreated with the herbicides.

REFERENCES

Alexander M (1967) The breakdown of pesticides in soils. In Brady N, ed. Agriculture and the quality of our environment, pp. 331-342. Amer. Ass. Adv. Sci., Washington D.C.

Bakalivanov D and Popova J (1978) Detoxication of the herbicide atrazine by adapted enzyme of soil microscopic fungi. IV Congress of microbiology, Sofia, p.365-368.

Burnside O (1965) Longevity of amiben, atrazine and 2,3,6-TBA in incubated soils, Weeds 13, 274.

Fryer P, Smith D, and Hance R (1980)Field experiments to investigate long term effects of repeated application of MCPA, tri-allate, simazine and linurone: Crop performance and residues 1969-78, Weed Research 20, 103-110.

Hiltbolt A (1974) Persistence of pesticides in soil. In Guenzi WD, ed. Pesticides in soils and water. Soil Sci. Soc. Amer.,Madison, Wisconsin.

Holly K and Roberts H (1963) Persistence of phytotoxic residue of triazine herbicides in soil, Weed Research 3, 1-10.

Kaufman D (1964) Microbial degradation of 2,2-dichlorpropionic acid in five soils, Canad. J. Microb. 10, 843-852.

Kaufman D, Kearney Ph and Sheets T (1965) Microbial degradation of simazine, Agr. Food Chem. 13 (3), 238-242.

Kirkland K and Fryer J (1966) Pre-treatment of soil with MCPA as a factor affecting persistence of a subsequent application. In Proc. 8th Brit. weed Contr. Conf., p. 616-621.

Kirkland K (1967) Inactivation of MCPA in soil, Weed Research 7, 364-367.

Kruglov J (1976) Soil microflora and herbicides. In Mouromzev G, ed. Agronomicheskaja mikrobiologia, p. 204-229, Kolos, Leningrad.

Pilnik V and Tarassov A (1965) Vestn. Selhos. Nauki 2, 42.

Sheets T (1970) Persistence of trazine herbicides in soils, Residue Reviews 32, 287-310.

Sikka H and Davis D (1966) Dissipation of atrazine from soil by corn sorghum and Johnsongrass, Weeds 14, 289.

Suess A (1970) Bayer Landwirt. Jahrbericht 3,4, 401-408.

DETERMINATION OF ZINC, IRON AND ALUMINIUM IN HALOPHYTIC SPECIES OF THE RIVER SADO ESTUARY - PORTUGAL,
BY ATOMIC ABSORPTION SPECTROMETRY F. REBOREDO - Bolseiro do INIC, Centro de Engenharia Biológica da
Universidade de Lisboa (BL1), Depart. Ciências do Ambiente - Universidade Nova de Lisboa, Quinta da
Torre. 2825, Monte da Caparica, Almada-Portugal.

ABSTRACT

In our research emphasis is placed upon
bioaccumulation of Zn, Fe and Al by salt marsh
vegetation of coastal areas of River Sado
estuary.

The differences in trace metal accumulation in
each species, collected at different sampling
points is discussed, considering the possible
influence of some local environmental factors,
and the position of the species in the salt
marsh.

The results show that roots of Halimione
portulacoides and Arthrocnemum perenne
accumulated large amounts of trace metals,
rather than above-ground tissues.

The highest mean values for each metal, were
always determined in A. perenne roots.

This suggest the importance of the root system
of the studied species in the recycling of
metals and in the decontamination of the
environment.

INTRODUCTION

The determination of heavy metals in biological
material has increased in the last ten years
(BOYDEN and ROMERIL, 1974, BADSHA and SAINSBURY,
1978; STONEBURNER and HARRISON, 1981), because
of the large "input" of metal pollutants to
the environment, as a consequence of indus-
trialization and urbanisation. Aquatic organisms have received particular attention
(HARDISTY et al., 1974) due to the high sensitivity of these species to pollution. The
transfer of pollutants through the food chain,
may reach human being and produce chronic and
acute diseases or even death (GOLDWATER, 1971).

Terrestrial organisms, are also tested in order to
know their sensitivity and accumulation degree,
specially vegetal species which are submitted to
phosphate fertilizers application (BEAUFAYS and
NANGNIOT, 1976).

Coastal vegetation has received particular attention of botanists, because of the distinct zonation
of the species and its relation with salt tolerance and response to flooding (ROZEMA, 1978).
Also the important role of marsh vegetation in
metal recycling has been shown by several authors
(BANUS et al, 1974; WINDOM, 1975).

The following points suggest the importance of
coastal marshes in this context(BEEFTINK, 1977):
1. Signalling the harmful effects of marine pollution on plants and animals, decontamination of
the ecosystem by absorption or destruction of
marine or estuarine pollutants and as a buffer
environment between the marine or estuarine and
terrestrial zones.
2. Estimation of the agricultural potentialities
3. As gene-reservoirs for further hybridization
and selection of new crops

Our study deals with trace metal accumulation -
Zn, Fe and Al in Halimione portulacoides (L) and
Arthrocnemum perenne (Miller) moss from a polluted
area of River Sado estuary. It is a preliminary
contribution to the knowledge of potential biological indicators in portuguese salt marshes.

EXPERIMENTAL

Apparatus. A Perkin-Elmer model 403 atomic
absorption spectrophotometer equipped with a
deuterium background corrector was used for flame
analysis, in the case of Zn and Fe determinations.
The conditions for analysis were those recommended

in the Perkin-Elmer Analytical Methods book.
A Pye Unicam SP 1900 atomic absorption spectrophotometer was used for Al analysis. In order
to remove interferences or suppress ionisation,
the samples and standards to be analysed for
aluminium were made up to 2000 $mg.l^{-1}$ in potassium chloride.

Reagents. Reagents pro-analysis were used in all
determinations
Nitric Acid 65%, sp. gr. 1.40
Perchloric Acid 70%, sp. gr. 1.67
Hydrochloric Acid 37%, sp. gr. 1,19
Standard Metal Solutions. Standards were prepared
by serial dilution of 1000 $mg.l^{-1}$ metal stock

Plant Part	Local Sampling	Fe	RSD(%)	Al	Mean µg/g D.W. RSD (%)	Zn	RSD(%)
R		718.2	13.0	913.0	7.5	30.5	12.2
S	1	154.6	12.3	128.6	6.8	23.8	10.5
L		527.3	5.6	292.4	15.8	40.0	6.3
R		1826.1	11.2	1866.7	19.1	85.1	13.1
S	2	919.5	9.0	610.5	10.1	54.9	18.6
L		1321.7	5.5	776.1	17.0	62.9	1.9
R		1700.0	4.6	1920.0	5.7	37.0	7.4
S	3	791.7	5.3	420.0	13.1	27.0	7.7
L		941.7	6.7	760.0	10.3	59.0	10.6

TABLE 1. Mean values* found in H. portulacoides
      R= Root   S= Stem   L= Leave
      RSD= Relative Standard Deviation

Plant Part	Local Sampling	Fe	RSD(%)	Al	Mean µg/g D.W. RSD(%)	Zn	RSD(%)
R.		1316.7	13.7	812.7	16.9	342.5	9.5
L.S.	1	445.4	8.5	168.7	17.6	15.5	4.6
S.S.		377.1	8.2	309.5	23.7	20.0	8.8
R.		3363.6	15.4	2866.7	5.3	160.6	5.1
L.S.	2	781.9	26.1	592.6	18.8	21.6	20.0
S.S.		808.3	8.6	711.1	8.6	24.2	23.6
R.		3990.0	9.4	3600.0	11.6	364.7	12.2
L.S.	3	642.9	5.6	710.0	11.6	9.0	15.2
S.S.		685.7	13.5	920.0	11.9	15.0	16.7

TABLE 2. Mean values* found in A. perenne
      R= Root   L.S.= Long erect stem
      S.S.= Short erect non-flowering
          and flowering stem
      RSD = Relative Standard Deviation

*each value is the mean of five replications

solutions. All standard solutions contain the same
reagents, as those added to the samples.
Vegetal Samples. Plant material was collected at
low tide from three points as indicated in Figure
1. After a careful rinsing by distilled water, the
samples were divided into leaves, stems and roots

in the case of H. portulacoides or long erect stems short erect non-flowering and flowering stems and roots in the case of A. perenne.

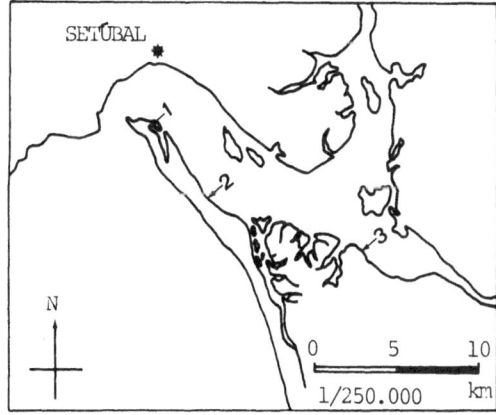

FIGURE 1 - Localization of the sampling
          points

Nitric Acid-Perchloric Acid digestion. For each gram of dry wheight, 20 ml of HNO_3 and 5 ml of $HClO_4$ were used. Digestions were carried out in the same way as that previously described for sediments (REBOREDO, 1981).

RESULTS AND DISCUSSION

Great differences were noted between metal concentrations in each species collected at the different sampling points.
The highest mean values of H. portulacoides and A. perenne were always found in samples from station 3, except for iron and zinc in the case of H. portulacoides. However, these levels do not differ greatly from those found in Station 2 (Table 1 and 2).
The lowest mean values were found in Station 1 except for Zn in the case of A. perenne (Table 1 and 2).
These differences can probably be explained by three factors.
1. In Station 1, the studied species live about 15 m off the low tide line.
2. In the other Stations, these species live nearer the low tide marsh and the sea purslane

H. portulacoides borders drainage channels.
3. Probably, soil water retention does not differ significantly in different sampling points.
However, the duration of soil waterlogging in Station 1 during low tide, would be shorter than in other stations.
Soil water would be lost by evaporation more easily, and so increase soil salinity which affects the uptake of trace metals by salt marsh vegetation.
Trace metals are greatly concentrated in roots in all species, rather than in the above-ground tissues. A similar finding was observed by BRETELER et al. (1981) for mercury in relation to Spartina alterniflora.
However, leaves of H. portulacoides from Station 1 and 3 exhibit the highest levels of zinc, which may constitute a metabolic requirement, while the other peak for this element was found in roots from Station 2. In spite of all species being able to take up large amounts of trace metals into their roots, only a small quantity of these elements is carried to other tissues.
Tables 3 and 4 present a whole plant correlation analysis for Zn, Fe and Al. Strong correlations occur among different plant parts, in the great majority of cases.
Although zinc levels were generally highly correlated, the correlation between long erect stems and short erect stems in A. perenne from Stations 1 and 2 was smaller.
Iron levels were also highly correlated in all tissues for the two species.
In spite of the generally high correlations for aluminium, weekly comparative correlations occur, in a few cases.
This suggests a close relationship between root metal concentrations and those in other tissues. There is probably a controlling mechanism for the distribution of the elements taken up by roots, according to the physiological state of the plants and their metabolic requirements.
These finding emphasise the importance of salt

marsh vegetation in the decontamination of this stressed environment and in the recycling of metal pollutants and agree with previousl works of BEEFTINK (1977) and WINDOM (1975).

H. portulacoides

		Local Sampling 1			2			3		
		Plant Part								
		R	S	L	R	S	L	R	S	L
Fe	R	-			-			-		
	S	.99	-		.89	-		.91	-	
	L	.95	.96	-	.89	.99	-	.84	.99	-
Al	R	-			-			-		
	S	.87	-		.59	-		.84	-	
	L	.95	.90	-	.78	.91	-	.67	.96	-
Zn	R	-			-			-		
	S	.93	-		.87	-		.76	-	
	L	.84	.88	-	.99	.86	-	.87	.90	-

TABLE 3. Correlation coefficients for Fe, Al and Zn, among different plant parts
R= Root   S=Stem   L= Leave

A. perenne

		Local Sampling 1			2			3		
		Plant Part								
		R	LS	SS	R	LS	SS	R	LS	SS
Fe	R	-			-			-		
	L.S.	.80	-		.90	-		.99	-	
	S.S.	.99	.80	-	.94	.91	-	.94	.97	-
Al	R	-			-			-		
	L.S.	.88	-		.96	-		.87	-	
	S.S.	.86	.94	-	.77	.71	-	.76	.67	-
Zn	R	-			-			-		
	L.S.	.74	-		.78	-		.84	-	
	S.S.	.95	.65	-	.96	.61	-	.82	.91	-

TABLE 4. Correlation coefficients for Fe, Al and Zn, among different plant parts
R= Root   L.S.= long erect stem
S.S.= Short erect non-flowering and flowering stem.

BIBLIOGRAPHY

Badsha KS and Sainsbury (1978) Some aspects of the biology and heavy metal accumulation of the fish Liparis liparis in the Severn estuary. Estuar. Coast. Mar. Sci. 7, 381-389.

Banus M Valiela I and Teal JM (1974) Export of lead from Salt Mashes. Mar. Pollut. Bull., 5, 6-9.

Beaufays JM and Nangniot P (1976) Etude Comparative du dosage de Cd dans les eaux, les engrais et les plantes par polarographie impulsionelle différentielle et spectrométrie d'absorption atomique. Analysis, 4, 193-199.

Beeftink WG (1977) "Salt-Marshes" 93-121 (Chapter 6) in "The Coastline" Ed. by RSK Barnes John Wiley and Sons Ltd. London.

Boyden CR and Romeril MG (1974) A trace metal problem in pond oyster culture. Mar.Pollut.Bull. 5, 74-74.

Breteler RJ Valiela I and Teal JM (1981) Bioavailability of mercury in several North-eastern U.S. Spartina ecosystems. Estuar. Coast. Shelf Sci., 12, 155-166.

Goldwater L (1971) Mercury in the environment. Sci. Amer., 244, 15-21.

Hardisty MW Huggins RJ Katar S. and Sainsbury M (1974) Ecological implications of heavy metal in fish from the Severn estuary Mar. Pollut. Bull., 5, 12-15.

Reboredo FHS (1981) Determinação de metais pesados em sedimentos do estuário do rio Tejo por espectrofotometria de absorção atómica por processo de chama e sem chama. Relatório Final de Estágio, Faculdade de Ciências de Lisboa. 54-54.

Rozema J (1978) On the ecology of some halophytes from a beach plain in the Netherlands. Ph. D. Thesis, Free Univ. of Amsterdam, 191.

Stoneburner DL and Harrison CS (1981) Heavy metal residues in sooty tern tissues from the Gulf of Mexico and North Central Pacific Ocean. Sci. Tot. Environ., 17, 51-58.

Windom HL (1975) Heavy metal fluxes through salt-marsh estuaries in "Estuarine Research" Vol.I, Ed. by LE Cronin Academic Press, London.